THE LYMPHATIC SYSTEM IN COLORECTAL CANCER

THE LYMPHATIC SYSTEM IN COLORECTAL CANCER

Basic Concepts, Pathology, Imaging, and Treatment Perspectives

---

Edited by

WIM P. CEELEN

Ghent University Hospital & Cancer Research Institute Ghent (CRIG), Ghent, Belgium

Academic Press is an imprint of Elsevier
125 London Wall, London EC2Y 5AS, United Kingdom
525 B Street, Suite 1650, San Diego, CA 92101, United States
50 Hampshire Street, 5th Floor, Cambridge, MA 02139, United States
The Boulevard, Langford Lane, Kidlington, Oxford OX5 1GB, United Kingdom

Copyright © 2022 Elsevier Inc. All rights reserved.

No part of this publication may be reproduced or transmitted in any form or by any means, electronic or mechanical, including photocopying, recording, or any information storage and retrieval system, without permission in writing from the publisher. Details on how to seek permission, further information about the Publisher's permissions policies and our arrangements with organizations such as the Copyright Clearance Center and the Copyright Licensing Agency, can be found at our website: www.elsevier.com/permissions.

This book and the individual contributions contained in it are protected under copyright by the Publisher (other than as may be noted herein).

Notices

Knowledge and best practice in this field are constantly changing. As new research and experience broaden our understanding, changes in research methods, professional practices, or medical treatment may become necessary.

Practitioners and researchers must always rely on their own experience and knowledge in evaluating and using any information, methods, compounds, or experiments described herein. In using such information or methods they should be mindful of their own safety and the safety of others, including parties for whom they have a professional responsibility.

To the fullest extent of the law, neither the Publisher nor the authors, contributors, or editors, assume any liability for any injury and/or damage to persons or property as a matter of products liability, negligence or otherwise, or from any use or operation of any methods, products, instructions, or ideas contained in the material herein.

British Library Cataloguing-in-Publication Data
A catalogue record for this book is available from the British Library

Library of Congress Cataloging-in-Publication Data
A catalog record for this book is available from the Library of Congress

ISBN: 978-0-12-824297-1

For Information on all Academic Press publications
visit our website at https://www.elsevier.com/books-and-journals

Publisher: Stacy Masucci
Acquisitions Editor: Rafael E. Teixeira
Editorial Project Manager: Sara Pianavilla
Production Project Manager: Omer Mukthar
Cover Designer: Christian J. Bilbow

Typeset by MPS Limited, Chennai, India

Contents

List of contributors xi
Preface xv

1
Basic concepts

1 Hypoxic signaling in lymphatic colorectal cancer metastasis 3
Luana Schito and Sergio Rey

1.1 Introduction 3
1.2 Hypoxic signaling in the tumor microenvironment 4
1.3 Hypoxia and metastatic dissemination 5
    1.3.1 EMT, cell motility, and invasion 6
    1.3.2 Lymphatic intravasation and extravasation 8
    1.3.3 Anoikis, mitochondrial hyperpolarization, and lymphatic dissemination 8
    1.3.4 Lymphatic niche formation and clonal expansion 9
1.4 Therapeutic perspectives 10
Acknowledgments 10
Conflicts of interest 10
References 10

2 Biomechanical aspects of the normal and cancer-associated lymphatic system 21
Wim P. Ceelen, Hooman Salavati, Ghazal Adeli Koudehi, Carlos Alejandro Silvera Delgado, Patrick Segers and Charlotte Debbaut

2.1 Introduction 21
2.2 Biomechanics of the normal lymphatic system 21
    2.2.1 Mechanical properties of the interstitium 21
    2.2.2 Interstitial flow and lymph formation 22
    2.2.3 Lymphatic endothelial cells and initial lymphatics 24
    2.2.4 Collecting lymphatics and lymphangions 26
    2.2.5 Lymph nodes 29
2.3 Biomechanics of lymphatic metastasis 30
    2.3.1 The biomechanical environment of cancer tissue 30
    2.3.2 Lymphangiogenesis in cancer 31
    2.3.3 Interstitial and lymphatic transport in cancer tissue 33
    2.3.4 Mechanisms of cancer cell invasion of initial lymphatics 33
    2.3.5 The metastatic lymph node 34
2.4 Conclusions and future perspectives 34
References 35

3 Mechanisms of lymphatic spread in colon cancer: insights from molecular and genetic studies 43
Mary Smithson and Karin Hardiman

3.1 Introduction 43
3.2 Mechanisms of the lymphatic system 43
    3.2.1 Lymph formation and movement 43
3.3 Tumor microenvironment 44
3.4 Colorectal cancer staging and treatment 45
    3.4.1 Colorectal cancer staging 45
    3.4.2 Colorectal cancer treatment 45
3.5 Evolutionary mechanisms of lymph node metastasis 46
    3.5.1 Linear models 46

3.6 Molecular markers of colorectal cancer spread 50
3.7 Molecular markers of metastases 50
3.8 Conclusions 52
Acknowledgments 52
Conflict of interest 52
References 52

4 Anatomy and embryology of the lymphatic system of the colon and rectum 57

Wouter Willaert

4.1 Introduction 57
4.2 Embryology of the lymphatic system 58
    4.2.1 Molecular mechanisms 58
    4.2.2 The lymph sacs 58
4.3 Anatomy of the lymphatic system 59
    4.3.1 Lymphatic capillaries 59
    4.3.2 Collecting lymphatic vessels and lymph nodes 60
    4.3.3 Lymphatic trunks and ducts 61
4.4 Anatomy of the lymphatic system of the colon and rectum 61
    4.4.1 Macroscopic anatomy of the colon and rectum 61
    4.4.2 Vascularization of the colon and rectum 62
    4.4.3 Lymphatic vessel anatomy in the mesocolon 62
    4.4.4 Lymphatic vessel anatomy in the mesorectum 63
    4.4.5 Lymphatic drainage of the colon and rectum 63
4.5 Conclusion 69
References 70

2
Pathology and imaging

5 Imaging of colorectal nodal disease 75

Lishan Cai, Zuhir Bodalal, Stefano Trebeschi, Selam Waktola, Tania C. Sluckin, Miranda Kusters, Monique Maas, Regina Beets-Tan and Sean Benson

5.1 Introduction 75

5.2 Staging, segmentation, and endoscopic detection 76
    5.2.1 Staging 76
    5.2.2 Segmentation and endoscopic detection 77
5.3 Metastasis classification and prediction 79
    5.3.1 Pathological classification 81
    5.3.2 Radiological LNM classification 82
5.4 Treatment response, recurrence, and survival 83
5.5 Summary and future directions 85
References 85

6 Tumor deposits in colorectal cancer 89

Nelleke Pietronella Maria Brouwer, Kai Francke and Iris D. Nagtegaal

6.1 Introduction 89
6.2 Definition of tumor deposits 89
6.3 The origins and biology of tumor deposits 91
    6.3.1 Where do tumor deposits come from? 91
    6.3.2 The biological mechanisms underlying tumor deposit development 92
6.4 Staging of tumor deposits 92
6.5 Prognostic value of tumor deposits 93
6.6 Tumor deposits and neoadjuvant therapies 94
6.7 Pathological assessment of tumor deposits: interobserver variation 96
    6.7.1 Interobserver agreement regarding tumor deposits in previous TNM editions 96
    6.7.2 Interobserver agreement regarding tumor deposits in the TNM 8th edition 96
    6.7.3 How to improve interobserver variation regarding tumor deposits? 97
6.8 Radiological assessment of tumor deposits: advances and challenges 98
    6.8.1 Tumor deposits in colon cancer, identification by CT 98
    6.8.2 Tumor deposits in rectal cancer, identification by MRI 99
    6.8.3 A radiological concept regarding the origin of tumor deposits 100
    6.8.4 The future of tumor deposits in radiology 100
References 101

7 Lymph node classification in colorectal cancer: tumor node metastasis versus the Japanese system 107

Kozo Kataoka, Yukihide Kanemitsu, Manabu Shiozawa and Masataka Ikeda

7.1 Japanese D3 lymphadenectomy 107
    7.1.1 Japanese lymph node classification 107
    7.1.2 Concept of Japanese lymphadenectomy 108
    7.1.3 Lateral lymph node dissection 110
    7.1.4 Japanese D3 versus complete mesocolic excision 111
7.2 Tumor node metastasis versus Japanese lymph node classification 112
7.3 Conclusion 113
References 113

8 Detection and significance of micrometastases and isolated tumor cells in lymph nodes of colorectal cancer resections 115

Anne Hoorens

8.1 Definition of micrometastases and isolated tumor cells 115
    8.1.1 AJCC/UICC definition 115
    8.1.2 Methods of detection 116
    8.1.3 Biological significance 118
    8.1.4 Reporting 119
    8.1.5 Summary 119
8.2 Micrometastases and isolated tumor cells in colorectal cancer 119
    8.2.1 Implications in colorectal cancer 119
    8.2.2 Occult disease in lymph nodes in colorectal cancer 120
    8.2.3 Recommendations for standardized histopathological analysis of lymph nodes 123
8.3 Micrometastases and isolated tumor cells in sentinel lymph-node biopsy in colorectal cancer 123
    8.3.1 Sentinel lymph-node mapping with ultra-staging 123
    8.3.2 Sentinel lymph-node biopsy in colorectal cancer 124
8.4 Micrometastases and isolated tumor cells after neoadjuvant therapy 126
8.5 Differential diagnosis of micrometastases and isolated tumor cells 127
8.6 Conclusion 127
References 128

9 Anatomical and temporal patterns of lymph node metastasis in colorectal cancer 131

Mathieu J.R. Struys and Wim P. Ceelen

9.1 Introduction 131
9.2 Mechanisms of lymphatic spread in colon cancer 132
9.3 Temporal patterns of metastasis in colon cancer 132
    9.3.1 Evidence from circulating and tissue biomarkers 133
    9.3.2 Evidence from genomic and phylogenetic studies 133
    9.3.3 Evidence from growth rate of primary and metastatic colorectal tumors 135
    9.3.4 Evidence from autopsy findings in metastatic colorectal cancer 136
    9.3.5 Evidence from clinical studies 136
9.4 Anatomical patterns of lymph node metastasis in colon cancer 141
9.5 Conclusion and implications for research 144
References 144

3
Treatment

10 Neoadjuvant treatment and lymph node metastasis in rectal cancer 155

Jesse P. Wright, Alexandra Elias and John R.T. Monson

10.1 Introduction 155
10.2 Importance of lymph node yield 156
10.3 Lymph node yield 157
10.4 Neoadjuvant therapy and lymph node yield 158
10.5 Lymph node ratio 160

10.6 Impact of nodal involvement with complete clinical response after neoadjuvant therapy 161
10.7 Effects of total neoadjuvant therapy 162
10.8 Conclusion 162
References 163
Further reading 166

11 Complete mesocolic excision in colon cancer 167
Alice C. Westwood, Jim P. Tiernan and Nicholas P. West

11.1 Introduction 167
11.2 Outline of the key anatomy 168
11.3 Principles of CME surgery 169
    11.3.1 Role of minimally invasive CME surgery 170
11.4 Oncological benefits of CME 171
    11.4.1 Which patients benefit from CME? 172
11.5 Potential limitations of CME and associated controversies 174
    11.5.1 Complications 174
    11.5.2 Technical difficulties 174
    11.5.3 CME in transverse colon cancer 175
11.6 Importance of pathological quality control in CME surgery 176
    11.6.1 Integrity of the mesocolon (plane of surgery) 177
    11.6.2 Distance between the tumor and the central arterial ligation point 179
    11.6.3 Length of bowel resected 181
    11.6.4 Lymph node yield 182
    11.6.5 Specimen photography 183
11.7 Conclusion 185
References 186

12 Japanese D3 dissection in cancer of the colon: technique and results 193
Yuichiro Tsukada and Masaaki Ito

12.1 Introduction 193
12.2 History of lymphadenectomy for colon cancer in Japan 194
    12.2.1 Japanese classification and Japanese guidelines 194
12.3 Basic principles of lymph node dissection in Japan 194
12.4 Changes in the recommended area of lymph node dissection in Japan 195
12.5 Current classifications of lymph node metastasis (N) and lymph node dissection (D) in Japan 196
    12.5.1 Basic principles of regional lymph node classification 196
12.6 Lymph node groups and station numbers 198
12.7 Classification of lymph node metastases (N) 198
12.8 Classification of lymph node dissection (D) 200
12.9 Technique of Japanese D3 dissection for colon cancer 200
12.10 Cecum cancer 200
12.11 Ascending colon cancer 200
12.12 Transverse colon cancer 202
12.13 Descending colon cancer 206
12.14 Sigmoid colon cancer 207
12.15 Preservation of the LCA for left-sided colon cancer 207
12.16 Outcomes of Japanese D3 dissection for colon cancer 208
    12.16.1 D3 lymphadenectomy versus D2 lymphadenectomy 208
12.17 Current recommendations of the Japanese guidelines, 2019 209
12.18 Comparison of Japanese D3 dissection with European CME with CVL 210
12.19 Future perspective 212
12.20 Summary 212
Acknowledgment 213
Disclosure 213
References 213

13 Management of para-aortic nodal disease in colon cancer 215
Alexander De Clercq and Gabrielle H. van Ramshorst

13.1 Introduction 215
13.2 Imaging and implications for prognosis 216
13.3 Differences between right-sided and left-sided para-aortic node involvement 216
13.4 Metachronous isolated lymph node metastasis 217
13.5 Synchronous metastases 218
13.6 Morbidity of surgery 220

13.7 The role of chemotherapy 221
13.8 Conclusion and future perspectives 222
Acknowledgment 223
References 223

14 Lateral lymph node dissection in rectal cancer 227

Tania C. Sluckin, Sanne-Marije J. A. Hazen, Takashi Akiyoshi and Miranda Kusters

14.1 Introduction 227
14.2 East versus West 228
14.3 Defining lateral nodal disease 231
14.4 The future 234
14.5 The lateral lymph node dissection 235
14.6 Risks 239
14.7 Procedural variation 239
14.8 Conclusion 241
References 241

15 Fluorescence-guided sentinel lymph node detection in colorectal cancer surgery 245

Ruben P.J. Meijer, Hidde A. Galema, Lorraine J. Lauwerends, Cornelis Verhoef, Jacobus Burggraaf, Stijn Keereweer, Merlijn Hutteman, Alexander L. Vahrmeijer and Denise E. Hilling

15.1 Concept of sentinel lymph node mapping 245
15.2 Fluorescence-guided surgery 246
15.3 Fluorescence-guided sentinel lymph node detection in colorectal cancer 247
15.4 Future perspectives 249
15.5 Conclusions 252
Disclosure 252
References 252

16 Systemic treatment of localized colorectal cancer 257

Dedecker Hans, Vandamme Timon, Teuwen Laure-Anne, Wuyts Laura, Prenen Hans, ten Tije, Albert Jan and Peeters Marc

16.1 Introduction 257
16.2 Relapse risk assessment in stage II disease 258
16.3 Adjuvant treatment in stage II disease 259
16.4 Time to treat 262
16.5 Adjuvant treatment in stage III disease 262
16.6 Neoadjuvant chemotherapy for locally advanced colon cancer 266
16.7 Nutrition and lifestyle modification reduce relapse risk 266
16.8 Adjuvant treatment in elderly 267
16.9 Dihydropyrimidine dehydrogenase-deficient patients 267
16.10 Future perspectives for (neo)adjuvant therapy in stage III colon cancer 268
16.11 Conclusion 268
References 269

17 Radiotherapy for metastatic nodal disease in colorectal cancer 273

Melissa A. Frick, Phoebe Loo, Lucas K. Vitzthum, Erqi L. Pollom and Daniel T. Chang

17.1 Introduction to radiation 273
17.2 Rationale for local treatment of metastatic colorectal cancer 274
17.3 Experience with radiation 276
  17.3.1 Radiation in the oligometastasic state 276
  17.3.2 Radiation for CRC oligometastases 276
  17.3.3 Radiation for colorectal nodal oligometastases 277
  17.3.4 Neoadjuvant radiotherapy 282
  17.3.5 Prognostic factors 283
  17.3.6 Future directions 283
17.4 Practical considerations in radiation and radiation techniques 286
  17.4.1 Patient selection 287
  17.4.2 Simulation 288
  17.4.3 Treatment planning 288
  17.4.4 Treatment delivery 289
  17.4.5 Treatment-related toxicity/dose constraints 289
  17.4.6 Follow-up 289
17.5 Case examples 289
  17.5.1 Case 1 289
  17.5.2 Case 2 292
References 294

Index 299

List of contributors

Takashi Akiyoshi Department of Gastroenterological Surgery, Cancer Institute Hospital of the Japanese Foundation for Cancer Research, Tokyo, Japan

Regina Beets-Tan Department of Radiology, The Netherlands Cancer Institute, Amsterdam, The Netherlands; GROW School for Oncology and Developmental Biology, Maastricht University, Maastricht, The Netherlands

Sean Benson Department of Radiology, The Netherlands Cancer Institute, Amsterdam, The Netherlands

Zuhir Bodalal Department of Radiology, The Netherlands Cancer Institute, Amsterdam, The Netherlands; GROW School for Oncology and Developmental Biology, Maastricht University, Maastricht, The Netherlands

Nelleke Pietronella Maria Brouwer Department of Pathology, Radboud University Medical Center, Nijmegen, The Netherlands

Jacobus Burggraaf Centre for Human Drug Research, Leiden, The Netherlands

Lishan Cai Department of Radiology, The Netherlands Cancer Institute, Amsterdam, The Netherlands; GROW School for Oncology and Developmental Biology, Maastricht University, Maastricht, The Netherlands

Wim P. Ceelen Department of GI Surgery, Ghent University Hospital and Department of Human Structure and Repair, Ghent University, Ghent, Belgium; Cancer Research Institute Ghent (CRIG), Ghent University, Belgium

Daniel T. Chang Department of Radiation Oncology, Stanford University School of Medicine, Palo Alto, CA, United States

Alexander De Clercq Department of Gastrointestinal Surgery, Ghent University Hospital, Ghent, Belgium

Charlotte Debbaut Cancer Research Institute Ghent (CRIG), Ghent University, Belgium; Institute for Biomedical Engineering and Technology (IBiTech)-Biofluid, Tissue and Solid Mechanics for Medical Applications (Biommeda), Ghent University, Belgium

Carlos Alejandro Silvera Delgado Institute for Biomedical Engineering and Technology (IBiTech)-Biofluid, Tissue and Solid Mechanics for Medical Applications (Biommeda), Ghent University, Belgium

Alexandra Elias Center for Colon and Rectal Surgery, Digestive Health and Surgery Institute, AdventHealth, Orlando, FL, United states

Kai Francke Department of Pathology, Radboud University Medical Center, Nijmegen, The Netherlands

Melissa A. Frick Department of Radiation Oncology, Stanford University School of Medicine, Palo Alto, CA, United States

Hidde A. Galema Department of Surgical Oncology and Gastrointestinal Surgery, Erasmus MC Cancer Institute, Rotterdam, The Netherlands; Department of Otorhinolaryngology, Head and Neck Surgery, Erasmus MC Cancer Institute, Rotterdam, The Netherlands

Dedecker Hans Department of Oncology, University of Antwerp, Antwerp, Belgium

Prenen Hans Department of Oncology, University of Antwerp, Antwerp, Belgium

Karin Hardiman Department of Surgery, Birmingham Veterans Affairs Medical Center, Birmingham, AL, United States

Sanne-Marije J.A. Hazen Department of Surgery, Cancer Center Amsterdam, Amsterdam University Medical Centers, Vrije Universiteit Amsterdam, Amsterdam, The Netherlands

Denise E. Hilling Department of Surgery, Leiden University Medical Center, Leiden, The Netherlands; Department of Surgical Oncology and Gastrointestinal Surgery, Erasmus MC Cancer Institute, Rotterdam, The Netherlands

Anne Hoorens Department of Pathology, Ghent University Hospital, Ghent University, Ghent, Belgium

Merlijn Hutteman Department of Surgery, Leiden University Medical Center, Leiden, The Netherlands

Masataka Ikeda Hyogo College of Medicine, Japan

Masaaki Ito Department of Colorectal Surgery, National Cancer Center Hospital East, Kashiwa, Japan

Albert Jan Department of Oncology, University of Antwerp, Antwerp, Belgium

Yukihide Kanemitsu National Cancer Center, Japan

Kozo Kataoka Hyogo College of Medicine, Japan

Stijn Keereweer Department of Surgical Oncology and Gastrointestinal Surgery, Erasmus MC Cancer Institute, Rotterdam, The Netherlands; Department of Otorhinolaryngology, Head and Neck Surgery, Erasmus MC Cancer Institute, Rotterdam, The Netherlands

Ghazal Adeli Koudehi Institute for Biomedical Engineering and Technology (IBiTech)-Biofluid, Tissue and Solid Mechanics for Medical Applications (Biommeda), Ghent University, Belgium

Miranda Kusters Department of Surgery, Cancer Center Amsterdam, Amsterdam University Medical Centers, Vrije Universiteit Amsterdam, Amsterdam, The Netherlands

Wuyts Laura Department of Oncology, University of Antwerp, Antwerp, Belgium

Teuwen Laure-Anne Department of Oncology, University of Antwerp, Antwerp, Belgium

Lorraine J. Lauwerends Department of Surgical Oncology and Gastrointestinal Surgery, Erasmus MC Cancer Institute, Rotterdam, The Netherlands; Department of Otorhinolaryngology, Head and Neck Surgery, Erasmus MC Cancer Institute, Rotterdam, The Netherlands

Phoebe Loo Department of Radiation Oncology, Stanford University School of Medicine, Palo Alto, CA, United States

Monique Maas Department of Radiology, The Netherlands Cancer Institute, Amsterdam, The Netherlands

Peeters Marc Department of Oncology, University of Antwerp, Antwerp, Belgium

Ruben P.J. Meijer Department of Surgery, Leiden University Medical Center, Leiden, The Netherlands; Centre for Human Drug Research, Leiden, The Netherlands

John R.T. Monson Center for Colon and Rectal Surgery, Digestive Health and Surgery Institute, AdventHealth, Orlando, FL, United states

Iris D. Nagtegaal Department of Pathology, Radboud University Medical Center, Nijmegen, The Netherlands

Erqi L. Pollom Department of Radiation Oncology, Stanford University School of Medicine, Palo Alto, CA, United States

Sergio Rey UCD School of Medicine, University College Dublin, Dublin, Ireland

Hooman Salavati Department of GI Surgery, Ghent University Hospital and Department of Human Structure and Repair, Ghent University, Ghent, Belgium; Cancer Research Institute Ghent (CRIG), Ghent University, Belgium; Institute for Biomedical Engineering and Technology (IBiTech)-Biofluid, Tissue and Solid Mechanics for Medical Applications (Biommeda), Ghent University, Belgium

Luana Schito UCD School of Medicine, University College Dublin, Dublin, Ireland

Patrick Segers Institute for Biomedical Engineering and Technology (IBiTech)-Biofluid, Tissue and Solid Mechanics for Medical Applications (Biommeda), Ghent University, Belgium

Manabu Shiozawa Kanagawa Cancer Center, Japan

Tania C. Sluckin Department of Surgery, Cancer Center Amsterdam, Amsterdam University Medical Centers, Vrije Universiteit Amsterdam, Amsterdam, The Netherlands; Department of Radiology, The Netherlands Cancer Institute, Amsterdam, The Netherlands

Mary Smithson Department of Surgery, University of Alabama Birmingham, Birmingham, AL, United States

Mathieu J.R. Struys Department of GI Surgery, Ghent University Hospital and Department of Human Structure and Repair, Ghent University, Ghent, Belgium

Jim P. Tiernan John Goligher Colorectal Unit, Leeds Teaching Hospitals NHS Trust, Leeds, United Kingdom

ten Tije Department of Oncology, University of Antwerp, Antwerp, Belgium

Vandamme Timon Department of Oncology, University of Antwerp, Antwerp, Belgium

Stefano Trebeschi Department of Radiology, The Netherlands Cancer Institute, Amsterdam, The Netherlands

Yuichiro Tsukada Department of Colorectal Surgery, National Cancer Center Hospital East, Kashiwa, Japan

Alexander L. Vahrmeijer Department of Surgery, Leiden University Medical Center, Leiden, The Netherlands

Gabrielle H. van Ramshorst Department of Gastrointestinal Surgery, Ghent University Hospital, Ghent, Belgium

Cornelis Verhoef Department of Surgical Oncology and Gastrointestinal Surgery, Erasmus MC Cancer Institute, Rotterdam, The Netherlands

Lucas K. Vitzthum Department of Radiation Oncology, Stanford University School of Medicine, Palo Alto, CA, United States

Selam Waktola Department of Radiology, The Netherlands Cancer Institute, Amsterdam, The Netherlands

Nicholas P. West Pathology & Data Analytics, Leeds Institute of Medical Research at St. James's, University of Leeds, Leeds, United Kingdom; Department of Histopathology, Leeds Teaching Hospitals NHS Trust, Leeds, United Kingdom

Alice C. Westwood Pathology & Data Analytics, Leeds Institute of Medical Research at St. James's, University of Leeds, Leeds, United Kingdom; Department of Histopathology, Leeds Teaching Hospitals NHS Trust, Leeds, United Kingdom

Wouter Willaert Department of Human Structure and Repair, Ghent University, Ghent, Belgium

Jesse P. Wright Center for Colon and Rectal Surgery, Digestive Health and Surgery Institute, AdventHealth, Orlando, FL, United states

Preface

With a global incidence rate of around 1.1 million new cases per year and an average mortality rate approaching 50%, colon cancer remains a considerable global health care burden. Surgery is the cornerstone of treatment in Stage I–III disease, and contrary to the rapid evolution of treatment modalities in Stage IV disease, the basic surgical approach changed little over the past century: a segment of colon is removed together with its supporting mesentery. The presence of metastatic lymph nodes (LN) is the strongest prognostic variable in patients with colon cancer. However, the biological significance of lymphatic metastasis in colon cancer remains incompletely understood. Specifically, it is not clear whether invaded nodes act as innocent bystanders in a cancer which spreads systemically from the start, or whether they are an essential and required step in the metastatic cascade. Recent data based on molecular and phylogenetic studies suggest that the former scenario is more likely. This uncertainty reflects on the current debate surrounding the ideal extent of lymphadenectomy in colon cancer. Proponents of extensive lymphadenectomy argue that removing a maximal number of LN not only improves staging but also prevents recurrent cancer in the remaining nodes, and may interrupt the metastatic cascade by removing the most distal cancer invaded LN. Others argue that the Halstedian, linear progression model of cancer spread does not reflect the biology of colon cancer, and that, therefore, extensive lymphadenectomy is unlikely to affect the risk of systemic relapse.

This volume is the first to offer a comprehensive, multidisciplinary overview of the lymphatic system in colon cancer, covering anatomy, physiology, molecular biology and genetics, imaging and staging, locoregional treatment, and systemic treatment. It will be an essential source of information for students, clinicians, and researchers with an interest in colon cancer. I sincerely thank all authors for their outstanding contributions, and hope that this volume will encourage further interest and research in this fascinating field.

Wim P. Ceelen
Ghent University Hospital & Cancer Research Institute Ghent (CRIG), Ghent, Belgium

SECTION 1

Basic Concepts

Hypoxic signaling in lymphatic colorectal cancer metastasis

Luana Schito and Sergio Rey
UCD School of Medicine, University College Dublin, Dublin, Ireland

1.1 Introduction

Colorectal cancer (CRC) is the second most deadly and third most frequent cancer in the world (Sung et al., 2021), typically evolving from benign adenomatous lesions in the colon or rectum to adenocarcinomas (Powell et al., 1992; Vogelstein et al., 1988) as a result of mutations in oncogenes and tumor suppressor genes acquired by colonic cells over a period of decades (Vogelstein & Kinzler, 2004). Depending on the stage, CRC can be locally confined to the colonic mucosa, invade the outermost colonic layers, and/or spread to nearby lymph nodes, tissues, or organs, eventually disseminating to distant sites, most notably the liver, lung, peritoneum, or distant lymph nodes, with rare bone, heart, pancreatic, and ovarian metastasis (Hugen et al., 2014; Lewis et al., 2006; Park et al., 2019). Lymphatic vessels as well as lymph nodes play a pivotal role in promoting CRC progression, acting as key routes for dissemination. Consistently, CRC cell invasion of intratumoral and peritumoral lymphatic vessels has been shown to positively correlate with the presence of CRC cells in lymph nodes (Liang et al., 2007; Lin et al., 2010), whose degree of involvement is in turn utilized as an important factor to determine CRC staging, therapeutic approach, and prognosis (Chang et al., 2007; Le Voyer et al., 2003; Parsons et al., 2011). Interestingly, recent studies have challenged the sequential model of cancer progression, in which primary CRC cell invasion leads to nodal metastasis followed by distant metastasis (Fidler, 2003; Langley & Fidler, 2011; Weinberg, 2008); indeed, phylogenetic tree reconstruction through somatic variants in CRC biopsies has uncovered that approximately two-thirds of CRC lymph node and distant metastases originate from independent subclones in the primary tumor, whereas the remaining third shares a common clonal origin (Naxerova et al., 2017), thereby suggesting that CRC cells are able to reach distant sites without passing through the lymph nodes, potentially explaining the apparent lack of a consistent therapeutic benefit of lymphadenectomy in CRC patients (Brouwer, 2020; Gervasoni et al., 2007). Furthermore, perineural and intramural vascular invasion have recently emerged as likely routes for CRC cell dissemination, also at odds with the sequential model (Knijn et al., 2016, 2018).

Hypoxia, a prevalent microenvironmental feature of CRC (Goethals et al., 2006), acts as a key clonal selection factor promoting CRC metastasis through the activation of an adaptive program that enhances motility and invasion upon subjacent colonic layers, nearby tissues, organs, and the vasculature, eventually leading to the colonization of distant sites, wherein a secondary (or multiple) tumor(s) can be established. As we will henceforth describe in detail, the hypoxic signal is chiefly transduced by a family of evolutionarily conserved heterodimeric proteins known as hypoxia-inducible factors (HIFs), exerting a central role in many aspects of cancer progression including tumor initiation, progression, metastasis, and resistance to chemotherapy and radiotherapy (Furlan et al., 2008; Schito & Semenza, 2016; Rey et al., 2017). Consistently, HIF paralog overexpression has been associated with an advanced CRC stage, regional and distant metastasis, as well as resistance to chemotherapy, while acting as a statistically independent adverse, prognostic factor (Baba et al., 2010; Cao et al., 2009; Rajaganeshan et al., 2009; Rasheed et al., 2009; Tang et al., 2018; Xu et al., 2019). Here, we summarize current knowledge of the cellular and molecular mechanisms promoting CRC lymphatic dissemination within the context of hypoxia and HIF signaling. Insights into potential HIF-targeting strategies aimed at blocking CRC dissemination are provided as well.

1.2 Hypoxic signaling in the tumor microenvironment

CRC tumorigenesis is a well-established sequential process (Fearon & Vogelstein, 1990), where mutations and epigenetic changes in oncogenes and tumor suppressors drive the stepwise progression of colonic, glandular epithelial cells from a nonmalignant to a cancerous phenotype with metastatic potential (Fearon & Vogelstein, 1990; Jung et al., 2020; Keum & Giovannucci, 2019). Sustained proliferation, a hallmark of tumorigenesis (Hanahan & Weinberg, 2011), is an energy-demanding process initially supported by local O_2 levels via diffusion. However, as tumor cells proliferate and their distance from the nearest capillary exceeds $\approx 100-200$ μm, diffusion is no longer sufficient to meet proliferative O_2 requirements, thereby resulting in hypoxic microenvironments. Cancer cells (CCs) typically respond to hypoxia by activating transcriptional and translational pathophysiological programs aimed at restoring O_2 supply while adjusting cellular metabolism in order to survive and metastasize (Elia et al., 2018; Schito & Rey, 2018; Schito & Semenza, 2016). Remodeling of the host vasculature, including the formation of blood (angiogenesis) and lymphatic (lymphangiogenesis) vessels from existing ones, represents an early event in CRC progression, presumably aimed at improving O_2 supply (blood vessels), while facilitating the entry of CRC cells into the blood and/or lymphatic circulation to promote metastasis (Bossi et al., 1995; Schito & Rey, 2020; Schito, 2019; Sundlisaeter, 2007). The resultant tumor-associated vasculature, abnormal in structure and function, contributes to cancer progression by paradoxically decreasing O_2 supply (blood vessels), while creating heterogeneous gradients of interstitial fluid pressure (blood and lymphatic vessels) that promote cell migration and invasion into regional and distant sites, while compromising the efficacy of drug delivery (Flessner et al., 2005; Martin et al., 2019; Padera et al., 2002;

Shieh, 2011). Therefore the ensuing physicochemical disruption of the microenvironment aggravates hypoxia while sustaining its signaling pathway activation in a feedforward manner, further promoting lymphangiogenesis (and angiogenesis), lymphatic permeability, and movement of CRC cells in and out of lymphovascular structures, all factors that coalesce in a prometastatic microenvironment in most solid cancers and, in particular, within the colonic wall, where there is limited space to accommodate interstitial edema. Importantly, CRC cells might rely on the co-option of immune and lymphatic endothelial cell (LEC) homeostatic mechanisms to further promote malignant dissemination (Maisel et al., 2017; Permanyer et al., 2018; Swartz, 2014).

HIFs are heterodimeric, basic helix−loop−helix transcription factors, consisting of an O_2-regulated subunit (HIF-1α, -2α, or -3α; henceforth referred to as HIFα) and a constitutively expressed HIF-1β subunit (Huang et al., 1996; Wang et al., 1995). In well-oxygenated cells (>5% O_2), human HIFα is hydroxylated at two proline and one asparagine residues ($Pro^{402}/Pro^{564}/Asn^{803}$ in HIF-1α or $Pro^{405}/Pro^{531}/Asn^{851}$ in HIF-2α) by prolyl-4-hydroxylases using O_2, Fe^{2+}, α-ketoglutarate, and ascorbate as cosubstrates (Bruick & McKnight, 2001; Epstein et al., 2001; Kaelin & Ratcliffe, 2008). This posttranslational modification allows HIFα proteins to be recognized by the von Hippel−Lindau (VHL) tumor suppressor, leading to ubiquitin-dependent proteasomal degradation (Cockman et al., 2000; Maxwell et al., 1999). By contrast, in hypoxic cells, HIFα hydroxylation is inhibited, leading to HIFα stabilization, dimerization with HIF-1β, translocation to the nucleus, and transcriptional target gene activation (Semenza et al., 1996). In addition, HIFα expression and transactivity can be regulated through O_2-independent mechanisms activated by chemotherapy (Cao et al., 2013; Samanta et al., 2014; Schito et al., 2020), growth factors, or mutations in oncogenes and tumor suppressors (Semenza, 2013b). A salient tumor suppressor is VHL, whose inactivation blocks nonhypoxic HIFα degradation, while promoting constitutive angiogenic signaling pathway activation, a central process underlying CRC progression, that can be targeted by therapeutics such as bevacizumab (Giles et al., 2006; Kuwai et al., 2004; Zhuang et al., 1996). Although HIF-1α and HIF-2α are similarly induced by hypoxia and are both able to recognize the same hypoxia response elements within the genomic sequence of target genes, they have been shown to display overlapping, yet distinct patterns of target gene activation in a cell-type dependent manner; in SW480 colon cancer cells, HIFα paralogs engage in a transcriptional counterbalance wherein HIF-1α promotes cellular proliferation, migration, and in vivo tumorigenicity, while HIF-2α inhibits these processes (Imamura et al., 2009). Of relevance, HIF-1α has been shown to support CRC progression by disrupting β-catenin→TCF4 signaling that is aberrantly activated by mutations in the β-catenin (*Ctnnb1*) or adenomatous polyposis *coli* (*Apc*) genes (Fodde et al., 2001; Morin et al., 1997; Kaidi et al., 2007). In line, β-catenin can augment HIF-1α transactivity resulting in increased CRC cell survival and adaptation to hypoxia (Kaidi et al., 2007).

1.3 Hypoxia and metastatic dissemination

The metastatic process involves multiple interrelated steps which can be organized for simplicity as follows: (1) loss of epithelial cell differentiation and gain of a motile, mesenchymal-like cell phenotype, a process known as epithelial—mesenchymal transition (EMT); (2) disruption and remodeling of the basement membrane and the extracellular matrix (ECM), leading to lymphovascular infiltration (invasion and intravasation); (3) survival under anchorage-independent conditions in the lymphatic and/or blood circulation (anoikis); (4) egress from the vasculature into secondary sites (extravasation); and (5) establishment of premetastatic niches. These steps involve autocrine and paracrine interactions between CCs and the stroma, resulting in the activation and/or recruitment of bone-marrow-derived cells, endothelial cells, pericytes, tumor-infiltrating immune cells, and cancer-associated fibroblasts, collectively supporting a prometastatic microenvironment [reviewed in Conti and Thomas (2011) and Dongre and Weinberg (2019)]. In line with this framework, several studies have shown that the tumor-to-stroma ratio acts as a strong independent prognostic factor in CRC patients (Park et al., 2014; Sandberg et al., 2018; West et al., 2010); in addition, metastatic CRC lymph nodes exhibit abnormally high levels of myofibroblasts scattered throughout, suggesting that CRC cells are able to recapitulate the primary tumor microenvironment closely resembling physiological stem cell niches (Yeung et al., 2013). In this chapter, emphasis is put upon discussing established and emerging cell-autonomous, HIFα-dependent transcriptional mechanisms underpinning lymphatic dissemination of CRC cells.

1.3.1 EMT, cell motility, and invasion

The EMT process consists in the loss of intercellular junctions and cell polarity, accompanied by cytoskeletal reorganization and gene expression changes that allow CRC cells to dislodge from the primary tumor, thus acquiring single-cell motility. Critically, repression of the transmembrane protein E-cadherin has been shown to be central for EMT, since it causes *adherens* junction loss and increased cell motility. In line, E-cadherin-dependent disruption of epithelial integrity causes downregulation of the transcriptional repressor REST, leading to de-repression of the hypoxia-inducible glycoprotein L1 cell adhesion molecule (L1CAM), known to mediate CRC cell blood vessel extravasation, metastasis, and chemoresistance (Fang et al., 2020; Ganesh et al., 2020; Zhang et al., 2012); importantly, E-cadherin-REST-dependent modulation of L1CAM expression enables a "regenerative" metastatic CRC phenotype, thought to represent a hijacked genetic wound healing program present in the normal, colonic mucosa (Ganesh et al., 2020). In CRC, EMT occurs at the invasive front, characterized by HIF-1α^+ migratory cells that have lost membranous E-cadherin expression and accumulated nuclear β-catenin (Brabletz et al., 2001; Rajaganeshan et al., 2009; Righi et al., 2015; Thiery et al., 2009); consistently, loss of E-cadherin correlates with CRC stage, lymph node involvement, liver metastasis, lymphatic and blood vessel intravasation (Mohri, 1997). HIF-1α activates EMT specification through transcriptional activation of genes encoding for the transcription factors SNAIL1, SNAIL2, ZEB1, TWIST, and TCF3 (Krishnamachary et al., 2006; Yang et al., 2008; Zhang et al., 2015), known to repress E-cadherin and other epithelial markers, while de-repressing mesenchymal markers, such as vimentin, N-cadherin, smooth muscle actin, fibronectin, and vitronectin (Batlle et al., 2000; Peinado et al., 2007; Toiyama et al., 2013). Interestingly, recent data have revealed that hypoxic CRC cells can additionally repress E-cadherin through Sox9→USP47-dependent stabilization of Snail (Choi et al., 2017), whose overexpression is associated with progression and mortality (Kim et al., 2014). Of relevance, the acquisition of mesenchymal and loss of epithelial markers in EMT do not necessarily occur as mutually exclusive events through a linear sequence, as suggested by data wherein CCs coexhibit epithelial and mesenchymal traits, thus indicating the existence of an "EMT hybrid state" [reviewed in Dongre and Weinberg (2019) and Pastushenko and Blanpain (2019)].

A key signaling pathway implicated in cytoskeleton remodeling, an early event in motility, is the hepatocyte growth factor (HGF)→c-Met axis, whose aberrant activation during CRC progression promotes cell motility, ECM degradation, tumor invasion, and nodal metastasis (Di Renzo et al., 1995; Takeuchi et al., 2003), in addition to its role in the maintenance of CRC stemness within the primary and nodal microenvironments (Vermeulen et al., 2010; Yeung et al., 2013). Importantly, in non-CRC cells, HIF-1α directly binds the *Met* promoter (Pennacchietti et al., 2003), whereas HGF is able to promote HIF-1α transcriptional activity, thereby potentiating cell motility and invasion (Pennacchietti et al., 2003; Tacchini et al., 2001). Notwithstanding, hypoxic activation of HGF→c-Met as a consequence of antiangiogenic therapies has been shown to promote hepatic and pulmonary metastases in experimental CRC models (Mira et al., 2017).

The interplay between cell−cell and cell−ECM interactions, as well as cytoskeleton reorganization, allows CCs to acquire motile and invasive properties. Integrins are a family of cell surface receptors that play a pivotal role in CRC progression by transducing ECM signals onto downstream effectors mediating cell adhesion, migration, and invasion, dependent on the activity of matrix metalloproteinases (MMPs) (Seguin et al., 2015). The hypoxic microenvironment stimulates the expression of integrin α_2, α_5, and β_1, which help CRC cells to bind collagen and fibronectin, consequently facilitating migration and invasion (Hongo et al., 2013). Tetraspanins, a family of membrane glycoproteins regulating several cell processes such as adhesion, spread, and migration, have been shown to be integrin-binding partners in several cancer types (Detchokul et al., 2014). Within the context of tumoral hypoxia, CRC cells suppress tetraspanin CD151 expression via HIF-1α facilitating detachment from surrounding cells and ECM, therein promoting metastasis (Chien et al., 2008; Semenza, 2008); consistent with this role, CD151 expression is significantly decreased in CRC biopsies, as compared to paired colonic tissue (Chien et al., 2008). Additionally, HIF-1α can exert control over the migratory and invasive abilities of CRC cells by directly regulating the expression of galectin-1, a member of the lectin family mediating cell−cell and cell−matrix interactions (Hittelet et al., 2003; Zhao et al., 2010); importantly, galectin-1 is overexpressed in both primary CRC and lymph node metastases, while correlating with progression (Sanjuán et al., 1997).

An additional factor enabling hypoxic CRC cells to penetrate the basement membrane is the urokinase-type plasminogen activator receptor (uPAR), a known HIF-1α target whose expression is found in primary CRCs, in stromal cells adjacent to the invasive front, and within lymph node metastases; importantly, uPAR expression correlates with patient mortality, while acting as an independent prognostic factor upon distant metastases (Yang et al., 2000; Krishnamachary et al., 2003; Boonstra et al., 2014; Pyke et al., 1991). In addition, CRC cells rely on the proteolytic activity of MMPs for remodeling or degradation of the basement membrane and ECM at the invasive front (Krishnamachary et al., 2003), a mechanism sustained under hypoxia via HIFα-dependent upregulation of MMP2, MMP9, and membrane type-1 (MT1)-MMP (Krishnamachary et al., 2003; Muñoz-Nájar et al., 2006; Petrella et al., 2005), predicting poor CRC clinical outcomes as well (Cho et al., 2007; Kanazawa et al., 2010; Salem et al., 2016; Zeng et al., 1996). In parallel, hypoxia has been described to alternatively enhance MMP2 and MMP9 expression through upregulation of SIRT1, an evolutionarily conserved histone and nonhistone deacetylase that promotes CRC cell migration and invasion (Yu et al., 2019).

1.3.2 Lymphatic intravasation and extravasation

Further to enhanced cell migration and invasiveness, metastasizing CRC cells are unlikely to reach regional and distant lymph nodes unless they acquire the ability to transmigrate into (intravasation) or out of (extravasation) a dense network of lymphatic vessels. These processes are in principle thought to be perpetuated by an underlying baseline distortion of tumor-associated lymphatic vessels, wherein enlarged LEC spaces and a discontinuous basement membrane increase permeability and facilitate CRC cell transit. Likewise, vascular permeability and intravasation can be further modulated by the HIFα-dependent induction of growth and motogenic factors such as platelet-derived growth factor (PDGF)B, vascular growth factor (VEGF)A, angiopoietin (ANGPT)2, and C-X-C motif chemokine ligand (CXCL) 12 (also known as SDF1) (Huang et al., 2009; Irigoyen et al., 2007; Semenza, 2013a; Schito et al., 2012), known to play key roles in the remodeling and invasion of tumor-associated lymphatic and/or blood vessels. A number of preclinical studies suggest that tumoral lymphangiogenesis, resulting from VEGFC- and VEGFD-dependent signaling upon VEGFR3, can promote lymphatic metastasis in autochthonous and immunocompromised mouse models of lung, prostate, pancreatic, *cervix*, and breast carcinomas (Burton et al., 2008; Chaudary et al., 2011; He et al., 2005; Mandriota et al., 2001; Skobe et al., 2001), whereas in human CRC biopsies, VEGFR3 expression has been associated with nodal metastases and poor prognosis (Garouniatis et al., 2013). As a result, these data might indicate that both increased lymphatic vessel density and permeability are responsible for nodal metastases. Notwithstanding, other hypoxia-dependent mechanisms not directly elicited by HIFα transactivity can support lymphatic metastasis; for instance, pancreatic, lung, and breast mouse tumors display hypoxic induction of VEGFC via internal ribosomal entry site-dependent mRNA translation, in a HIFα-independent manner (Morfoisse et al., 2014). From the translational standpoint, it is plausible that HIFα blockade might not offer durable clinical benefits upon metastatic control unless supplemented by the inhibition of VEGFC→VEGFR3 axis.

1.3.3 Anoikis, mitochondrial hyperpolarization, and lymphatic dissemination

Preclinical data from hematogeneous metastatic mouse models indicate that CCs halt within the microvasculature of secondary sites shortly after intravasation (Weiss et al., 1988), reflecting the amount of cellular stress imposed on circulating epithelial CCs. Likewise, a seminal study by Fidler showed that ≤0.2% of melanoma cells can survive for ≥14 days in the circulation and establish lung nodules, indicating that extravasation and proliferation at the secondary site are critical steps limiting the metastatic process (Fidler, 1970; Massagué &

Obenauf, 2016). This harsh microenvironmental challenge selects for metastasizing CCs that are able to survive within the lymphatic and blood circulation by inhibiting pathways that normally induce apoptosis upon loss of anchorage to the ECM (anoikis). Importantly, the anatomical features of blood vessels within the target organs (i.e., bone marrow, liver, brain, and lungs) and the molecular interactions between CCs and ECs (Massagué & Obenauf, 2016; Shibue & Weinberg, 2011) are two main determinants of microvascular CC arrest. Mechanistically, anchorage of CCs to the luminal surface of ECs typically resembles the "rolling over" process observed in activated leukocytes during margination, a critical step preceding extravasation whenever these cells pursue pathogenic microorganisms (Simon & Green, 2005). Furthermore, one study revealed that inhibition of cellular respiration and oxidative phosphorylation can lead to mitochondrial hyperpolarization, increased invasiveness, and survival in nonadherent two-dimensional and three-dimensional cell culture conditions, thereby suggesting anoikis blockade in hypoxic CRCs (Heerdt et al., 2006). Of note, similar mechanisms seem to be at play in gastric CCs and during mammary gland morphogenesis, wherein ROS-dependent HIFα stabilization and EGFR→MAPK→ERK signaling pathway can promote anoikis resistance under hypoxia (Rohwer et al., 2008; Whelan et al., 2010, 2013).

1.3.4 Lymphatic niche formation and clonal expansion

The metastatic niche hypothesis postulates that primary tumor cells are able to precondition distant site microenvironments favoring metastatic colonization, proliferation, and growth [reviewed in Peinado et al. (2017)]. These "premetastatic niches" depend on tumor-secreted factors, such as angiogenic factors, cytokines, and chemoattractants that enable the mobilization and recruitment of tissue resident or circulating supporting cells into premetastatic sites, ultimately leading to the generation of a permissive niche (Hiratsuka et al., 2006; Kaplan et al., 2005; Peinado et al., 2017). Specifically, the CXCL1→CXCR2 axis, significantly upregulated in CRCs, has been implicated in myeloid-derived suppressor cell (MDSC) homing to the colonic mucosa and tumors (Katoh et al., 2013; Ogata et al., 2010; Rubie et al., 2008) and found to facilitate hepatic premetastatic niche formation during CRC progression (Wang et al., 2017). Interestingly, this process is regulated by tumor-derived VEGFA that induces CXCL1 in tumor-associated macrophages, thereby recruiting $CXCR2^+$ MDSCs from the circulation into the premetastatic liver (Wang et al., 2017). Taken together, these data are consistent with a role of hypoxic signaling in HIFα→VEGFA→CXCL1→CXCR2 premetastatic niche formation.

Lymphatic remodeling in draining lymph nodes has been shown to precede CC arrival via lymphangiogenic factors released by primary CCs and/or the tumor stroma, likely representing conditioning steps for premetastatic "lymphovascular niches" (Commerford et al., 2018; Hirakawa et al., 2007; Ogawa et al., 2014; Olmeda et al., 2017). Indeed, the configuration of a premetastatic niche in tumor-draining lymph nodes relies upon LEC-derived CXCL12 secretion and recruitment of $CXCR4^+$ cells [reviewed in Rey and Semenza (2010)]. These cellular processes provide a molecular mechanism implying a direct role for hypoxia-induced HIFα→CXCL12→CXCR4 signaling in both the acquisition of invasive abilities in CRC cells (Hongo et al., 2013), and the establishment of a permissive premetastatic niche for lymphatic CRC cell dissemination.

1.4 Therapeutic perspectives

Integration of the experimental and clinical data heretofore summarized indicates remarkable similarities among the molecular mechanisms co-opted by CRC cells during lymphatic and hematogenous dissemination, while suggesting hypoxia and HIFα signaling as common drivers underlying these processes. As a result, HIFα targeting might serve a dual effect upon CRC progression and dissemination by blocking shared lymphatic and hematogenous pathways such as VEGFA, CXCL12, or PDGFB, and/or specific lymphatic signaling such as VEGFC/D→VEGFR3 (Schito et al., 2012; Schito & Rey, 2020; Stacker et al., 2014). Interestingly, recent preclinical studies suggest low-dose metronomic chemotherapy as a feasible strategy to forestall lymphatic metastatic signaling and dissemination in hypoxic CRCs by replacing conventional maximum-tolerated dose approaches, known to worsen intratumoral hypoxia and induce HIFα (Schito et al., 2020). Whether HIFα targeting through this modality should be attempted on its own or in combination with specific HIFα inhibitors remains as a central outstanding question in the field.

Acknowledgments

The authors regret that several relevant papers could not be cited due to space limitations. LS and SR are Fellows of the UCD Conway Institute of Biomolecular and Biomedical Research, supported by the UCD Ad Astra Fellows Programme, University College Dublin, Ireland.

Conflicts of interest

The authors declare that no conflicts of interest exist.

References

Baba, Y., Nosho, K., Shima, K., Irahara, N., Chan, A. T., Meyerhardt, J. A., Chung, D. C., Giovannucci, E. L., Fuchs, C. S., & Ogino, S. (2010). HIF1A overexpression is associated with poor prognosis in a cohort of 731 colorectal cancers. *American Journal of Pathology*, 176(5), 2292–2301. Available from https://doi.org/10.2353/ajpath.2010.090972.

Batlle, E., Sancho, E., Francí, C., Domínguez, D., Monfar, M., Baulida, J., & De Herreros, A. G. (2000). The transcription factor Snail is a repressor of E-cadherin gene expression in epithelial tumour cells. *Nature Cell Biology*, 2(2), 84–89. Available from https://doi.org/10.1038/35000034.

Boonstra, M. C., Verbeek, F. P. R., Mazar, A. P., Prevoo, H. A. J. M., Kuppen, P. J. K., Van de Velde, C. J. H., Vahrmeijer, A. L., & Sier, C. F. M. (2014). Expression of uPAR in tumor-associated stromal cells is associated with colorectal cancer patient prognosis: A TMA study. *BMC Cancer*, 14(1). Available from https://doi.org/10.1186/1471-2407-14-269.

Bossi, P., Coggi, G., Viale, G., Viale, G., Alfano, R. M., Lee, A. K. C., Lee, A. K. C., & Bosari, S. (1995). Angiogenesis in colorectal tumors: Microvessel quantitation in adenomas and carcinomas with clinicopathological correlations. *Cancer Research*, 55(21), 5049–5053.

Brabletz, T., Jung, A., Reu, S., Porzner, M., Hlubek, F., Kunz-Schughart, L. A., Knuechel, R., & Kirchner, T. (2001). Variable β-catenin expression in colorectal cancers indicates tumor progression driven by the tumor environment. *Proceedings of the National Academy of Sciences of the United States of America*, 98(18), 10356–10361. Available from https://doi.org/10.1073/pnas.171610498.

Brouwer, N. P. M. (2020). More extensive lymphadenectomy in colon cancer; how far are we willing to go for a biomarker? *Techniques in Coloproctology*, *24*, 761−764.

Bruick, R. K., & McKnight, S. L. (2001). A conserved family of prolyl-4-hydroxylases that modify HIF. *Science (New York, N.Y.)*, *294*(5545), 1337−1340. Available from https://doi.org/10.1126/science.1066373.

Burton, J. B., Priceman, S. J., Sung, J. L., Brakenhielm, E., Dong, S. A., Pytowski, B., Alitalo, K., & Wu, L. (2008). Suppression of prostate cancer nodal and systemic metastasis by blockade of the lymphangiogenic axis. *Cancer Research*, *68*(19), 7828−7837. Available from https://doi.org/10.1158/0008-5472.CAN-08-1488.

Cao, D., Hou, M., Guan, Y. S., Jiang, M., Yang, Y., & Gou, H. F. (2009). Expression of HIF-1alpha and VEGF in colorectal cancer: Association with clinical outcomes and prognostic implications. *BMC Cancer*, *9*. Available from https://doi.org/10.1186/1471-2407-9-432.

Cao, Y., Eble, J. M., Moon, E., Yuan, H., Weitzel, D. H., Landon, C. D., Nien, C. Y.-C., Hanna, G., Rich, J. N., Provenzale, J. M., & Dewhirst, M. W. (2013). Tumor cells upregulate normoxic HIF-1α in response to doxorubicin. *Cancer Research*, *73*(20), 6230−6242. Available from https://doi.org/10.1158/0008-5472.CAN-12-1345.

Chang, G. J., Rodriguez-Bigas, M. A., Skibber, J. M., & Moyer, V. A. (2007). Lymph node evaluation and survival after curative resection of colon cancer: Systematic review. *Journal of the National Cancer Institute*, *99*(6), 433−441. Available from https://doi.org/10.1093/jnci/djk092.

Chaudary, N., Milosevic, M., & Hill, R. P. (2011). Suppression of vascular endothelial growth factor receptor 3 (VEGFR3) and vascular endothelial growth factor C (VEGFC) inhibits hypoxia-induced lymph node metastasis in cervix cancer. *Gynecologic Oncology*, *123*(2), 393−400. Available from https://doi.org/10.1016/j.ygyno.2011.07.006.

Chien, C. W., Lin, S. C., Lai, Y. Y., Lin, B. W., Lin, S. C., Lee, J. C., & Tsai, S. J. (2008). Regulation of CD151 by hypoxia controls cell adhesion and metastasis in colorectal cancer. *Clinical Cancer Research*, *14*(24), 8043−8051. Available from https://doi.org/10.1158/1078-0432.CCR-08-1651.

Cho, Y. B., Lee, W. Y., Song, S. Y., Shin, H. J., Yun, S. H., & Chun, H. K. (2007). Matrix metalloproteinase-9 activity is associated with poor prognosis in T3-T4 node-negative colorectal cancer. *Human Pathology*, *38*(11), 1603−1610. Available from https://doi.org/10.1016/j.humpath.2007.03.018.

Choi, B. J., Park, S. A., Lee, S. Y., Cha, Y. N., & Surh, Y. J. (2017). Hypoxia induces epithelial-mesenchymal transition in colorectal cancer cells through ubiquitin-specific protease 47-mediated stabilization of snail: A potential role of Sox9. *Scientific Reports*, *7*(1). Available from https://doi.org/10.1038/s41598-017-15139-5.

Cockman, M. E., Masson, N., Mole, D. R., Jaakkola, P., Chang, G. W., Clifford, S. C., Maher, E. R., Pugh, C. W., Ratcliffe, P. J., & Maxwell, P. H. (2000). Hypoxia inducible factor-α binding and ubiquitylation by the von Hippel-Lindau tumor suppressor protein. *Journal of Biological Chemistry*, *275*(33), 25733−25741. Available from https://doi.org/10.1074/jbc.M002740200.

Commerford, C. D., Dieterich, L. C., He, Y., Hell, T., Montoya-Zegarra, J. A., Noerrelykke, S. F., Russo, E., Röcken, M., & Detmar, M. (2018). Mechanisms of tumor-induced lymphovascular niche formation in draining lymph nodes. *Cell Reports*, *25*(13), 3554−3563. Available from https://doi.org/10.1016/j.celrep.2018.12.002, e4.

Conti, J., & Thomas, G. (2011). The role of tumour stroma in colorectal cancer invasion and metastasis. *Cancers*, *3*(2), 2160−2168. Available from https://doi.org/10.3390/cancers3022160.

Detchokul, S., Williams, E. D., Parker, M. W., & Frauman, A. G. (2014). Tetraspanins as regulators of the tumour microenvironment: Implications for metastasis and therapeutic strategies. *British Journal of Pharmacology*, *171*(24), 5462−5490. Available from https://doi.org/10.1111/bph.12260.

Di Renzo, M. F., Olivero, M., Giacomini, A., Porte, H., Chastre, E., Mirossay, L., Nordlinger, B., Bretti, S., Bottardi, S., & Giordano, S. (1995). Overexpression and amplification of the met/HGF receptor gene during the progression of colorectal cancer. *Clinical Cancer Research: An Official Journal of the American Association for Cancer Research*, *1*(2), 147−154.

Dongre, A., & Weinberg, R. A. (2019). New insights into the mechanisms of epithelial−mesenchymal transition and implications for cancer. *Nature Reviews. Molecular Cell Biology*, *20*(2), 69−84. Available from https://doi.org/10.1038/s41580-018-0080-4.

Elia, I., Doglioni, G., & Fendt, S. M. (2018). Metabolic hallmarks of metastasis formation. *Trends in Cell Biology*, *28*(8), 673−684. Available from https://doi.org/10.1016/j.tcb.2018.04.002.

Epstein, A. C. R., Gleadle, J. M., McNeill, L. A., Hewitson, K. S., O'Rourke, J., Mole, D. R., Mukherji, M., Metzen, E., Wilson, M. I., Dhanda, A., Tian, Y. M., Masson, N., Hamilton, D. L., Jaakkola, P., Barstead, R., Hodgkin, J., Maxwell, P. H., Pugh, C. W., Schofield, C. J., & Ratcliffe, P. J. (2001). C. elegans EGL-9 and mammalian homologs define a family of dioxygenases that regulate HIF by prolyl hydroxylation. *Cell*, *107*(1), 43−54. Available from https://doi.org/10.1016/S0092-8674(01)00507-4.

Fang, Q.-X., Zheng, X.-C., & Zhao, H.-J. (2020). L1CAM is involved in lymph node metastasis via ERK1/2 signaling in colorectal cancer. *American Journal of Translational Research, 12*(3), 837−846.

Fearon, E. R., & Vogelstein, B. (1990). A genetic model for colorectal tumorigenesis. *Cell, 61*(5), 759−767. Available from https://doi.org/10.1016/0092-8674(90)90186-I.

Fidler, I. J. (1970). Metastasis: Quantitative analysis of distribution and fate of tumor emboli labeled with 125I-5-Iodo-2′-deoxyuridine. *Journal of the National Cancer Institute, 45*(4), 773−782. Available from https://doi.org/10.1093/jnci/45.4.773.

Fidler, I. J. (2003). The pathogenesis of cancer metastasis: The "seed and soil" hypothesis revisited. *Nature Reviews Cancer, 3*(6), 453−458. Available from https://doi.org/10.1038/nrc1098.

Flessner, M. F., Choi, J., Credit, K., Deverkadra, R., & Henderson, K. (2005). Resistance of tumor interstitial pressure to the penetration of intraperitoneally delivered antibodies into metastatic ovarian tumors. *Clinical Cancer Research, 11*(8), 3117−3125. Available from https://doi.org/10.1158/1078-0432.CCR-04-2332.

Fodde, R., Smits, R., & Clevers, H. (2001). APC, signal transduction and genetic instability in colorectal cancer. *Nature Reviews Cancer, 1*(1), 55−67. Available from https://doi.org/10.1038/35094067.

Furlan, D., Sahnane, N., Carnevali, I., Cerutti, R., Bertoni, F., Kwee, I., Uccella, S., Bertolini, V., Chiaravalli, A. M., & Capella, C. (2008). Up-regulation of the hypoxia-inducible factor-1 transcriptional pathway in colorectal carcinomas. *Human Pathology, 39*(10), 1483−1494. Available from https://doi.org/10.1016/j.humpath.2008.02.013.

Ganesh, K., Basnet, H., Kaygusuz, Y., Laughney, A. M., He, L., Sharma, R., O'Rourke, K. P., Reuter, V. P., Huang, Y.-H., Turkekul, M., Emrah, E., Masilionis, I., Manova-Todorova, K., Weiser, M. R., Saltz, L. B., Garcia-Aguilar, J., Koche, R., Lowe, S. W., Pe'er, D., ... Massagué, J. (2020). L1CAM defines the regenerative origin of metastasis-initiating cells in colorectal cancer. *Nature Cancer, 1*(1), 28−45. Available from https://doi.org/10.1038/s43018-019-0006-x.

Garouniatis, A., Zizi-Sermpetzoglou, A., Rizos, S., Kostakis, A., Nikiteas, N., & Papavassiliou, A. G. (2013). Vascular endothelial growth factor receptors 1,3 and caveolin-1 are implicated in colorectal cancer aggressiveness and prognosis - Correlations with epidermal growth factor receptor, CD44v6, focal adhesion kinase, and c-Met. *Tumor Biology, 34*(4), 2109−2117. Available from https://doi.org/10.1007/s13277-013-0776-1.

Gervasoni, J. E., Sbayi, S., & Cady, B. (2007). Role of lymphadenectomy in surgical treatment of solid tumors: An update on the clinical data. *Annals of Surgical Oncology, 14*(9), 2443−2462. Available from https://doi.org/10.1245/s10434-007-9360-5.

Giles, R. H., Lolkema, M. P., Snijckers, C. M., Belderbos, M., Van Der Groep, P., Mans, D. A., Van Beest, M., Van Noort, M., Goldschmeding, R., Van Diest, P. J., Clevers, H., & Voest, E. E. (2006). Interplay between VHL/HIF1α and Wnt/β-catenin pathways during colorectal tumorigenesis. *Oncogene, 25*(21), 3065−3070. Available from https://doi.org/10.1038/sj.onc.1209330.

Goethals, L., Debucquoy, A., Perneel, C., Geboes, K., Ectors, N., De Schutter, H., Penninckx, F., McBride, W. H., Begg, A. C., & Haustermans, K. M. (2006). Hypoxia in human colorectal adenocarcinoma: Comparison between extrinsic and potential intrinsic hypoxia markers. *International Journal of Radiation Oncology, Biology, Physics, 65*(1), 246−254. Available from https://doi.org/10.1016/j.ijrobp.2006.01.007.

Hanahan, D., & Weinberg, R. A. (2011). Hallmarks of cancer: The next generation. *Cell, 144*(5), 646−674. Available from https://doi.org/10.1016/j.cell.2011.02.013.

He, Y., Rajantie, I., Pajusola, K., Jeltsch, M., Holopainen, T., Yla-Herttuala, S., Harding, T., Jooss, K., Takahashi, T., & Alitalo, K. (2005). Vascular endothelial cell growth factor receptor 3-mediated activation of lymphatic endothelium is crucial for tumor cell entry and spread via lymphatic vessels. *Cancer Research, 65*(11), 4739−4746. Available from https://doi.org/10.1158/0008-5472.CAN-04-4576.

Heerdt, B. G., Houston, M. A., & Augenlicht, L. H. (2006). Growth properties of colonic tumor cells are a function of the intrinsic mitochondrial membrane potential. *Cancer Research, 66*(3), 1591−1596. Available from https://doi.org/10.1158/0008-5472.CAN-05-2717.

Hirakawa, S., Brown, L. F., Kodama, S., Paavonen, K., Alitalo, K., & Detmar, M. (2007). VEGF-C-induced lymphangiogenesis in sentinel lymph nodes promotes tumor metastasis to distant sites. *Blood, 109*(3), 1010−1017. Available from https://doi.org/10.1182/blood-2006-05-021758.

Hiratsuka, S., Watanabe, A., Aburatani, H., & Maru, Y. (2006). Tumour-mediated upregulation of chemoattractants and recruitment of myeloid cells predetermines lung metastasis. *Nature Cell Biology, 8*(12), 1369−1375. Available from https://doi.org/10.1038/ncb1507.

References

Hittelet, A., Legendre, H., Nagy, N., Bronckart, Y., Pector, J. C., Salmon, I., Yeaton, P., Gabius, H. J., Kiss, R., & Camby, I. (2003). Upregulation of galectins-1 and -3 in human colon cancer and their role in regulating cell migration. *International Journal of Cancer, 103*(3), 370–379. Available from https://doi.org/10.1002/ijc.10843.

Hongo, K., Tsuno, N. H., Kawai, K., Sasaki, K., Kaneko, M., Hiyoshi, M., Murono, K., Tada, N., Nirei, T., Sunami, E., Takahashi, K., Nagawa, H., Kitayama, J., & Watanabe, T. (2013). Hypoxia enhances colon cancer migration and invasion through promotion of epithelial-mesenchymal transition. *Journal of Surgical Research, 182*(1), 75–84. Available from https://doi.org/10.1016/j.jss.2012.08.034.

Huang, L. E., Arany, Z., Livingston, D. M., & Bunn, H. F. (1996). Activation of hypoxia-inducible transcription factor depends primarily upon redox-sensitive stabilization of its α subunit. *Journal of Biological Chemistry, 271*(50), 32253–32259. Available from https://doi.org/10.1074/jbc.271.50.32253.

Huang, Y., Song, N., Ding, Y., Yuan, S., Li, X., Cai, H., Shi, H., & Luo, Y. (2009). Pulmonary vascular destabilization in the premetastatic phase facilitates lung metastasis. *Cancer Research, 69*(19), 7529–7537. Available from https://doi.org/10.1158/0008-5472.CAN-08-4382.

Hugen, N., van de Velde, C. J. H., de Wilt, J. H. W., & Nagtegaal, I. D. (2014). Metastatic pattern in colorectal cancer is strongly influenced by histological subtype. *Annals of Oncology: Official Journal of the European Society for Medical Oncology, 25*(3), 651–657. Available from https://doi.org/10.1093/annonc/mdt591.

Imamura, T., Kikuchi, H., Herraiz, M. T., Park, D. Y., Mizukami, Y., Mino-Kenduson, M., Lynch, M. P., Rueda, B. R., Benita, Y., Xavier, R. J., & Chung, D. C. (2009). HIF-1α and HIF-2α have divergent roles in colon cancer. *International Journal of Cancer, 124*(4), 763–771. Available from https://doi.org/10.1002/ijc.24032.

Irigoyen, M., Ansó, E., Martínez, E., Garayoa, M., Martínez-Irujo, J. J., & Rouzaut, A. (2007). Hypoxia alters the adhesive properties of lymphatic endothelial cells. A transcriptional and functional study. *Biochimica et Biophysica Acta - Molecular Cell Research, 1773*(6), 880–890. Available from https://doi.org/10.1016/j.bbamcr.2007.03.001.

Jung, G., Hernández-Illán, E., Moreira, L., Balaguer, F., & Goel, A. (2020). Epigenetics of colorectal cancer: Biomarker and therapeutic potential. *Nature Reviews Gastroenterology and Hepatology, 17*(2), 111–130. Available from https://doi.org/10.1038/s41575-019-0230-y.

Kaelin, W. G., & Ratcliffe, P. J. (2008). Oxygen sensing by metazoans: The central role of the HIF hydroxylase pathway. *Molecular Cell, 30*(4), 393–402. Available from https://doi.org/10.1016/j.molcel.2008.04.009.

Kaidi, A., Williams, A. C., & Paraskeva, C. (2007). Interaction between β-catenin and HIF-1 promotes cellular adaptation to hypoxia. *Nature Cell Biology, 9*(2), 210–217. Available from https://doi.org/10.1038/ncb1534.

Kanazawa, A., Oshima, T., Yoshihara, K., Tamura, S., Yamada, T., Inagaki, D., Sato, T., Yamamoto, N., Shiozawa, M., Morinaga, S., Akaike, M., Kunisaki, C., Tanaka, K., Masuda, M., & Imada, T. (2010). Relation of MT1-MMP gene expression to outcomes in colorectal cancer. *Journal of Surgical Oncology, 102*(6), 571–575. Available from https://doi.org/10.1002/jso.21703.

Kaplan, R. N., Riba, R. D., Zacharoulis, S., Bramley, A. H., Vincent, L., Costa, C., MacDonald, D. D., Jin, D. K., Shido, K., Kerns, S. A., Zhu, Z., Hicklin, D., Wu, Y., Port, J. L., Altorki, N., Port, E. R., Ruggero, D., Shmelkov, S. V., Jensen, K. K., & Lyden, D. (2005). VEGFR1-positive haematopoietic bone marrow progenitors initiate the pre-metastatic niche. *Nature, 438*(7069), 820–827. Available from https://doi.org/10.1038/nature04186.

Katoh, H., Wang, D., Daikoku, T., Sun, H., Dey, S. K., & DuBois, R. N. (2013). CXCR2-expressing myeloid-derived suppressor cells are essential to promote colitis-associated tumorigenesis. *Cancer Cell, 24*(5), 631–644. Available from https://doi.org/10.1016/j.ccr.2013.10.009.

Keum, N. N., & Giovannucci, E. (2019). Global burden of colorectal cancer: emerging trends, risk factors and prevention strategies. *Nature Reviews Gastroenterology and Hepatology, 16*(12), 713–732. Available from https://doi.org/10.1038/s41575-019-0189-8.

Kim, Y. H., Kim, G., Kwon, C. I., Kim, J. W., Park, P. W., & Hahm, K. B. (2014). TWIST1 and SNAI1 as markers of poor prognosis in human colorectal cancer are associated with the expression of ALDH1 and TGF-β1. *Oncology Reports, 31*(3), 1380–1388. Available from https://doi.org/10.3892/or.2014.2970.

Knijn, N., Mogk, S. C., Teerenstra, S., Simmer, F., & Nagtegaal, I. D. (2016). Perineural invasion is a strong prognostic factor in colorectal cancer. *American Journal of Surgical Pathology, 40*(1), 103–112. Available from https://doi.org/10.1097/PAS.0000000000000518.

Knijn, N., van Exsel, U. E. M., de Noo, M. E., & Nagtegaal, I. D. (2018). The value of intramural vascular invasion in colorectal cancer – A systematic review and meta-analysis. *Histopathology, 72*(5), 721–728. Available from https://doi.org/10.1111/his.13404.

Krishnamachary, B., Berg-Dixon, S., Kelly, B., Agani, F., Feldser, D., Ferreira, G., Iyer, N., LaRusch, J., Pak, B., Taghavi, P., & Semenza, G. L. (2003). Regulation of colon carcinoma cell invasion by hypoxia-inducible factor 1. *Cancer Research*, *63*(5), 1138−1143.

Krishnamachary, B., Zagzag, D., Nagasawa, H., Rainey, K., Okuyama, H., Baek, J. H., & Semenza, G. L. (2006). Hypoxia-inducible factor-1-dependent repression of E-cadherin in von Hippel-Lindau tumor suppressor-null renal cell carcinoma mediated by TCF3, ZFHX1A, and ZFHX1B. *Cancer Research*, *66*(5), 2725−2731. Available from https://doi.org/10.1158/0008-5472.CAN-05-3719.

Kuwai, T., Kitadai, Y., Tanaka, S., Hiyama, T., Tanimoto, K., & Chayama, K. (2004). Mutation of the von Hippel-Lindau (VHL) gene in human colorectal carcinoma: Association with cytoplasmic accumulation of hypoxia-inducible factor (HIF)-1α. *Cancer Science*, *95*(2), 149−153. Available from https://doi.org/10.1111/j.1349-7006.2004.tb03196.x.

Langley, R. R., & Fidler, I. J. (2011). The seed and soil hypothesis revisited-The role of tumor-stroma interactions in metastasis to different organs. *International Journal of Cancer*, *128*(11), 2527−2535. Available from https://doi.org/10.1002/ijc.26031.

Le Voyer, T. E., Sigurdson, E. R., Hanlon, A. L., Mayer, R. J., Macdonald, J. S., Catalano, P. J., & Haller, D. G. (2003). Colon cancer survival is associated with increasing number of lymph nodes analyzed: A secondary survey of intergroup trial INT-0089. *Journal of Clinical Oncology*, *21*(15), 2912−2919. Available from https://doi.org/10.1200/JCO.2003.05.062.

Lewis, M. R., Deavers, M. T., Silva, E. G., & Malpica, A. (2006). Ovarian involvement by metastatic colorectal adenocarcinoma: Still a diagnostic challenge. *The American Journal of Surgical Pathology*, *30*(2), 177−184. Available from https://doi.org/10.1097/01.pas.0000176436.26821.8a.

Liang, P., Nakada, I., Hong, J.-W., Tabuchi, T., Motohashi, G., Takemura, A., Nakachi, T., Kasuga, T., & Tabuchi, T. (2007). Prognostic significance of immunohistochemically detected blood and lymphatic vessel invasion in colorectal carcinoma: Its impact on prognosis. *Annals of Surgical Oncology*, *14*(2), 470−477. Available from https://doi.org/10.1245/s10434-006-9189-3.

Lin, M., Ma, S.-P., Lin, H.-Z., Ji, P., Xie, D., & Yu, J.-X. (2010). Intratumoral as well as peritumoral lymphatic vessel invasion correlates with lymph node metastasis and unfavourable outcome in colorectal cancer. *Clinical & Experimental Metastasis*, *27*(3), 123−132. Available from https://doi.org/10.1007/s10585-010-9309-0.

Maisel, K., Sasso, M. S., Potin, L., & Swartz, M. A. (2017). Exploiting lymphatic vessels for immunomodulation: Rationale, opportunities, and challenges. *Advanced Drug Delivery Reviews*, *114*, 43−59. Available from https://doi.org/10.1016/j.addr.2017.07.005.

Mandriota, S. J., Jussila, L., Jeltsch, M., Compagni, A., Baetens, D., Prevo, R., Banerji, S., Huarte, J., Montesano, R., Jackson, D. G., Orci, L., Alitalo, K., Christofori, G., & Pepper, M. S. (2001). Vascular endothelial growth factor-C-mediated lymphangiogenesis promotes tumour metastasis. *EMBO Journal*, *20*(4), 672−682. Available from https://doi.org/10.1093/emboj/20.4.672.

Martin, J. D., Seano, G., & Jain, R. K. (2019). Normalizing Function of tumor vessels: Progress, opportunities, and challenges. *Annual Review of Physiology*, *81*, 505−534. Available from https://doi.org/10.1146/annurev-physiol-020518-114700.

Massagué, J., & Obenauf, A. C. (2016). Metastatic colonization by circulating tumour cells. *Nature*, *529*(7586), 298−306. Available from https://doi.org/10.1038/nature17038.

Maxwell, P. H., Wiesener, M. S., Chang, G. W., Clifford, S. C., Vaux, E. C., Cockman, M. E., Wykoff, C. C., Pugh, C. W., Maher, E. R., & Ratcliffe, P. J. (1999). The tumour suppressor protein VHL targets hypoxia-inducible factors for oxygen-dependent proteolysis. *Nature*, *399*(6733), 271−275. Available from https://doi.org/10.1038/20459.

Mira, A., Morello, V., Céspedes, M. V., Perera, T., Comoglio, P. M., Mangues, R., & Michieli, P. (2017). Stroma-derived HGF drives metabolic adaptation of colorectal cancer to angiogenesis inhibitors. *Oncotarget*, *8*(24), 38193−38213. Available from https://doi.org/10.18632/oncotarget.16942.

Mohri, Y. (1997). Prognostic significance of E-cadherin expression in human colorectal cancer tissue. *Surgery Today*, *27*(7), 606−612. Available from https://doi.org/10.1007/BF02388215.

Morfoisse, F., Kuchnio, A., Frainay, C., Gomez-Brouchet, A., Delisle, M. B., Marzi, S., Helfer, A. C., Hantelys, F., Pujol, F., Guillermet-Guibert, J., Bousquet, C., Dewerchin, M., Pyronnet, S., Prats, A. C., Carmeliet, P., & Garmy-Susini, B. (2014). Hypoxia induces VEGF-C expression in metastatic tumor cells via a HIF-1α-independent translation-mediated mechanism. *Cell Reports*, *6*(1), 155−167. Available from https://doi.org/10.1016/j.celrep.2013.12.011.

Morin, P. J., Sparks, A. B., Korinek, V., Barker, N., Clevers, H., Vogelstein, B., & Kinzler, K. W. (1997). Activation of beta-catenin-Tcf signaling in colon cancer by mutations in beta-catenin or APC. *Science (New York, N.Y.)*, 275 (5307), 1787–1790.

Muñoz-Nájar, U. M., Neurath, K. M., Vumbaca, F., & Claffey, K. P. (2006). Hypoxia stimulates breast carcinoma cell invasion through MT1-MMP and MMP-2 activation. *Oncogene*, 25(16), 2379–2392. Available from https://doi.org/10.1038/sj.onc.1209273.

Naxerova, K., Reiter, J. G., Brachtel, E., Lennerz, J. K., Van De Wetering, M., Rowan, A., Cai, T., Clevers, H., Swanton, C., Nowak, M. A., Elledge, S. J., & Jain, R. K. (2017). Origins of lymphatic and distant metastases in human colorectal cancer. *Science (New York, N.Y.)*, 357(6346). Available from https://doi.org/10.1126/science.aai8515.

Ogata, H., Sekikawa, A., Yamagishi, H., Ichikawa, K., Tomita, S., Imura, J., Ito, Y., Fujita, M., Tsubaki, M., Kato, H., Fujimori, T., & Fukui, H. (2010). GROα promotes invasion of colorectal cancer cells. *Oncology Reports*, 24(6), 1479–1486. Available from https://doi.org/10.3892/or-00001008.

Ogawa, F., Amano, H., Eshima, K., Ito, Y., Matsui, Y., Hosono, K., Kitasato, H., Iyoda, A., Iwabuchi, K., Kumagai, Y., Satoh, Y., Narumiya, S., & Majima, M. (2014). Prostanoid induces premetastatic niche in regional lymph nodes. *Journal of Clinical Investigation*, 124(11), 4882–4894. Available from https://doi.org/10.1172/JCI73530.

Olmeda, D., Cerezo-Wallis, D., Riveiro-Falkenbach, E., Pennacchi, P. C., Contreras-Alcalde, M., Ibarz, N., Cifdaloz, M., Catena, X., Calvo, T. G., Cañón, E., Alonso-Curbelo, D., Suarez, J., Osterloh, L., Graña, O., Mulero, F., Megías, D., Cañamero, M., Martínez-Torrecuadrada, J. L., Mondal, C., & Soengas, M. S. (2017). Whole-body imaging of lymphovascular niches identifies pre-metastatic roles of midkine. *Nature*, 546(7660), 676–680. Available from https://doi.org/10.1038/nature22977.

Padera, T. P., Kadambi, A., Di Tomaso, E., Mouta Carreira, C., Brown, E. B., Boucher, Y., Choi, N. C., Mathisen, D., Wain, J., Mark, E. J., Munn, L. L., & Jain, R. K. (2002). Lymphatic metastasis in the absence of functional intratumor lymphatics. *Science (New York, N.Y.)*, 296(5574), 1883–1886. Available from https://doi.org/10.1126/science.1071420.

Park, J. H., Richards, C. H., McMillan, D. C., Horgan, P. G., & Roxburgh, C. S. D. (2014). The relationship between tumour stroma percentage, the tumour microenvironment and survival in patients with primary operable colorectal cancer. *Annals of Oncology*, 25(3), 644–651. Available from https://doi.org/10.1093/annonc/mdt593.

Park, S., Ahn, H. K., Lee, D. H., Jung, Y., Jeong, J.-W., Nam, S., & Lee, W.-S. (2019). Systematic mutation analysis in rare colorectal cancer presenting ovarian metastases. *Scientific Reports*, 9(1), 16990. Available from https://doi.org/10.1038/s41598-019-53182-6.

Parsons, H. M., Tuttle, T. M., Kuntz, K. M., Begun, J. W., McGovern, P. M., & Virnig, B. A. (2011). Association between lymph node evaluation for colon cancer and node positivity over the past 20 years. *JAMA - Journal of the American Medical Association*, 306(10), 1089–1097. Available from https://doi.org/10.1001/jama.2011.1285.

Pastushenko, I., & Blanpain, C. (2019). EMT transition states during tumor progression and metastasis. *Trends in Cell Biology*, 29(3), 212–226. Available from https://doi.org/10.1016/j.tcb.2018.12.001.

Peinado, H., Olmeda, D., & Cano, A. (2007). Snail, ZEB and bHLH factors in tumour progression: An alliance against the epithelial phenotype? *Nature Reviews Cancer*, 7(6), 415–428. Available from https://doi.org/10.1038/nrc2131.

Peinado, H., Zhang, H., Matei, I. R., Costa-Silva, B., Hoshino, A., Rodrigues, G., Psaila, B., Kaplan, R. N., Bromberg, J. F., Kang, Y., Bissell, M. J., Cox, T. R., Giaccia, A. J., Erler, J. T., Hiratsuka, S., Ghajar, C. M., & Lyden, D. (2017). Pre-metastatic niches: Organ-specific homes for metastases. *Nature Reviews Cancer*, 17(5), 302–317. Available from https://doi.org/10.1038/nrc.2017.6.

Pennacchietti, S., Michieli, P., Galluzzo, M., Mazzone, M., Giordano, S., & Comoglio, P. M. (2003). Hypoxia promotes invasive growth by transcriptional activation of the met protooncogene. *Cancer Cell*, 3(4), 347–361. Available from https://doi.org/10.1016/S1535-6108(03)00085-0.

Permanyer, M., Bošnjak, B., & Förster, R. (2018). Dendritic cells, T cells and lymphatics: Dialogues in migration and beyond. *Current Opinion in Immunology*, 53, 173–179. Available from https://doi.org/10.1016/j.coi.2018.05.004.

Petrella, B. L., Lohi, J., & Brinckerhoff, C. E. (2005). Identification of membrane type-1 matrix metalloproteinase as a target of hypoxia-inducible factor-2α in von Hippel-Lindau renal cell carcinoma. *Oncogene*, 24(6), 1043–1052. Available from https://doi.org/10.1038/sj.onc.1208305.

Powell, S. M., Zilz, N., Beazer-Barclay, Y., Bryan, T. M., Hamilton, S. R., Thibodeau, S. N., Vogelstein, B., & Kinzler, K. W. (1992). APC mutations occur early during colorectal tumorigenesis. *Nature*, 359(6392), 235–237. Available from https://doi.org/10.1038/359235a0.

Pyke, C., Kristensen, P., Ralfkiaer, E., Grondahl-Hansen, J., Eriksen, J., Blasi, F., & Dano, K. (1991). Urokinase-type plasminogen activator is expressed in stromal cells and its receptor in cancer cells at invasive foci in human colon adenocarcinomas. *American Journal of Pathology*, *138*(5), 1059−1067.

Rajaganeshan, R., Prasad, R., Guillou, P. J., Scott, N., Poston, G., & Jayne, D. G. (2009). Expression patterns of hypoxic markers at the invasive margin of colorectal cancers and liver metastases. *European Journal of Surgical Oncology*, *35*(12), 1286−1294. Available from https://doi.org/10.1016/j.ejso.2009.05.008.

Rasheed, S., Harris, A. L., Tekkis, P. P., Turley, H., Silver, A., McDonald, P. J., Talbot, I. C., Glynne-Jones, R., Northover, J. M. A., & Guenther, T. (2009). Hypoxia-inducible factor-1alpha and -2alpha are expressed in most rectal cancers but only hypoxia-inducible factor-1alpha is associated with prognosis. *British Journal of Cancer*, *100*(10), 1666−1673. Available from https://doi.org/10.1038/sj.bjc.6605026.

Rey, S., & Semenza, G. L. (2010). Hypoxia-inducible factor-1-dependent mechanisms of vascularization and vascular remodelling. *Cardiovascular Research*, *86*(2), 236−242. Available from https://doi.org/10.1093/cvr/cvq045.

Rey, S., Schito, L., Wouters, B. G., Eliasof, S., & Kerbel, R. S. (2017). Targeting hypoxia-inducible factors for antiangiogenic cancer therapy. *Trends in Cancer*, *3*(7), 529−541. Available from https://doi.org/10.1016/j.trecan.2017.05.002.

Righi, A., Sarotto, I., Casorzo, L., Cavalchini, S., Frangipane, E., & Risio, M. (2015). Tumour budding is associated with hypoxia at the advancing front of colorectal cancer. *Histopathology*, *66*(7), 982−990. Available from https://doi.org/10.1111/his.12602.

Rohwer, N., Welzel, M., Daskalow, K., Pfander, D., Wiedenmann, B., Detjen, K., & Cramer, T. (2008). Hypoxia-inducible factor 1α mediates anoikis resistance via suppression of α5 integrin. *Cancer Research*, *68*(24), 10113−10120. Available from https://doi.org/10.1158/0008-5472.CAN-08-1839.

Rubie, C., Frick, V., Wagner, M., Schuld, J., Gräber, S., Brittner, B., Bohle, R. M., & Schilling, M. K. (2008). ELR + CXC chemokine expression in benign and malignant colorectal conditions. *BMC Cancer*, *8*. Available from https://doi.org/10.1186/1471-2407-8-178.

Salem, N., Kamal, I., Al-Maghrabi, J., Abuzenadah, A., Peer-Zada, A. A., Qari, Y., Al-Ahwal, M., Al-Qahtani, M., & Buhmeida, A. (2016). High expression of matrix metalloproteinases: MMP-2 and MMP-9 predicts poor survival outcome in colorectal carcinoma. *Future Oncology*, *12*(3), 323−331. Available from https://doi.org/10.2217/fon.15.325.

Samanta, D., Gilkesa, D. M., Chaturvedia, P., Xiang, L., & Semenza, G. L. (2014). Hypoxia-inducible factors are required for chemotherapy resistance of breast cancer stem cells. *Proceedings of the National Academy of Sciences of the United States of America*, *111*(50), E5429−E5438. Available from https://doi.org/10.1073/pnas.1421438111.

Sandberg, T. P., Oosting, J., van Pelt, G. W., Mesker, W. E., Tollenaar, R. A. E. M., & Morreau, H. (2018). Molecular profiling of colorectal tumors stratified by the histological tumor-stroma ratio - Increased expression of galectin-1 in tumors with high stromal content. *Oncotarget*, *9*(59), 31502−31515. Available from https://doi.org/10.18632/oncotarget.25845.

Sanjuán, X., Fernández, P. L., Castells, A., Castronovo, V., van den Brule, F., Liu, F. T., Cardesa, A., & Campo, E. (1997). Differential expression of galectin 3 and galectin 1 in colorectal cancer progression. *Gastroenterology*, *113*(6), 1906−1915.

Schito, L. (2019). Hypoxia-dependent angiogenesis and lymphangiogenesis in cancer. In Gilkes, D. (Ed.), *Advances in Experimental Medicineand Biology* (vol. 1136, pp. 71−85). New York LLC: Springer. Available from https://doi.org/10.1007/978-3-030-12734-3_5.

Schito, L., & Rey, S. (2018). Cell-autonomous metabolic reprogramming in hypoxia. *Trends in Cell Biology*, *28*(2), 128−142. Available from https://doi.org/10.1016/j.tcb.2017.10.006.

Schito, L., & Rey, S. (2020). Hypoxia: Turning vessels into vassals of cancer immunotolerance. *Cancer Letters*, *487*, 74−84. Available from https://doi.org/10.1016/j.canlet.2020.05.015.

Schito, L., & Semenza, G. L. (2016). Hypoxia-inducible factors: Master regulators of cancer progression. *Trends in Cancer*, *2*(12), 758−770. Available from https://doi.org/10.1016/j.trecan.2016.10.016.

Schito, L., Rey, S., Tafani, M., Zhang, H., Wong, C. C. L., Russo, A., Russo, M. A., & Semenza, G. L. (2012). Hypoxia-inducible factor 1-dependent expression of platelet-derived growth factor B promotes lymphatic metastasis of hypoxic breast cancer cells. *Proceedings of the National Academy of Sciences of the United States of America*, *109*(40), E2707−E2716. Available from https://doi.org/10.1073/pnas.1214019109.

References

Schito, L., Rey, S., Xu, P., Man, S., Cruz-Muñoz, W., & Kerbel, R. S. (2020). Metronomic chemotherapy offsets HIFα induction upon maximum-tolerated dose in metastatic cancers. *EMBO Molecular Medicine, 12*(9). Available from https://doi.org/10.15252/emmm.201911416.

Seguin, L., Desgrosellier, J. S., Weis, S. M., & Cheresh, D. A. (2015). Integrins and cancer: Regulators of cancer stemness, metastasis, and drug resistance. *Trends in Cell Biology, 25*(4), 234–240. Available from https://doi.org/10.1016/j.tcb.2014.12.006.

Semenza, G. L. (2008). Does loss of CD151 expression promote the metastasis of hypoxic colon cancer cells? *Clinical Cancer Research, 14*(24), 7969–7970. Available from https://doi.org/10.1158/1078-0432.CCR-08-2417.

Semenza, G. L. (2013a). Cancer-stromal cell interactions mediated by hypoxia-inducible factors promote angiogenesis, lymphangiogenesis, and metastasis. *Oncogene, 32*(35), 4057–4063. Available from https://doi.org/10.1038/onc.2012.578.

Semenza, G. L. (2013b). HIF-1 mediates metabolic responses to intratumoral hypoxia and oncogenic mutations. *Journal of Clinical Investigation, 123*(9), 3664–3671. Available from https://doi.org/10.1172/JCI67230.

Semenza, G. L., Jiang, B. H., Leung, S. W., Passantino, R., Concordat, J. P., Maire, P., & Giallongo, A. (1996). Hypoxia response elements in the aldolase A, enolase 1, and lactate dehydrogenase a gene promoters contain essential binding sites for hypoxia-inducible factor 1. *Journal of Biological Chemistry, 271*(51), 32529–32537. Available from https://doi.org/10.1074/jbc.271.51.32529.

Shibue, T., & Weinberg, R. A. (2011). Metastatic colonization: Settlement, adaptation and propagation of tumor cells in a foreign tissue environment. *Seminars in Cancer Biology, 21*(2), 99–106. Available from https://doi.org/10.1016/j.semcancer.2010.12.003.

Shieh, A. C. (2011). Biomechanical forces shape the tumor microenvironment. *Annals of Biomedical Engineering, 39*(5), 1379–1389. Available from https://doi.org/10.1007/s10439-011-0252-2.

Simon, S. I., & Green, C. E. (2005). Molecular mechanics and dynamics of leukocyte recruitment during inflammation. *Annual Review of Biomedical Engineering, 7*, 151–185. Available from https://doi.org/10.1146/annurev.bioeng.7.060804.100423.

Skobe, M., Hawighorst, T., Jackson, D. G., Prevo, R., Janes, L., Velasco, P., Riccardi, L., Alitalo, K., Claffey, K., & Detmar, M. (2001). Induction of tumor lymphangiogenesis by VEGF-C promotes breast cancer metastasis. *Nature Medicine, 7*(2), 192–198. Available from https://doi.org/10.1038/84643.

Stacker, S. A., Williams, S. P., Karnezis, T., Shayan, R., Fox, S. B., & Achen, M. G. (2014). Lymphangiogenesis and lymphatic vessel remodelling in cancer. *Nature Reviews Cancer, 14*(3), 159–172. Available from https://doi.org/10.1038/nrc3677.

Sundlisaeter, E. (2007). Lymphangiogenesis in colorectal cancer—prognostic and therapeutic aspects. *Int J Cancer, 121*, 1401–1409.

Sung, H., Ferlay, J., Siegel, R. L., Laversanne, M., Soerjomataram, I., Jemal, A., & Bray, F. (2021). Global cancer statistics 2020: GLOBOCAN Estimates of incidence and mortality worldwide for 36 cancers in 185 countries. *Cancer Journal for Clinicians, 71*(3), 209–249. Available from https://doi.org/10.3322/caac.21660.

Swartz, M. A. (2014). Immunomodulatory roles of lymphatic vessels in cancer progression. *Cancer Immunology Research, 2*(8), 701–707. Available from https://doi.org/10.1158/2326-6066.CIR-14-0115.

Tacchini, L., Dansi, P., Matteucci, E., & Desiderio, M. A. (2001). Hepatocyte growth factor signalling stimulates hypoxia inducible factor-1 (HIF-1) activity in HepG2 hepatoma cells. *Carcinogenesis, 22*(9), 1363–1371. Available from https://doi.org/10.1093/carcin/22.9.1363.

Takeuchi, H., Bilchik, A., Saha, S., Turner, R., Wiese, D., Tanaka, M., Kuo, C., Wang, H.-J., & Hoon, D. S. B. (2003). c-MET expression level in primary colon cancer: A predictor of tumor invasion and lymph node metastases. *Clinical Cancer Research: An Official Journal of the American Association for Cancer Research, 9*(4), 1480–1488.

Tang, Y. A., Chen, Y. F., Bao, Y., Mahara, S., Yatim, S. M. J. M., Oguz, G., Lee, P. L., Feng, M., Cai, Y., Tan, E. Y., Fong, S. S., Yang, Z. H., Lan, P., Wu, X. J., & Yu, Q. (2018). Hypoxic tumor microenvironment activates GLI2 via HIF-1α and TGF-β2 to promote chemoresistance in colorectal cancer. *Proceedings of the National Academy of Sciences of the United States of America, 115*(26), E5990–E5999. Available from https://doi.org/10.1073/pnas.1801348115.

Thiery, J. P., Acloque, H., Huang, R. Y. J., & Nieto, M. A. (2009). Epithelial-mesenchymal transitions in development and disease. *Cell, 139*(5), 871–890. Available from https://doi.org/10.1016/j.cell.2009.11.007.

Toiyama, Y., Yasuda, H., Saigusa, S., Tanaka, K., Inoue, Y., Goel, A., & Kusunoki, M. (2013). Increased expression of slug and vimentin as novel predictive biomarkers for lymph node metastasis and poor prognosis in colorectal cancer. *Carcinogenesis, 34*(11), 2548–2557. Available from https://doi.org/10.1093/carcin/bgt282.

Vermeulen, L., De Sousa, E., Melo, F., Van Der Heijden, M., Cameron, K., De Jong, J. H., Borovski, T., Tuynman, J. B., Todaro, M., Merz, C., Rodermond, H., Sprick, M. R., Kemper, K., Richel, D. J., Stassi, G., & Medema, J. P. (2010). Wnt activity defines colon cancer stem cells and is regulated by the microenvironment. *Nature Cell Biology*, *12*(5), 468–476. Available from https://doi.org/10.1038/ncb2048.

Vogelstein, B., & Kinzler, K. W. (2004). Cancer genes and the pathways they control. *Nature Medicine*, *10*(8), 789–799.

Vogelstein, B., Fearon, E. R., Hamilton, S. R., Kern, S. E., Preisinger, A. C., Leppert, M., Nakamura, Y., White, R., Smits, A. M., & Bos, J. L. (1988). Genetic alterations during colorectal-tumor development. *The New England Journal of Medicine*, *319*(9), 525–532. Available from https://doi.org/10.1056/NEJM198809013190901.

Wang, D., Sun, H., Wei, J., Cen, B., & DuBois, R. N. (2017). CXCL1 is critical for premetastatic niche formation and metastasis in colorectal cancer. *Cancer Research*, *77*(13), 3655–3665. Available from https://doi.org/10.1158/0008-5472.CAN-16-3199.

Wang, G. L., Jiang, B. H., Rue, E. A., & Semenza, G. L. (1995). Hypoxia-inducible factor 1 is a basic-helix-loop-helix-PAS heterodimer regulated by cellular O_2 tension. *Proceedings of the National Academy of Sciences of the United States of America*, *92*(12), 5510–5514. Available from https://doi.org/10.1073/pnas.92.12.5510.

Weinberg, R. A. (2008). Mechanisms of malignant progression. *Carcinogenesis*, *29*(6), 1092–1095. Available from https://doi.org/10.1093/carcin/bgn104.

Weiss, L., Orr, F. W., & Hohn, K. V. (1988). Interactions of cancer cells with the microvasculature during metastasis. *FASEB Journal*, *2*(1), 12–21. Available from https://doi.org/10.1096/fasebj.2.1.3275560.

West, N. P., Dattani, M., McShane, P., Hutchins, G., Grabsch, J., Mueller, W., Treanor, D., Quirke, P., & Grabsch, H. (2010). The proportion of tumour cells is an independent predictor for survival in colorectal cancer patients. *British Journal of Cancer*, *102*(10), 1519–1523. Available from https://doi.org/10.1038/sj.bjc.6605674.

Whelan, K. A., Caldwell, S. A., Shahriari, K. S., Jackson, S. R. E., Franchetti, L. D., Johannes, G. J., & Reginato, M. J. (2010). Hypoxia suppression of Bim and Bmf blocks anoikis and luminal clearing during mammary morphogenesis. *Molecular Biology of the Cell*, *21*(22), 3829–3837. Available from https://doi.org/10.1091/mbc.E10-04-0353.

Whelan, K. A., Schwab, L. P., Karakashev, S. V., Franchetti, L., Johannes, G. J., Seagroves, T. N., & Reginato, M. J. (2013). The oncogene HER2/neu (ERBB2) requires the hypoxia-inducible factor HIF-1 for mammary tumor growth and anoikis resistance. *Journal of Biological Chemistry*, *288*(22), 15865–15877. Available from https://doi.org/10.1074/jbc.M112.426999.

Xu, K., Zhan, Y., Yuan, Z., Qiu, Y., Wang, H., Fan, G., Wang, J., Li, W., Cao, Y., Shen, X., Zhang, J., Liang, X., & Yin, P. (2019). Hypoxia induces drug resistance in colorectal cancer through the HIF-1α/miR-338-5p/IL-6 feedback loop. *Molecular Therapy*, *27*(10), 1810–1824. Available from https://doi.org/10.1016/j.ymthe.2019.05.017.

Yang, J. L., Seetoo, D. Q., Wang, Y., Ranson, M., Berney, C. R., Ham, J. M., Russell, P. J., & Crowe, P. J. (2000). Urokinase-type plasminogen activator and its receptor in colorectal cancer: Independent prognostic factors of metastasis and cancer-specific survival and potential therapeutic targets. *International Journal of Cancer*, *89*(5), 431–439.

Yang, M. H., Wu, M. Z., Chiou, S. H., Chen, P. M., Chang, S. Y., Liu, C. J., Teng, S. C., & Wu, K. J. (2008). Direct regulation of TWIST by HIF-1α promotes metastasis. *Nature Cell Biology*, *10*(3), 295–305. Available from https://doi.org/10.1038/ncb1691.

Yeung, T. M., Buskens, C., Wang, L. M., Mortensen, N. J., & Bodmer, W. F. (2013). Myofibroblast activation in colorectal cancer lymph node metastases. *British Journal of Cancer*, *108*(10), 2106–2115. Available from https://doi.org/10.1038/bjc.2013.209.

Yu, S., Zhou, R., Yang, T., Liu, S., Cui, Z., Qiao, Q., & Zhang, J. (2019). Hypoxia promotes colorectal cancer cell migration and invasion in a SIRT1-dependent manner. *Cancer Cell International*, *19*(1). Available from https://doi.org/10.1186/s12935-019-0819-9.

Zeng, Z. S., Huang, Y., Cohen, A. M., & Guillem, J. G. (1996). Prediction of colorectal cancer relapse and survival via tissue RNA levels of matrix metalloproteinase-9. *Journal of Clinical Oncology*, *14*(12), 3133–3140. Available from https://doi.org/10.1200/JCO.1996.14.12.3133.

Zhang, W., Shi, X., Peng, Y., Wu, M., Zhang, P., Xie, R., Wu, Y., Yan, Q., Liu, S., & Wang, J. (2015). HIF-1α promotes epithelial-mesenchymal transition and metastasis through direct regulation of ZEB1 in colorectal cancer. *PLoS One*, *10*(6). Available from https://doi.org/10.1371/journal.pone.0129603.

Zhang, H., Wong, C. C. L., Wei, H., Gilkes, D. M., Korangath, P., Chaturvedi, P., Schito, L., Chen, J., Krishnamachary, B., Winnard, P. T., Jr., Raman, V., Zhen, L., Mitzner, W. A., Sukumar, S., & Semenza, G. L. (2012). HIF-1-dependent expression of angiopoietin-like 4 and L1CAM mediates vascular metastasis of hypoxic breast cancer cells to the lungs. *Oncogene*, *31*(14), 1757–1770. Available from https://doi.org/10.1038/onc.2011.365.

Zhao, X. Y., Chen, T. T., Xia, L., Guo, M., Xu, Y., Yue, F., Jiang, Y., Chen, G. Q., & Zhao, K. W. (2010). Hypoxia inducible factor-1 mediates expression of galectin-1: The potential role in migration/invasion of colorectal cancer cells. *Carcinogenesis*, *31*(8), 1367–1375. Available from https://doi.org/10.1093/carcin/bgq116.

Zhuang, Z., Emmert-Buck, M. R., Roth, M. J., Gnarra, J., Linehan, W. M., Liotta, L. A., & Lubensky, I. A. (1996). Von Hippel-Lindau disease gene deletion detected in microdissected sporadic human colon carcinoma specimens. *Human Pathology*, *27*(2), 152–156.

CHAPTER 2

Biomechanical aspects of the normal and cancer-associated lymphatic system

Wim P. Ceelen[1,2], Hooman Salavati[1,2,3], Ghazal Adeli Koudehi[3], Carlos Alejandro Silvera Delgado[3], Patrick Segers[3] and Charlotte Debbaut[2,3]

[1]Department of GI Surgery, Ghent University Hospital and Department of Human Structure and Repair, Ghent University, Ghent, Belgium [2]Cancer Research Institute Ghent (CRIG), Ghent University, Belgium [3]Institute for Biomedical Engineering and Technology (IBiTech)-Biofluid, Tissue and Solid Mechanics for Medical Applications (Biommeda), Ghent University, Belgium

2.1 Introduction

The development of cancer is an intrinsically dynamic process, which evolves in time and space. Biophysical phenomena such as fluid and cellular transport, growth-induced forces and pressures exerted and sensed, and other mechanical cues are increasingly recognized as essential drivers of the malignant phenotype, although they have traditionally received much less attention compared to the deciphering of the molecular and biological background of cancer. During the process of lymphatic metastasis, the fluid flow mechanics of the lymphatic circulatory system can be coopted to improve the efficiency of cancer cell transit from the primary tumor, extravasation, and metastatic seeding (Follain et al., 2020). Here, we review the biomechanical aspects of the normal lymphatic system and discuss the relevance of biophysical mechanisms in the process of lymphatic metastasis, with a focus on colorectal cancer (CRC).

2.2 Biomechanics of the normal lymphatic system

2.2.1 Mechanical properties of the interstitium

Vascularized tissue consists of cell populations and blood vessels embedded in a scaffold of structural proteins, glycoproteins, proteoglycans, and polysaccharides with distinct chemical and mechanical properties (Hynes, 2009). This extracellular matrix (ECM) or interstitium provides structural tissue support, determines the macroscopic mechanical tissue properties, and controls cellular activities including proliferation, differentiation, adhesion, migration, and survival. From a continuum perspective, tissue interstitium can be seen as a poroelastic medium, characterized by intrinsically coupled fluid and solid stresses (Wiig & Swartz, 2012). The tissue pressure gradients can cause interstitial fluid pressure (IFP) variations and flow, which exert both tangential (shear) stress and normal (drag) stresses on cells and ECM components. In addition, biomechanical stresses in the interstitium can also arise from intrinsic forces due to fibroblast contraction or tissue growth. Interstitial stress results in strain or deformation, the degree of which is determined by tissue properties such as matrix stiffness. Material stiffness is typically quantified by Young's modulus E, the slope of the stress versus strain relationship in the region of linear elastic behavior (typically for small deformations). Since most tissues are nonlinearly elastic or viscoelastic and often anisotropic due to the presence of fibrous constituents with dominant directions, Young's modulus is typically measured within a small linear range of stress/strain or described more accurately using different, more complex models taking into account the nonlinearities and anisotropy. Experimentally, tissue compliance, which is the inverse of stiffness, is usually measured. Several imaging-based techniques such as MR elastography and ultrasonic shear wave elastography are emerging tools for the clinical measurement of soft tissue stiffness (Bercoff et al., 2004; Fovargue et al., 2018).

2.2.2 Interstitial flow and lymph formation

Tissue fluid homeostasis is maintained by the balance between capillary filtration, venous reabsorption (90%), and lymphatic absorption (10%) (Huxley & Scallan, 2011). The lymphatic system, which is present in most organs of the body, collects and returns to the venous circulation an average of 5–6 L of extravasated fluid, peptides, proteins, long-chain fatty acids, fat-soluble vitamins, and immune cells per day. The composition and physicochemical properties of lymph depend on the organ or tissue it drains from, and on disease states such as inflammation and cancer (Hansen et al., 2015). The oxygen content of lymph is comparable to that of venous blood (partial oxygen pressure of 8–35 mmHg) (Hangai-Hoger et al., 2004). Lymph draining from the small intestine is termed chyle, which depending on dietary load contains 14–210 mmol/L of total fat, including neutral fat, free fatty acids, phospholipids, sphingomyelin, and cholesterol (Sriram et al., 2016). Lymph flow from the human large intestine amounts to approximately 0.015 mL/min/100 g (Alexander et al., 2010). The composition of lymph changes significantly in cancer tissue: in an animal model of breast cancer, proteins involved in cadherin-mediated endocytosis, acute phase response, junction signaling, gap junction, VEGF (vascular endothelial growth factor) signaling, and PI3K/AKT signaling pathways were overrepresented in the lymph from metastatic tumor-bearing animals compared to the lymph from nonmetastatic tumor-bearing animals (Mohammed et al., 2020).

Interstitial fluid flow with lymphatic drainage results from a pressure difference between the blood and lymphatic capillaries. The corresponding IFP gradient ∇P (representing the pressure difference over length) results in a fluid velocity vector $\vec{v}$, the magnitude of which depends on the hydraulic conductivity (K) of the interstitium, as described by Darcy's law (Eq. 2.1):

$$\vec{v} = -K\nabla P = -\frac{k}{\mu}\nabla P \, (\text{Darcy's law}) \quad (2.1)$$

The hydraulic conductivity K is determined by bulk tissue properties (such as material density, matrix fiber orientation, and electrochemical charge) and the interstitial fluid dynamic viscosity (μ). As shown in Eq. (2.1), the hydraulic conductivity can also be expressed as the tissue permeability (k) divided by the fluid dynamic viscosity (μ). Darcy's law is a first-order differential equation and does not account for viscous stresses. Although the latter effect might be limited, this can be addressed by using the Brinkman equation, adding a second-order viscous term to Darcy's law (rearranged form in Eq. 2.2) to describe the relation between interstitial pressure gradients and flow velocities as shown in Eq. (2.3) (Brinkman, 1949). Reported values of interstitial fluid velocity in normal tissue are in the range of 0.1–0.4 μm/s (Chary & Jain, 1989).

$$\nabla P = -\frac{\mu}{k}\vec{v} \, (\text{rearranged format of Darcy's law}) \quad (2.2)$$

$$\nabla P = -\frac{\mu}{k}\vec{v} + \mu\nabla^2\vec{v} \, (\text{Brinkman equation}) \quad (2.3)$$

Fluid extravasation from interstitial capillaries is determined by properties of the capillary wall, hydrostatic pressure, and protein concentrations in the blood and interstitium (leading to osmotic pressure differences) and described by the well-known Starling equation (Eq. 2.4) (Levick & Michel, 2010):

$$J_v = L_p \frac{S}{V}(\Delta P - \sigma\Delta\pi) \quad (2.4)$$

with J_v the fluid flux, L_p the permeability of the capillary wall, S/V the surface-area-to-volume ratio, ΔP the local hydrostatic pressure difference across the lymphatic capillary wall, $\Delta\pi$ the local osmotic pressure difference across the lymphatic capillary wall, and σ the capillary osmotic reflection coefficient.

However, the Starling principle has recently been revised based on several observations, among them the (incorrect) assumption by Starling that the microvasculature is impermeable to plasma proteins, whereas in fact postcapillary venules do allow protein plasma leakage (Levick & Michel, 2010). Lymphatic flow velocity is low, in the order of 0.02–1 mm/s (on average, but with peak values up to 9 mm/s), with Reynolds number far below one ($3.3 \cdot 10^{-5}$–$2 \cdot 10^{-3}$) (Follain et al., 2020). Such low velocities and Reynolds numbers mean that the flow in lymphatic vessels with a diameter smaller than 100 μm and in the thoracic duct is always laminar with low shear stresses exerted on the lymphatic endothelium (0.064–1.2 Pa).

Importantly, while the geometry of conducting channels and tissue conductivity determine the magnitude of interstitial fluid flow, the reverse phenomenon also exists: interstitial flow has important effects on tissue morphogenesis and function, cell migration and differentiation, and matrix remodeling (Rutkowski & Swartz, 2007). The mechanisms that govern the effect of interstitial flow can be purely mechanical, involving shear stress on the cell surface, pressure forces pushing on cells, or tethering forces on cell–matrix connections. Also, changes in interstitial flow may affect the spatial distribution and gradients of cytokines and chemokines such as VEGF-C (Goldman et al., 2007). In addition, increased flow or pressure may mobilize interstitial stores of transforming growth factor β (TGF-β), which drives myofibroblast differentiation and matrix alignment (Ng et al., 2005).

2.2.3 Lymphatic endothelial cells and initial lymphatics

The lymphatic collecting system originates in the interstitium as initial lymphatic vessels, also termed lymphatic capillaries or terminal lymphatics (Fig. 2.1). These microvessels are noncontractile, permeable, and considerably larger (10–60 μm) than surrounding blood capillaries. They consist of a single layer of oak leaf-shaped lymphatic endothelial cells (LECs) without pericytes or smooth muscle media and have little or no basement membrane (Azzali, 2003). LECs are a distinct endothelial cell lineage that expresses the homeobox transcription factor prospero-related homeobox 1 (PROX1) and receptor tyrosine kinase—VEGF receptor 3 (VEGF-R3) (Cho et al., 2019). The latter is considered the primary mechanosensory trigger in LECs and responds to mechanical forces including shear stress and interstitial pressure by tyrosine phosphorylation mediated by β1 integrin (Baeyens et al., 2015). Experiments in the mammalian embryo showed indeed that increasing the interstitial fluid volume (and thus IFP) elongates the LECs, increases VEGF-R3 phosphorylation, and leads to β1 integrin-dependent LEC proliferation (Planas-Paz et al., 2012). LECs are coated on the luminal side with the glycocalyx, a matrix of glycoproteins and glycosaminoglycans (GAGs) with a thickness between 50 and 500 nm and which extends over the intercellular clefts (Arkill et al., 2011). In addition, the majority of LECs express the membrane mucin-like glycoprotein podoplanin, which plays a crucial role in the development of the lymphatic vascular system (Ji et al., 2010; Quintanilla et al., 2019). Recently, telocytes were identified as a novel interstitial cell type displaying telopodes (long and thin prolongations) which intimately surround the abluminal side of LECs, suggesting a possible role of telocytes in the regulation of lymphatic capillary function (Rosa et al., 2020).

Lymphangiogenesis may be stimulated by contact with biomaterials, and there is increasing interest in the use of implantable scaffolds or fibers to induce lymphangiogenesis in peripheral tissue to improve lymph flow (Sestito & Thomas, 2019). As an example, a nanofibrillar collagen scaffold (BioBridge, Fibralign Corporation, Stanford, United States) enhanced lymphangiogenesis and the generation of lymphatic collecting vessels. Furthermore, it improved lymphatic function and fluid clearance even without the addition of exogenous mediators such as VEGF-C in a porcine lymphedema model (Huang et al., 2013; Rochlin et al., 2020).

In the small intestine, lymphatic capillaries, or lacteals, are located exclusively in intestinal villi, whereas collecting lymphatic vessels are present in the mesentery (Unthank & Bohlen, 1988). The initial lymphatics are irregularly shaped but may efficiently increase in volume when the tissue is stretched (Ikomi & Hiruma, 2020). LECs form overlapping structures that may act as functional "button" like unidirectional valves (Baluk et al., 2007; Gerli et al., 2000; Leak, 1976). These primary valves create 2–3-μm diameter pores allowing interstitial fluid and molecules to enter into the vessel lumen and prevent leakage of

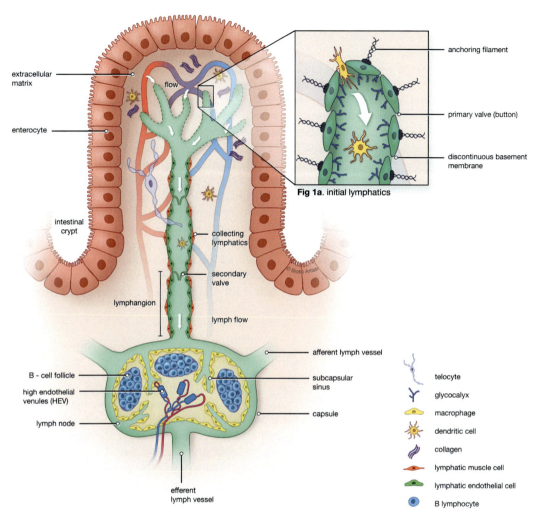

FIGURE 2.1 The normal colonic lymphatic system. Lymph formation is driven by hydrostatic and osmotic pressure gradients between the vascular bed and the initial lymphatics. Lymphatic drainage of large solutes and cells is facilitated by primary valves or buttons, formed by overlapping LECs. In the mesentery, lymph is transported in the collecting lymphatics toward mesenteric lymph nodes. The functional unit of the collecting lymphatics is the lymphangion, bounded by secondary valves which ensure unidirectional transport. See the text for details. *LECs*, Lymphatic endothelial cells.

lymph in the opposite direction. In addition to paracellular diffusion and convection of fluid, cells, and macromolecules, active transcellular transport contributes to initial lymph formation, and solute uptake occurs both in caveolae- and clathrin-coated vesicles (Dixon et al., 2009; Triacca et al., 2017).

Fluid transport in these initial lymphatics requires only small and transient pressure gradients. Only a few experimental studies have attempted to measure the magnitude of this pressure gradient. A gradient of only 0.08 mmHg/mm was obtained in rat hindlimb skin, well in line with the theoretical estimation suggesting that a gradient of 0.09 mmHg/mm is sufficient to drive even a large filtrate from capillaries to the initial lymphatics (Schmid-Schönbein, 1990; Zhang et al., 2000). Computational simulations predicted a volume flow through the primary valves of 0.114 nL/μm per cycle for a transmural pressure varying between 1.0 and 8.0 mmHg at 0.4 Hz (Galie & Spilker, 2009). Currently, the exact mechanisms driving lymphatic fluid transport within the interstitium are not well defined. Possible mechanisms include traction exerted by anchoring filaments that connect elastic fibers in the ECM and the outside aspect of the initial lymphatics, and intermittent pressure induced by surrounding muscle contraction (e.g., diaphragm and calf muscles), bowel peristalsis in case of intestinal lymphatics, and arterial pulsation. A second hypothesis explaining lymph flow in tissue is the "retrograde pump," which relates interstitial lymph flow to the generation of suction pressures caused by contraction and relaxation in the downstream collecting vessels (Schmid-Schönbein, 1990). Based on vessel geometries per branch order from rat mesenteric initial lymphatic networks, pressure drops were estimated to be in the range of 0.31–2.57 mmHg, sufficient for fluid flow through an initial lymphatic network (Sloas et al., 2016). Interestingly, subcutaneous normal lymphatics are often arranged in a hexagonal geometry, which—based on simulations—seems more efficient at draining fluid from the interstitial space compared to other geometries (a square grid or series of parallel tubes) for a given interstitial fluid load (Roose & Swartz, 2012).

2.2.4 Collecting lymphatics and lymphangions

The collecting lymph vessels are lined by lymphatic smooth muscle cells, which drive lymph transport by spontaneous cyclic contractions. In the rat mesentery the average diameter of collecting lymphatics is 91 μm, volume flow rate 13.95 μL/h, and Reynolds number 0.045 (laminar flow) (Dixon et al., 2006). Lymphatic vessels exhibit a highly nonlinear pressure–diameter behavior: at low pressures, the vessels are highly compliant, but above a threshold of about 3.5 mmHg stiffness increases exponentially (Moore & Bertram, 2018).

Transport is unidirectional due to the presence of one-way valves. Segments of collecting lymphatic vessels between valves are termed lymphangions (Fig. 2.1), which constitute the contractile compartment that propels lymph fluid to the next compartment. The organization of lymphatic vessels into a series of lymphangions separated by lymphatic valves plays a major role in moving lymph forward toward the large lymphatic trunks and assists lymph moving against gravity in lymphatic vessels located inferior to the heart. Lymph flow in collecting vessels passes through lymph nodes and is, therefore, classified as pre- or postnodal (or afferent or efferent). Large collecting vessels are innervated and seem to be under the control of pacemakers (Muthuchamy & Zawieja, 2008; von der Weid et al., 1996). These pacemakers generate voltage-gated Ca^{2+} channel-induced action potentials, which provide the depolarization and calcium influx required for the initiation of phasic constrictions of the lymphatic chambers, with resulting contractions that propagate at a speed of 4–8 mm/s (Von Der Weid, 2019; Zawieja et al., 1993). Electrical conduction is probably transmitted via gap junctions between muscle cells formed of connexin proteins (Scallan et al., 2016). The contraction of a rat mesenteric lymphangion typically lasts for 2 s, but the period of relaxation between contractions is highly variable. A 0.5 Hz sinusoid represents the 2 s contraction reasonably well, and with the measured microvessel diameters and lymph properties, a Womersley number value of up to 0.1 can be calculated, for which quasi-steady flow is a good approximation (Moore & Bertram, 2018). The resulting lymphatic pump can be described using properties similar to the heart pump such as lymphatic systole, diastole, contraction frequency, ejection fraction, stroke volume, and lymph flow. In the rat mesentery, collecting lymphatics contracted with a frequency of 6.4 ± 0.61 beats/min and ejected approximately 67% of their end-diastolic volume (Table 2.1) (Benoit et al., 1989). The contraction frequency of lymphangions is affected by temperature, intralymphatic pressure, shear stress, chemical modulators (prostaglandins, norepinephrine, acetylcholine, substance P), sodium load, and fluid osmolarity (Mizuno et al., 2015; Solari et al., 2017; Solari et al., 2018). The temperature-dependent regulation of contractile activity was recently shown to be mediated by transient expression of receptor potential channels of the vanilloid 4 subfamily (TRPV4) (Solari et al., 2020). Also, the presence of low-density lipoproteins was shown to be positively correlated with the contraction frequency of rat diaphragm lymphatic vessels (Solari, Marcozzi, Bistoletti, et al., 2020). In addition, studies in the rat mesenterial collecting lymphatic system demonstrated that an acute lipid load reduces pump function, in terms of contraction frequency and amplitude, with concomitant increased postprandial flow rate and viscosity (Kassis et al., 2016).

Experiments in isolated collecting lymphatics have shown how lymphangions adapt to changes in pressure and flow. Generally, an elevated transmural pressure will result in an increased frequency and force of phasic contractions, up to a maximal value after which phasic contractions become weaker against the increasing load (Davis et al., 2009). Also, several studies showed that increased inflow *pressure* enhances the pump function of chains of lymphangions by the Frank–Starling mechanism (Scallan et al., 2012). However, increased *flow* leads to flow-mediated NO-dependent vasodilation and inhibition of the lymphatic pump, which is considered an energy-conserving mechanism when lymph formation is high (Gashev et al., 2004). Recent studies suggest that NO production by the endothelial cells is mediated, at least in part, by mechanosensing and -transduction properties of the endothelial glycocalyx (Zeng et al., 2018). On the other hand, low lymph flow-induced shear stress induces the expression of genes required for lymphatic valve development and maturation (Sweet et al., 2015). How exactly mechanosensing of shear stress is linked to valve development is not completely elucidated. Recent data suggest a role for the mechanically activated ion channel PIEZO1 (Beech & Kalli, 2019; Choi et al.,

TABLE 2.1 Pump properties of the lymphatic system.

Contraction frequency	6.4 ± 0.61 beats/min
End-diastolic diameter	$69.2 \pm 6.5\ \mu m$
End systolic diameter	$39.4 \pm 5.6\ \mu m$
Peak pressure	9.9 ± 2.5 mmHg
Ejection fraction	0.67 ± 0.05

Values from Benoit, J. N., Zawieja, D. C., Goodman, A. H., & Granger, H. J. (1989). Characterization of intact mesenteric lymphatic pump and its responsiveness to acute edemagenic stress. The American Journal of Physiology, 257(6 Pt 2), H2059–H2069. https://doi.org/10.1152/ajpheart.1989.257.6.H2059.

2019a). Furthermore, observations using microfluidic devices suggest that spatial gradients in shear stress demarcate the locations of future valve formation and implicate E-selectin as a component of a mechanosensory process for detecting these gradients (Michalaki et al., 2020). In collecting vessels the lymphatic endothelium is exposed to high luminal flow, which generates two types of shear stresses: laminar pulsatile shear stress in the linear part of lymphangions and the luminal side of the valve leaflets, and reciprocating shear stress in the valve sinus, where lymph flow patterns are more complex due to mixed forward and retrograde fluid movements (Zawieja, 2009). The endothelial wall in collecting lymphatic vessels is permeable for macromolecules, and the extent of macromolecular passage across the lymphatic lining is influenced by hydrostatic pressure and by shear forces (Breslin & Kurtz, 2009; Scallan & Huxley, 2010).

Next to experimental data, computer simulations offer unique freedom and control (e.g. to investigate the impact of certain parameters) which is an added value compared to experimental/clinical settings. Computational fluid dynamics (CFD) modeling can offer valuable and complementary information, despite the simplifications and limitations of computational methods. For instance, Fig. 2.2 represents a CFD model of lymph propulsion in a 2D axisymmetric model of a collecting lymphatic vessel with three lymphangions

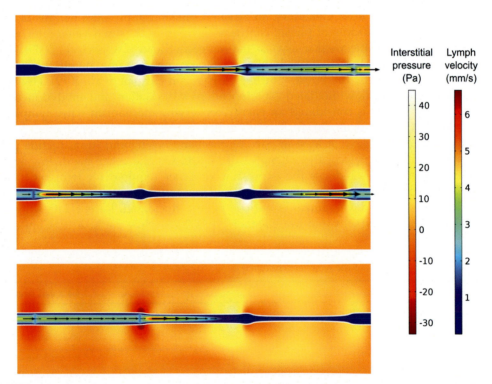

FIGURE 2.2 2D axisymmetric model of a collecting lymphatic vessel with three lymphangions in series showing the sequential contraction of three lymphangions in series. The velocity contours of lymph flow inside the lymphangions and the pressure contours of the interstitium are presented.

in series, immersed in a poroelastic interstitium. The poroelastic interstitium was defined using Biot's theory of poroelasticity, where the solid structure of the interstitium was assumed to be linear elastic, while the interstitial fluid inside the porous domain was governed by Darcy's law (Eq. 2.1). As such, the coupled interaction between the fluid and the deformation of the porous medium was determined. Using this computational model, the impact of different parameters including the mechanical properties of different domains, the timing of the contractions, and the boundary conditions could be evaluated in a very controlled manner. The sequential contractions of the lymphangions, shown in Fig. 2.2, reduced the vessel diameter down to 70% of its relaxing state which resulted in intralymphangion pressure gradients of up to 120 Pa, and lymph velocity magnitudes of up to 6.6 mm/s. Another consequence of the lymphangion contractions was the deformations in the poroelastic interstitium leading to interstitial pressure gradients.

2.2.5 Lymph nodes

Afferent lymphatics allow homing of immune cells to lymph nodes. Lymph fluid containing the local immune signature enters the lymph node through afferent lymphatic vessels and is channeled around the node through conduit channels (Fig. 2.1). These channels are lined with macrophages and dendritic cells that remove microorganisms and antigens and present them to T and B lymphocytes (Roozendaal et al., 2008). It was recently shown that activated T cells enter lymph nodes at high frequencies, come to an instantaneous arrest mediated passively by the mechanical 3D-sieve barrier of the node subcapsular sinus, and subsequently migrate randomly on the sinus floor independent of both chemokines and integrins (Martens et al., 2020). The composition of lymph is significantly altered after passage through the lymph node station. A proteomics-based comparison of plasma and prenodal lymph from human volunteers showed that the common proteome between the two fluids (144 out of 253 proteins) was dominated by complement activation and blood coagulation components, transporters, and protease inhibitors (Clement et al., 2013). The enriched proteome of human lymph (72 proteins) consisted of products derived from the ECM, apoptosis, and cellular catabolism.

Micro-CT imaging in the mouse showed that 75% of the vessel wall surface area in lymph nodes consists of capillaries distributed around the periphery of the node (Jafarnejad et al., 2019). Hemodynamic estimations based on 3D models of the microvascular network resulted in an arteriolar inlet pressure of 30 mmHg, venule exit pressure of 10 mmHg (resulting in a pressure drop of 20 mmHg), the total flow of 0.22 μL/min, and a transmural flow of 10 nL/min. A computational model of the mouse popliteal lymph node suggested that about 90% of the lymph takes a peripheral path via the subcapsular and medullary sinuses, while fluid perfusing deeper into the paracortex is sequestered by parenchymal blood vessels (Jafarnejad et al., 2015). Fluid absorption by these blood vessels is driven mainly by oncotic pressure differences between lymph and blood, although the magnitude of fluid transfer is highly dependent on blood vessel surface area. Also, model predictions showed that the hydraulic conductivity of the medulla should be at least three orders of magnitude larger than that of the paracortex to ensure physiologic pressures across the lymph node.

Lymph nodes are packed with lymphocytes, which cause tension on the fibroblastic reticular cell network and node capsule. This pressure increases its resistance to swelling and the influx rate of lymphocytes. It was shown that the organized growth of the fibroblastic reticular cell network is able to release tension, limit the swelling rate of the lymph node, and allow increased lymphocyte influx. Halting of lymph node swelling coincides with a thickening of the node's capsule, which lowers tension on the fibroblastic reticular cell network to form a new force equilibrium (Assen, 2019).

2.3 Biomechanics of lymphatic metastasis

Cancer mortality is caused by the ability of circulating tumor cells (CTCs) to take advantage of the circulatory system to create distant metastases. Using a computational model of the blood circulation, it was estimated that on average, 40% of the variation in the metastatic distribution can be attributed to blood circulation (36% in CRC) (Font-Clos et al., 2020). While blood-borne metastasis is relatively inefficient, with only a small fraction of disseminated CTCs being able to initiate distant organ metastasis, it has been shown that cancer cells can resist mechanical destruction in the circulation via a RhoA/actomyosin-dependent process of mechano-adaptation (Moose et al., 2020). In most solid cancers the presence of lymphatic vessel invasion and lymph node metastasis are considered adverse prognostic parameters. However, the exact underlying mechanisms are not fully elucidated, since it is increasingly realized that hematogenous distant organ metastasis occurs early and is independent from lymphatic metastasis.

2.3.1 The biomechanical environment of cancer tissue

Both the structural and biomechanical microenvironments differ considerably between normal and malignant tissue. One of the most striking features is the pathological *stiffness* of cancer tissue, which may often be clinically appreciated, and is increasingly used as a diagnostic biophysical biomarker using elastography (Chantarojanasiri & Kongkam, 2017; Pepin & McGee, 2018). Increased stiffness is caused by ECM remodeling due to fibroblast activation, which in some malignancies such as pancreatic cancer leads to a dense, impermeable, and desmoplastic stroma (Thomas & Radhakrishnan, 2019). Increased interstitial flow in the tumor stroma imposes tension on the matrix fibers to which cancer-associated fibroblasts (CAFs) respond with myofibroblast-like differentiation, TGF-β production, matrix synthesis and alignment, and consequent stiffening of the matrix, creating a positive feedback loop (Shieh et al., 2011; Swartz & Lund, 2012). Importantly, matrix stiffness affects the behavior of metastatic cancer cells, driven by their mechanosensing apparatus. Liu et al. (2020) showed that the metastatic pattern of prostate cancer cells depended on the stiffness of the artificial constructs in which they were grown: low stiffness promoted cell migration and proliferation by inducing cluster formation characterized by high expression of CD44.

In addition, the tumor microenvironment is characterized by pathologically elevated fluid and solid pressures. Elevated IFP is caused by collapsed primary lymphatics, dilated and hyperpermeable capillaries, increased angiogenesis at the tumor rim, and increased deposition of matrix components (Bockelmann & Schumacher, 2019). Solid stress can be schematically divided into three categories: (1) *external stress* is caused by the growing tumor pushing against and deforming the surrounding normal tissue, which exerts a reciprocal force on the tumor; (2) *swelling stress* resulting from electrostatic repulsive forces between negatively charged hyaluronic acid (HA) chains and from proliferation and swelling of cancer cells; and (3) *residual stress*, which remains in the tumor when it is removed from its surroundings (Stylianopoulos, 2017).

Computational models show that the absence of a functioning lymphatic system within the tumor matrix has a significant effect on IFP pressure profiles in the tumor (Fig. 2.3).

2.3.2 Lymphangiogenesis in cancer

Metastasis to the locoregional lymphatic basin is a key prognostic feature of most human solid cancers (Fig. 2.4). Lymphatic spread is a dynamic process during tumorigenesis, which involves lymphangiogenesis, lymphatic remodeling, and increased interstitial and lymphatic flow. Tumor-associated lymphangiogenesis is mainly driven by the VEGF-C–VEGF-R3 and VEGF-D–VEGF-R3 axes (Tammela & Alitalo, 2010). Interestingly, intratumor lymphatics are usually collapsed and seem to play no role in the process of lymphatic metastasis, in contrast to lymphatics in the tumor margin (Padera et al., 2002). In animal models, increased numbers of draining lymphatics were observed in the

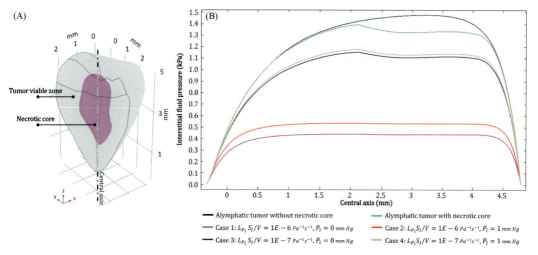

FIGURE 2.3 The impact of lymphatic drainage on the IFP profile. (A) The realistic 3D geometry of a peritoneal tumor of a mouse reconstructed from MRI imaging data. The tumor interstitium was divided into a viable zone and a necrotic core region. (B) IFP profiles plotted along the central axis of the tumor [shown in (A)] showing the impact of lymph vessel properties on the IFP profile. The lymphatic drainage rate ($L_{p_l}S_l/V$) noticeably has a significant effect on the IFP profile, even with minimal lymphatic functionality (Cases 3 and 4). Comparison between the IFP profiles of Cases 1 and 2 as well as Cases 3 and 4 show that the effective lymphatic pressure (P_l) also slightly affects the IFP profile. Interestingly, with a functional lymphatic system (Cases 1 and 2), the presence of a necrotic core does not impact the IFP profile visibly. *IFP*, Interstitial fluid pressure.

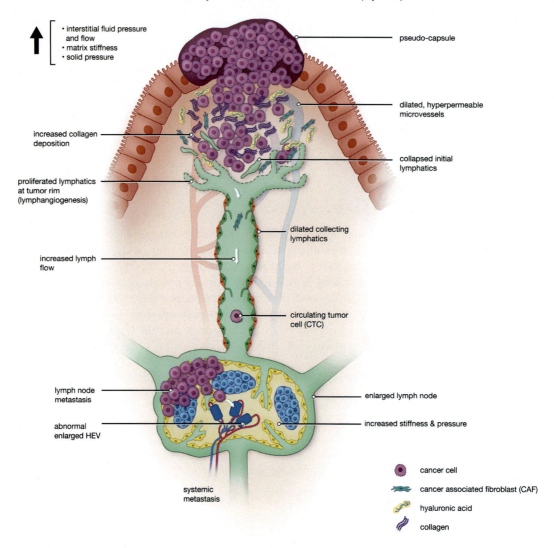

FIGURE 2.4 The lymphatic system in cancer. A combination of rapid growth, fibroblast activation and matrix deposition, hyperpermeable and dilated blood vessels, and collapsed initial lymphatics results in pathologically elevated fluid and solid stress inside the tumor microenvironment. The process of tumor-induced lymphangiogenesis results in lymphatic vessel proliferation at the rim of the tumor. Tumor-associated collecting lymphatics are dilated, and lymph flow is increased. Locoregional lymph nodes are enlarged and display increased stiffness and pressure. Nodes may be invaded by metastatic cancer cells, and distant metastasis is possible through cancer cell invasion of abnormal and enlarged HEVs. See the text for details. *HEVs*, High endothelial venules.

peritumor tissue. These vessels displayed a retrograde draining pattern, indicating possible dysfunction of the intraluminal valves (Isaka et al., 2004). In several human cancers, increased lymphatic vessel density and expression of the lymphangiogenic VEGF-C and VEGF-D are correlated with poor prognosis, invasion, and metastasis (Schmid-Schönbein,

1990; Zhang et al., 2000). Interestingly, however, recent studies have shown that tumor lymphangiogenesis is actually critical in the efficacy of immunotherapies, and inhibiting lymphangiogenesis was shown to invalidate the tumor response to photodynamic therapy and checkpoint blockade. Therefore therapies targeting lymphangiogenesis may have unintended consequences (Fankhauser et al., 2017; Muchowicz et al., 2017).

2.3.3 Interstitial and lymphatic transport in cancer tissue

The pressure difference with the surrounding normal tissue leads to a gradient that drives interstitial fluid flow outward toward the lymphatics (Baxter & Jain, 1989). Secretion of VEGF-C and VEGF-D by cancer and immune cell populations leads to morphological functional changes in the collecting lymphatics, which undergo increases in diameter and volumetric flow, thus facilitating the transport of cancer cells to regional nodes and distant organs (Hoshida et al., 2006; Karnezis et al., 2012; Solari et al., 2017; Sriram et al., 2016). Transcriptional profiling of collecting lymphatics draining to sentinel lymph nodes suggested a role for VEGF-D-induced and prostaglandin-dependent collecting lymphatic vessel dilation, an effect blocked by nonsteroidal anti-inflammatory drugs (NSAIDs) (Karnezis et al., 2012).

The lymph fluid velocity in tumors is estimated to be one order of magnitude higher than the interstitial fluid velocity, while increased flow in peritumor lymphatic vessels leads to a 200-fold increase in the accumulation of cancer cells in tumor-draining lymph nodes (TDLNs) and a 4-fold increase in the rate of metastasis formation (Berk et al., 1996; Hoshida et al., 2006). Lymphatic flow properties affect the behavior of disseminated cancer cells. Using a microfluidics model, Lee et al. (2017) showed that the low shear stress prevalent in initial lymphatics drives motility, adhesion, and invasion of prostate cancer cells through a Yes-associated protein 1-dependent mechanism. Similarly, microfluid flows and resulting shear stress were shown to induce epithelial-to-mesenchymal transition in ovarian and breast cancer cells, providing them with a motile and aggressive phenotype (Beech & Kalli, 2019; Choi et al., 2019b). Recently, the glycocalyx components such as GAGs including heparan sulfate and HA were identified as mechanosensors for interstitial flow on cancer cells, mediated by Glypican-1 which transduces the signal downstream (Moran et al., 2019).

2.3.4 Mechanisms of cancer cell invasion of initial lymphatics

A recent computational model suggests that tumor cells collectively invade the interstitial stroma, guided by activated CAFs, which create a "track" by ECM remodeling, and guide cancer cells to the initial lymphatics (Waldeland et al., 2020). Furthermore, cancer cells were observed in vitro to be directed toward lymphatic channels by gradients of chemotactic molecules such as CCR7, an effect that was shown to increase dramatically when a slow interstitial flow (0.2 μm/s) is imposed (Shields et al., 2007). Also, cancer cell invasion is facilitated by the immunosuppressive tumor microenvironment created by LECs through inhibiting dendritic cell maturation and limiting cytotoxic lymphocyte function through the production of inhibitory ligand PD-L1, inducible nitric oxide synthase, indoleamine 2,3-dioxygenase, and TGF-β (Garnier et al., 2019). Interestingly, lymphatic microvessel invasion may be facilitated by the adoption of an amoeboid cancer cell phenotype (Huang et al., 2013; Rochlin et al., 2020). CRC cells adapt their mechanical properties to facilitate lymphatic metastasis. In vitro biomechanical measurements showed that, compared to CRC cells from the primary tumor, cells from lymph node metastatic origin displayed a rounder shape, significantly lower Young's modulus (increased deformability), and higher nonspecific adhesion force (Palmieri et al., 2015). Recent data show that the lymphatic environment may confer a survival advantage to CTCs. In a melanoma model, exposure to the lymphatic environment protected cancer cells from ferroptosis and increased their ability to survive during subsequent hematogenous metastasis (Ubellacker et al., 2020).

2.3.5 The metastatic lymph node

Tumor-draining metastatic lymph nodes undergo a "priming" process, accompanied by biochemical and biomechanical changes, even before cancer cells arrive. Studies in mouse melanoma models showed that TDLNs exhibit significant morphological and functional changes even before apparent metastasis: nodes were enlarged and showed increased collagen and HA content, higher tissue elasticity and viscoelasticity, and increased intranodal pressures compared to non–TDLNs (Rohner et al., 2015). Nathanson et al. (2014) measured intranodal pressure in sentinel LNs of breast cancer patients and found significantly higher values in metastatic nodes compared to normal nodes (22.0 ± 1.3 mmHg vs 9.3 ± 0.7 mmHg, $P < .001$).

Also, increased angiogenesis and lymph flow were observed in TDLNs, before the appearance of metastatic cells (Harrell et al., 2007). This priming of TDLNs as a premetastatic niche is characterized by sprouting and proliferation and an altered transcriptional profile of LECs, while upregulation of integrin αIIb mediates LEC adhesion to fibrinogen (Commerford et al., 2018). In an experimental CRC model, tumor cells were shown to induce lymphatic network remodeling in lymph nodes by the secretion of exosomal interferon regulatory factor 2, which induces the release of VEGF-C by macrophages (Sun et al., 2019).

The composition of tumor-draining lymph differs considerably from normal lymph. Broggi et al. (2019) compared postoperative lymphatic exudate of metastatic melanoma patients after lymphadenectomy with lymph from healthy volunteers. They found that tumor-draining lymph was significantly enriched in tumor-derived factors, especially extracellular vesicles containing melanoma-associated proteins and miRNAs, with unique protein signatures reflecting early versus the advanced metastatic spread.

2.4 Conclusions and future perspectives

There is increasing interest in the role of the physical microenvironment in tumor development, progression, and metastasis (Nia et al., 2020). Although the role of mechanical drivers in the pathophysiology of lymphatic metastasis has only begun to be unraveled, the available data suggest that lymphatic seeding is accompanied and facilitated by altered biomechanical properties such as lymphatic fluid flow, microvessel and lymph node pressures, and nodal tissue stiffness. Expanding our insights into the biophysical microenvironment of cancer-associated lymphatic systems may advance the field of "immuno-engineering", allowing to target cancer-associated lymphatic systems by using biomaterials such as implantable scaffolds, injectable hydrogels, microfluidic medical devices, and particle-based drug delivery systems (Cote et al., 2019; Manspeaker & Thomas, 2020). Progress in this field will require a multidisciplinary effort from (bio)medical specialists, engineers and physicists, and cancer biologists. One of the prerequisites is the availability of improved computational models and in vivo and in vitro model systems that accurately recapitulate the biomechanical properties of the normal and cancer-associated lymphatic system. A promising platform is the construction of engineered microfluidic systems (lymphatic metastasis on a chip) to understand mechanisms such as cancer cell—LEC crosstalk, transendothelial migration, and lymphangiogenesis under controlled pressure and flow conditions (Greenlee & King, 2020). These high throughput devices may allow to identify novel therapeutic targets and enable precision prevention and treatment of lymphatic metastasis in clinical practice.

References

Alexander, J. S., Ganta, V. C., Jordan, P. A., & Witte, M. H. (2010). Gastrointestinal lymphatics in health and disease. *Pathophysiology: The Official Journal of the International Society for Pathophysiology/ISP, 17*(4), 315–335. Available from https://doi.org/10.1016/j.pathophys.2009.09.003.

Arkill, K. P., Knupp, C., Michel, C. C., Neal, C. R., Qvortrup, K., Rostgaard, J., & Squire, J. M. (2011). Similar endothelial glycocalyx structures in microvessels from a range of mammalian tissues: Evidence for a common filtering mechanism? *Biophysical Journal, 101*(5), 1046–1056. Available from https://doi.org/10.1016/j.bpj.2011.07.036.

Assen, F. P. (2019). *Lymph node mechanics: Deciphering the interplay between stroma contractility, morphology and lymphocyte trafficking.* PhD thesis, Institute of Science and Technology (IST), Austria, 2019. Available from https://doi.org/10.15479/AT:ISTA:6947.

Azzali, G. (2003). Transendothelial transport and migration in vessels of the apparatus lymphaticus peripherus absorbens (ALPA). *International Review of Cytology, 230*, 41–87. Available from https://doi.org/10.1016/s0074-7696(03)30002-6.

Baeyens, N., Nicoli, S., Coon, B. G., Ross, T. D., Van den Dries, K., Han, J., Lauridsen, H. M., Mejean, C. O., Eichmann, A., Thomas, J. L., Humphrey, J. D., & Schwartz, M. A. (2015). Vascular remodeling is governed by a VEGFR3-dependent fluid shear stress set point. *Elife, 4*. Available from https://doi.org/10.7554/eLife.04645.

Baluk, P., Fuxe, J., Hashizume, H., Romano, T., Lashnits, E., Butz, S., Vestweber, D., Corada, M., Molendini, C., Dejana, E., & McDonald, D. M. (2007). Functionally specialized junctions between endothelial cells of lymphatic vessels. *The Journal of Experimental Medicine, 204*(10), 2349–2362. Available from https://doi.org/10.1084/jem.20062596.

Baxter, L. T., & Jain, R. K. (1989). Transport of fluid and macromolecules in tumors. I. Role of interstitial pressure and convection. *Microvascular Research, 37*(1), 77–104. Available from https://doi.org/10.1016/0026-2862(89)90074-5.

Beech, D. J., & Kalli, A. C. (2019). Force sensing by piezo channels in cardiovascular health and disease. *Arteriosclerosis, Thrombosis, and Vascular Biology, 39*(11), 2228–2239. Available from https://doi.org/10.1161/atvbaha.119.313348.

Benoit, J. N., Zawieja, D. C., Goodman, A. H., & Granger, H. J. (1989). Characterization of intact mesenteric lymphatic pump and its responsiveness to acute edemagenic stress. *The American Journal of Physiology, 257*(6 Pt 2), H2059–H2069. Available from https://doi.org/10.1152/ajpheart.1989.257.6.H2059.

Bercoff, J., Tanter, M., & Fink, M. (2004). Supersonic shear imaging: A new technique for soft tissue elasticity mapping. *IEEE Transactions on Ultrasonics, Ferroelectrics, and Frequency Control, 51*(4), 396–409. Available from https://doi.org/10.1109/TUFFC.2004.1295425.

Berk, D. A., Swartz, M. A., Leu, A. J., & Jain, R. K. (1996). Transport in lymphatic capillaries. II. Microscopic velocity measurement with fluorescence photobleaching. *The American Journal of Physiology, 270*(1 Pt 2), H330–H337. Available from https://doi.org/10.1152/ajpheart.1996.270.1.H330.

Bockelmann, L. C., & Schumacher, U. (2019). Targeting tumor interstitial fluid pressure: Will it yield novel successful therapies for solid tumors? *Expert Opinion on Therapeutic Targets, 23*(12), 1005–1014. Available from https://doi.org/10.1080/14728222.2019.1702974.

Breslin, J. W., & Kurtz, K. M. (2009). Lymphatic endothelial cells adapt their barrier function in response to changes in shear stress. *Lymphatic Research and Biology, 7*(4), 229–237. Available from https://doi.org/10.1089/lrb.2009.0015.

Brinkman, H. C. (1949). A calculation of the viscous force exerted by a flowing fluid on a dense swarm of particles. *Flow, Turbulence and Combustion, 1*(1), 27. Available from https://doi.org/10.1007/BF02120313.

Broggi, M. A. S., Maillat, L., Clement, C. C., Bordry, N., Corthésy, P., Auger, A., Matter, M., Hamelin, R., Potin, L., Demurtas, D., Romano, E., Harari, A., Speiser, D. E., Santambrogio, L., & Swartz, M. A. (2019). Tumor-associated factors are enriched in lymphatic exudate compared to plasma in metastatic melanoma patients. *The Journal of Experimental Medicine, 216*(5), 1091–1107. Available from https://doi.org/10.1084/jem.20181618.

Chantarojanasiri, T., & Kongkam, P. (2017). Endoscopic ultrasound elastography for solid pancreatic lesions. *World Journal Gastrointestinal Endoscopy, 9*(10), 506–513. Available from https://doi.org/10.4253/wjge.v9.i10.506.

Chary, S. R., & Jain, R. K. (1989). Direct measurement of interstitial convection and diffusion of albumin in normal and neoplastic tissues by fluorescence photobleaching. *Proceedings of the National Academy of Sciences of the United States of America, 86*(14), 5385–5389. Available from https://doi.org/10.1073/pnas.86.14.5385.

Cho, H., Kim, J., Ahn, J. H., Hong, Y. K., Mäkinen, T., Lim, D. S., & Koh, G. Y. (2019). YAP and TAZ negatively regulate Prox1 during developmental and pathologic lymphangiogenesis. *Circulation Research, 124*(2), 225–242. Available from https://doi.org/10.1161/circresaha.118.313707.

Choi, H. Y., Yang, G. M., Dayem, A. A., Saha, S. K., Kim, K., Yoo, Y., Hong, K., Kim, J. H., Yee, C., Lee, K. M., & Cho, S. G. (2019a). Hydrodynamic shear stress promotes epithelial-mesenchymal transition by downregulating ERK and GSK3β activities. *Breast Cancer Research: BCR, 21*(1), 6. Available from https://doi.org/10.1186/s13058-018-1071-2.

Choi, D., Park, E., Jung, E., Cha, B., Lee, S., Yu, J., Kim, P. M., Lee, S., Hong, Y. J., Koh, C. J., Cho, C. W., Wu, Y. F., Jeon, N. L., Wong, A. K., Shin, L., Kumar, S. R., Bermejo-Moreno, I., Srinivasan, R. S., Cho, I. T., & Hong, Y. K. (2019b). Piezo1 incorporates mechanical force signals into the genetic program that governs lymphatic valve development and maintenance. *JCI Insight, 4*(5), 15. Available from https://doi.org/10.1172/jci.insight.125068.

Clement, C. C., Aphkhazava, D., Nieves, E., Callaway, M., Olszewski, W., Rotzschke, O., & Santambrogio, L. (2013). Protein expression profiles of human lymph and plasma mapped by 2D-DIGE and 1D SDS-PAGE coupled with nanoLC-ESI-MS/MS bottom-up proteomics. *Journal of Proteomics, 78*, 172–187. Available from https://doi.org/10.1016/j.jprot.2012.11.013.

Commerford, C. D., Dieterich, L. C., He, Y., Hell, T., Montoya-Zegarra, J. A., Noerrelykke, S. F., Russo, E., Röcken, M., & Detmar, M. (2018). Mechanisms of tumor-induced lymphovascular niche formation in draining lymph nodes. *Cell Reports, 25*(13), 3554–3563. Available from https://doi.org/10.1016/j.celrep.2018.12.002, e4.

Cote, B., Rao, D., Alany, R. G., Kwon, G. S., & Alani, A. W. G. (2019). Lymphatic changes in cancer and drug delivery to the lymphatics in solid tumors. *Advanced Drug Delivery Reviews, 144*, 16–34. Available from https://doi.org/10.1016/j.addr.2019.08.009.

Davis, M. J., Davis, A. M., Lane, M. M., Ku, C. W., & Gashev, A. A. (2009). Rate-sensitive contractile responses of lymphatic vessels to circumferential stretch. *The Journal of Physiology, 587*(1), 165–182. Available from https://doi.org/10.1113/jphysiol.2008.162438.

Dixon, J. B., Greiner, S. T., Gashev, A. A., Cote, G. L., Moore, J. E., & Zawieja, D. C. (2006). Lymph flow, shear stress, and lymphocyte velocity in rat mesenteric prenodal lymphatics. *Microcirculation (New York, N.Y.: 1994), 13*(7), 597–610. Available from https://doi.org/10.1080/10739680600893909.

Dixon, J. B., Raghunathan, S., & Swartz, M. A. (2009). A tissue-engineered model of the intestinal lacteal for evaluating lipid transport by lymphatics. *Biotechnology and Bioengineering, 103*(6), 1224–1235. Available from https://doi.org/10.1002/bit.22337.

References

Fankhauser, M., Broggi, M. A. S., Potin, L., Bordry, N., Jeanbart, L., Lund, A. W., Da Costa, E., Hauert, S., Rincon-Restrepo, M., Tremblay, C., Cabello, E., Homicsko, K., Michielin, O., Hanahan, D., Speiser, D. E., & Swartz, M. A. (2017). Tumor lymphangiogenesis promotes T cell infiltration and potentiates immunotherapy in melanoma. *Science Translational Medicine, 9*(407). Available from https://doi.org/10.1126/scitranslmed.aal4712.

Follain, G., Herrmann, D., Harlepp, S., Hyenne, V., Osmani, N., Warren, S. C., Timpson, P., & Goetz, J. G. (2020). Fluids and their mechanics in tumour transit: Shaping metastasis. *Nature Reviews Cancer, 20*(2), 107–124. Available from https://doi.org/10.1038/s41568-019-0221-x.

Font-Clos, F., Zapperi, S., & La Porta, C. A. M. (2020). Blood flow contributions to cancer metastasis. *IScience, 23*(5)101073. Available from https://doi.org/10.1016/j.isci.2020.101073.

Fovargue, D., Nordsletten, D., & Sinkus, R. (2018). Stiffness reconstruction methods for MR elastography. *NMR in Biomedicine, 31*(10), e3935. Available from https://doi.org/10.1002/nbm.3935.

Galie, P., & Spilker, R. L. (2009). A two-dimensional computational model of lymph transport across primary lymphatic valves. *Journal of Biomechanical Engineering, 131*(11), 111004. Available from https://doi.org/10.1115/1.3212108.

Garnier, L., Gkountidi, A. O., & Hugues, S. (2019). Tumor-associated lymphatic vessel features and immunomodulatory functions. *Frontiers in Immunology, 10*, 720. Available from https://doi.org/10.3389/fimmu.2019.00720.

Gashev, A. A., Davis, M. J., Delp, M. D., & Zawieja, D. C. (2004). Regional variations of contractile activity in isolated rat lymphatics. *Microcirculation (New York, N.Y.: 1994), 11*(6), 477–492. Available from https://doi.org/10.1080/10739680490476033.

Gerli, R., Solito, R., Weber, E., & Agliano, M. (2000). Specific adhesion molecules bind anchoring filaments and endothelial cells in human skin initial lymphatics. *Lymphology, 33*(4), 148–157.

Goldman, J., Conley, K. A., Raehl, A., Bondy, D. M., Pytowski, B., Swartz, M. A., Rutkowski, J. M., Jaroch, D. B., & Ongstad, E. L. (2007). Regulation of lymphatic capillary regeneration by interstitial flow in skin. *American Journal of Physiology. Heart and Circulatory Physiology, 292*(5), H2176–H2183. Available from https://doi.org/10.1152/ajpheart.01011.2006.

Greenlee, J. D., & King, M. R. (2020). Engineered fluidic systems to understand lymphatic cancer metastasis. *Biomicrofluidics, 14*(1), 011502. Available from https://doi.org/10.1063/1.5133970.

Hangai-Hoger, N., Cabrales, P., Briceño, J. C., Tsai, A. G., & Intaglietta, M. (2004). Microlymphatic and tissue oxygen tension in the rat mesentery. *American Journal of Physiology. Heart and Circulatory Physiology, 286*(3), H878–H883. Available from https://doi.org/10.1152/ajpheart.00913.2003.

Hansen, K. C., D'Alessandro, A., Clement, C. C., & Santambrogio, L. (2015). Lymph formation, composition and circulation: A proteomics perspective. *International Immunology, 27*(5), 219–227. Available from https://doi.org/10.1093/intimm/dxv012.

Harrell, M. I., Iritani, B. M., & Ruddell, A. (2007). Tumor-induced sentinel lymph node lymphangiogenesis and increased lymph flow precede melanoma metastasis. *The American Journal of Pathology, 170*(2), 774–786. Available from https://doi.org/10.2353/ajpath.2007.060761.

Hoshida, T., Isaka, N., Hagendoorn, J., Di Tomaso, E., Chen, Y. L., Pytowski, B., Fukumura, D., Padera, T. P., & Jain, R. K. (2006). Imaging steps of lymphatic metastasis reveals that vascular endothelial growth factor-C increases metastasis by increasing delivery of cancer cells to lymph nodes: Therapeutic implications. *Cancer Research, 66*(16), 8065–8075. Available from https://doi.org/10.1158/0008-5472.Can-06-1392.

Huang, N. F., Okogbaa, J., Lee, J. C., Jha, A., Zaitseva, T. S., Paukshto, M. V., Sun, J. S., Punjya, N., Fuller, G. G., & Cooke, J. P. (2013). The modulation of endothelial cell morphology, function, and survival using anisotropic nanofibrillar collagen scaffolds. *Biomaterials, 34*(16), 4038–4047. Available from https://doi.org/10.1016/j.biomaterials.2013.02.036.

Huxley, V. H., & Scallan, J. (2011). Lymphatic fluid: Exchange mechanisms and regulation. *The Journal of Physiology, 589*(Pt 12), 2935–2943. Available from https://doi.org/10.1113/jphysiol.2011.208298.

Hynes, R. O. (2009). The extracellular matrix: Not just pretty fibrils. *Science (New York, N.Y.), 326*(5957), 1216–1219. Available from https://doi.org/10.1126/science.1176009.

Ikomi, F., & Hiruma, S. (2020). Relationship between shape of peripheral initial lymphatics and efficiency of mechanical stimulation-induced lymph formation. *Microcirculation (New York, N.Y.: 1994), 14*. Available from https://doi.org/10.1111/micc.12606.

Isaka, N., Padera, T. P., Hagendoorn, J., Fukumura, D., & Jain, R. K. (2004). Peritumor lymphatics induced by vascular endothelial growth factor-C exhibit abnormal function. *Cancer Research, 64*(13), 4400–4404. Available from https://doi.org/10.1158/0008-5472.Can-04-0752.

Jafarnejad, M., Ismail, A. Z., Duarte, D., Vyas, C., Ghahramani, A., Zawieja, D. C., Lo Celso, C., Poologasundarampillai, G., & Moore, J. E., Jr. (2019). Quantification of the whole lymph node vasculature based on tomography of the vessel corrosion casts. *Scientific Reports, 9*(1), 13380. Available from https://doi.org/10.1038/s41598-019-49055-7.

Jafarnejad, M., Woodruff, M. C., Zawieja, D. C., Carroll, M. C., & Moore, J. E., Jr. (2015). Modeling lymph flow and fluid exchange with blood vessels in lymph nodes. *Lymphatic Research and Biology, 13*(4), 234–247. Available from https://doi.org/10.1089/lrb.2015.0028.

Ji, R. C., Eshita, Y., Xing, L., & Miura, M. (2010). Multiple expressions of lymphatic markers and morphological evolution of newly formed lymphatics in lymphangioma and lymph node lymphangiogenesis. *Microvascular Research, 80*(2), 195–201. Available from https://doi.org/10.1016/j.mvr.2010.04.002.

Karnezis, T., Shayan, R., Caesar, C., Roufail, S., Harris, N. C., Ardipradja, K., Zhang, Y. F., Williams, S. P., Farnsworth, R. H., Chai, M. G., Rupasinghe, T. W., Tull, D. L., Baldwin, M. E., Sloan, E. K., Fox, S. B., Achen, M. G., & Stacker, S. A. (2012). VEGF-D promotes tumor metastasis by regulating prostaglandins produced by the collecting lymphatic endothelium. *Cancer Cell, 21*(2), 181–195. Available from https://doi.org/10.1016/j.ccr.2011.12.026.

Kassis, T., Yarlagadda, S. C., Kohan, A. B., Tso, P., Breedveld, V., & Dixon, J. B. (2016). Postprandial lymphatic pump function after a high-fat meal: A characterization of contractility, flow, and viscosity. *American Journal of Physiology. Gastrointestinal and Liver Physiology, 310*(10), G776–G789. Available from https://doi.org/10.1152/ajpgi.00318.2015.

Leak, L. V. (1976). The structure of lymphatic capillaries in lymph formation. *Federation Proceedings, 35*(8), 1863–1871.

Lee, H. J., Diaz, M. F., Price, K. M., Ozuna, J. A., Zhang, S., Sevick-Muraca, E. M., Hagan, J. P., & Wenzel, P. L. (2017). Fluid shear stress activates YAP1 to promote cancer cell motility. *Nature Communications, 8*, 14122. Available from https://doi.org/10.1038/ncomms14122.

Levick, J. R., & Michel, C. C. (2010). Microvascular fluid exchange and the revised Starling principle. *Cardiovascular Research, 87*(2), 198–210. Available from https://doi.org/10.1093/cvr/cvq062.

Liu, Z. X., Wang, L. J., Xu, H., Du, Q. Q. G., Li, L., Wang, L., Zhang, E. S., Chen, G. S., & Wang, Y. (2020). Heterogeneous responses to mechanical force of prostate cancer cells inducing different metastasis patterns. *Advanced Science, 7*(15), 11. Available from https://doi.org/10.1002/advs.201903583.

Manspeaker, M. P., & Thomas, S. N. (2020). Lymphatic immunomodulation using engineered drug delivery systems for cancer immunotherapy. *Advanced Drug Delivery Reviews, 160*, 19–35. Available from https://doi.org/10.1016/j.addr.2020.10.004.

Martens, R., Permanyer, M., Werth, K., Yu, K., Braun, A., Halle, O., Halle, S., Patzer, G. E., Bosnjak, B., Kiefer, F., Janssen, A., Friedrichsen, M., Poetzsch, J., Kohli, K., Lueder, Y., Jauregui, R. G., Eckert, N., Worbs, T., Galla, M., & Forster, R. (2020). Efficient homing of T cells via afferent lymphatics requires mechanical arrest and integrin-supported chemokine guidance. *Nature Communications, 11*(1), 16. Available from https://doi.org/10.1038/s41467-020-14921-w.

Michalaki, E., Surya, V. N., Fuller, G. G., & Dunn, A. R. (2020). Perpendicular alignment of lymphatic endothelial cells in response to spatial gradients in wall shear stress. *Communications Biology, 3*(1), 9. Available from https://doi.org/10.1038/s42003-019-0732-8.

Mizuno, R., Isshiki, M., Ono, N., Nishimoto, M., & Fujita, T. (2015). A high salt diet alters pressure-induced mechanical activity of the rat lymphatics with enhancement of myogenic characteristics. *Lymphatic Research and Biology, 13*(1), 2–9. Available from https://doi.org/10.1089/lrb.2014.0028.

Mohammed, S. I., Torres-Luquis, O., Zhou, W., Lanman, N. A., Espina, V., & Liotta, L. (2020). Tumor-draining lymph secretome en route to the regional lymph node in breast cancer metastasis. *Breast Cancer, 12*, 57–67. Available from https://doi.org/10.2147/bctt.S236168.

Moore, J. E., Jr., & Bertram, C. D. (2018). Lymphatic system flows. *Annual Review of Fluid Mechanics, 50*, 459–482. Available from https://doi.org/10.1146/annurev-fluid-122316-045259.

Moose, D. L., Krog, B. L., Kim, T. H., Zhao, L., Williams-Perez, S., Burke, G., Rhodes, L., Vanneste, M., Breheny, P., Milhem, M., Stipp, C. S., Rowat, A. C., & Henry, M. D. (2020). Cancer cells resist mechanical destruction in circulation via RhoA/actomyosin-dependent mechano-adaptation. *Cell Reports, 30*(11), 3864–3874. Available from https://doi.org/10.1016/j.celrep.2020.02.080, e6.

Moran, H., Cancel, L. M., Mayer, M. A., Qazi, H., Munn, L. L., & Tarbell, J. M. (2019). The cancer cell glycocalyx proteoglycan Glypican-1 mediates interstitial flow mechanotransduction to enhance cell migration and metastasis. *Biorheology, 56*(2–3), 151–161. Available from https://doi.org/10.3233/bir-180203.

Muchowicz, A., Wachowska, M., Stachura, J., Tonecka, K., Gabrysiak, M., Wołosz, D., Pilch, Z., Kilarski, W. W., Boon, L., Klaus, T. J., & Golab, J. (2017). Inhibition of lymphangiogenesis impairs antitumour effects of photodynamic therapy and checkpoint inhibitors in mice. *European Journal of Cancer, 83*, 19–27. Available from https://doi.org/10.1016/j.ejca.2017.06.004.

Muthuchamy, M., & Zawieja, D. (2008). Molecular regulation of lymphatic contractility. *Annals of the New York Academy of Sciences, 1131*, 89–99. Available from https://doi.org/10.1196/annals.1413.008.

Nathanson, S. D., Shah, R., Chitale, D. A., & Mahan, M. (2014). Intraoperative clinical assessment and pressure measurements of sentinel lymph nodes in breast cancer. *Annals of Surgical Oncology, 21*(1), 81–85. Available from https://doi.org/10.1245/s10434-013-3249-2.

Ng, C. P., Hinz, B., & Swartz, M. A. (2005). Interstitial fluid flow induces myofibroblast differentiation and collagen alignment in vitro. *Journal of Cell Science, 118*(Pt 20), 4731–4739. Available from https://doi.org/10.1242/jcs.02605.

Nia, H. T., Munn, L. L., & Jain, R. K. (2020). Physical traits of cancer. *Science (New York, N.Y.), 370*(6516). Available from https://doi.org/10.1126/science.aaz0868.

Padera, T. P., Kadambi, A., di Tomaso, E., Carreira, C. M., Brown, E. B., Boucher, Y., Choi, N. C., Mathisen, D., Wain, J., Mark, E. J., Munn, L. L., & Jain, R. K. (2002). Lymphatic metastasis in the absence of functional intratumor lymphatics. *Science (New York, N.Y.), 296*(5574), 1883–1886. Available from https://doi.org/10.1126/science.1071420.

Palmieri, V., Lucchetti, D., Maiorana, A., Papi, M., Maulucci, G., Calapà, F., Ciasca, G., Giordano, R., Sgambato, A., & De Spirito, M. (2015). Mechanical and structural comparison between primary tumor and lymph node metastasis cells in colorectal cancer. *Soft Matter, 11*(28), 5719–5726. Available from https://doi.org/10.1039/c5sm01089f.

Pepin, K. M., & McGee, K. P. (2018). Quantifying tumor stiffness with magnetic resonance elastography: The role of mechanical properties for detection, characterization, and treatment stratification in oncology. *Topics in Magnetic Resonance Imaging: TMRI, 27*(5), 353–362. Available from https://doi.org/10.1097/rmr.0000000000000181.

Planas-Paz, L., Strilić, B., Goedecke, A., Breier, G., Fässler, R., & Lammert, E. (2012). Mechanoinduction of lymph vessel expansion. *The EMBO Journal, 31*(4), 788–804. Available from https://doi.org/10.1038/emboj.2011.456.

Quintanilla, M., Montero-Montero, L., Renart, J., & Martín-Villar, E. (2019). Podoplanin in inflammation and cancer. *International Journal Molecular Sciences, 20*(3). Available from https://doi.org/10.3390/ijms20030707.

Rochlin, D. H., Inchauste, S., Zelones, J., & Nguyen, D. H. (2020). The role of adjunct nanofibrillar collagen scaffold implantation in the surgical management of secondary lymphedema: Review of the literature and summary of initial pilot studies. *Journal of Surgical Oncology, 121*(1), 121–128. Available from https://doi.org/10.1002/jso.25576.

Rohner, N. A., McClain, J., Tuell, S. L., Warner, A., Smith, B., Yun, Y., Mohan, A., Sushnitha, M., & Thomas, S. N. (2015). Lymph node biophysical remodeling is associated with melanoma lymphatic drainage. *The FASEB Journal, 29*(11), 4512–4522. Available from https://doi.org/10.1096/fj.15-274761.

Roose, T., & Swartz, M. A. (2012). Multiscale modeling of lymphatic drainage from tissues using homogenization theory. *Journal of Biomechanics, 45*(1), 107–115. Available from https://doi.org/10.1016/j.jbiomech.2011.09.015.

Roozendaal, R., Mebius, R. E., & Kraal, G. (2008). The conduit system of the lymph node. *International Immunology, 20*(12), 1483–1487. Available from https://doi.org/10.1093/intimm/dxn110.

Rosa, I., Marini, M., Sgambati, E., Ibba-Manneschi, L., & Manetti, M. (2020). Telocytes and lymphatic endothelial cells: Two immunophenotypically distinct and spatially close cell entities. *Acta Histochemica, 122*(3), 151530. Available from https://doi.org/10.1016/j.acthis.2020.151530.

Rutkowski, J. M., & Swartz, M. A. (2007). A driving force for change: Interstitial flow as a morphoregulator. *Trends in Cell Biology, 17*(1), 44–50. Available from https://doi.org/10.1016/j.tcb.2006.11.007.

Scallan, J. P., & Huxley, V. H. (2010). In vivo determination of collecting lymphatic vessel permeability to albumin: A role for lymphatics in exchange. *The Journal of Physiology, 588*(Pt 1), 243–254. Available from https://doi.org/10.1113/jphysiol.2009.179622.

Scallan, J. P., Wolpers, J. H., Muthuchamy, M., Zawieja, D. C., Gashev, A. A., & Davis, M. J. (2012). Independent and interactive effects of preload and afterload on the pump function of the isolated lymphangion. *American Journal of Physiology. Heart and Circulatory Physiology, 303*(7), H809–H824. Available from https://doi.org/10.1152/ajpheart.01098.2011.

Scallan, J. P., Zawieja, S. D., Castorena-Gonzalez, J. A., & Davis, M. J. (2016). Lymphatic pumping: Mechanics, mechanisms and malfunction. *The Journal of Physiology, 594*(20), 5749–5768. Available from https://doi.org/10.1113/jp272088.

Schmid-Schönbein, G. W. (1990). Microlymphatics and lymph flow. *Physiological Reviews, 70*(4), 987–1028. Available from https://doi.org/10.1152/physrev.1990.70.4.987.

Sestito, L. F., & Thomas, S. N. (2019). Biomaterials for modulating lymphatic function in immunoengineering. *ACS Pharmacology & Translational Science., 2*(5), 293–310. Available from https://doi.org/10.1021/acsptsci.9b00047.

Shieh, A. C., Rozansky, H. A., Hinz, B., & Swartz, M. A. (2011). Tumor cell invasion is promoted by interstitial flow-induced matrix priming by stromal fibroblasts. *Cancer Research, 71*(3), 790–800. Available from https://doi.org/10.1158/0008-5472.Can-10-1513.

Shields, J. D., Fleury, M. E., Yong, C., Tomei, A. A., Randolph, G. J., & Swartz, M. A. (2007). Autologous chemotaxis as a mechanism of tumor cell homing to lymphatics via interstitial flow and autocrine CCR7 signaling. *Cancer Cell, 11*(6), 526–538. Available from https://doi.org/10.1016/j.ccr.2007.04.020.

Sloas, D. C., Stewart, S. A., Sweat, R. S., Doggett, T. M., Alves, N. G., Breslin, J. W., Gaver, D. P., & Murfee, W. L. (2016). Estimation of the pressure drop required for lymph flow through initial lymphatic networks. *Lymphatic Research and Biology, 14*(2), 62–69. Available from https://doi.org/10.1089/lrb.2015.0039.

Solari, E., Marcozzi, C., Bartolini, B., Viola, M., Negrini, D., & Moriondo, A. (2020). Acute exposure of collecting lymphatic vessels to low-density lipoproteins increases both contraction frequency and lymph flow: An in vivo mechanical insight. *Lymphatic Research and Biology, 18*(2), 146–155. Available from https://doi.org/10.1089/lrb.2019.0040.

Solari, E., Marcozzi, C., Bistoletti, M., Baj, A., Giaroni, C., Negrini, D., & Moriondo, A. (2020). TRPV4 channels' dominant role in the temperature modulation of intrinsic contractility and lymph flow of rat diaphragmatic lymphatics. *American Journal of Physiology. Heart and Circulatory Physiology, 319*(2), H507–H518. Available from https://doi.org/10.1152/ajpheart.00175.2020.

Solari, E., Marcozzi, C., Negrini, D., & Moriondo, A. (2017). Temperature-dependent modulation of regional lymphatic contraction frequency and flow. *American Journal of Physiology-Heart and Circulatory Physiology, 313*(5), H879–H889. Available from https://doi.org/10.1152/ajpheart.00267.2017.

Solari, E., Marcozzi, C., Negrini, D., & Moriondo, A. (2018). Fluid osmolarity acutely and differentially modulates lymphatic vessels intrinsic contractions and lymph flow. *Frontiers in Physiology, 9*, 15. Available from https://doi.org/10.3389/fphys.2018.00871.

Sriram, K., Meguid, R. A., & Meguid, M. M. (2016). Nutritional support in adults with chyle leaks. *Nutrition (Burbank, Los Angeles County, California), 32*(2), 281–286. Available from https://doi.org/10.1016/j.nut.2015.08.002.

Stylianopoulos, T. (2017). The solid mechanics of cancer and strategies for improved therapy. *Journal of Biomechanical Engineering, 139*(2). Available from https://doi.org/10.1115/1.4034991.

Sun, B., Zhou, Y., Fang, Y., Li, Z., Gu, X., & Xiang, J. (2019). Colorectal cancer exosomes induce lymphatic network remodeling in lymph nodes. *International Journal of Cancer, 145*(6), 1648–1659. Available from https://doi.org/10.1002/ijc.32196.

Swartz, M. A., & Lund, A. W. (2012). Lymphatic and interstitial flow in the tumour microenvironment: Linking mechanobiology with immunity. *Nature Reviews. Cancer, 12*(3), 210–219. Available from https://doi.org/10.1038/nrc3186.

Sweet, D. T., Jiménez, J. M., Chang, J., Hess, P. R., Mericko-Ishizuka, P., Fu, J., Xia, L., Davies, P. F., & Kahn, M. L. (2015). Lymph flow regulates collecting lymphatic vessel maturation in vivo. *The Journal of Clinical Investigation, 125*(8), 2995–3007. Available from https://doi.org/10.1172/jci79386.

Tammela, T., & Alitalo, K. (2010). Lymphangiogenesis: Molecular mechanisms and future promise. *Cell, 140*(4), 460–476. Available from https://doi.org/10.1016/j.cell.2010.01.045.

Thomas, D., & Radhakrishnan, P. (2019). Tumor-stromal crosstalk in pancreatic cancer and tissue fibrosis. *Molecular Cancer, 18*(1), 14. Available from https://doi.org/10.1186/s12943-018-0927-5.

Triacca, V., Güç, E., Kilarski, W. W., Pisano, M., & Swartz, M. A. (2017). Transcellular pathways in lymphatic endothelial cells regulate changes in solute transport by fluid stress. *Circulation Research, 120*(9), 1440–1452. Available from https://doi.org/10.1161/circresaha.116.309828.

Ubellacker, J. M., Tasdogan, A., Ramesh, V., Shen, B., Mitchell, E. C., Martin-Sandoval, M. S., Gu, Z., McCormick, M. L., Durham, A. B., Spitz, D. R., Zhao, Z., Mathews, T. P., & Morrison, S. J. (2020). Lymph protects metastasizing melanoma cells from ferroptosis. *Nature, 585*(7823), 113–118. Available from https://doi.org/10.1038/s41586-020-2623-z.

Unthank, J. L., & Bohlen, H. G. (1988). Lymphatic pathways and role of valves in lymph propulsion from small intestine. *The American Journal of Physiology, 254*(3 Pt 1), G389–G398. Available from https://doi.org/10.1152/ajpgi.1988.254.3.G389.

von der Weid, P. Y. (2019). Lymphatic vessel pumping. *Advances in Experimental Medicine and Biology*, *1124*, 357−377. Available from https://doi.org/10.1007/978-981-13-5895-1_15.

von der Weid, P. Y., Crowe, M. J., & Van Helden, D. F. (1996). Endothelium-dependent modulation of pacemaking in lymphatic vessels of the guinea-pig mesentery. *The Journal of Physiology*, *493*(Pt 2), 563−575. Available from https://doi.org/10.1113/jphysiol.1996.sp021404.

Waldeland, J. O., Polacheck, W. J., & Evje, S. (2020). Collective tumor cell migration in the presence of fibroblasts. *Journal of Biomechanics*, *100*, 15. Available from https://doi.org/10.1016/j.jbiomech.2019.109568.

Wiig, H., & Swartz, M. A. (2012). Interstitial fluid and lymph formation and transport: Physiological regulation and roles in inflammation and cancer. *Physiological Reviews*, *92*(3), 1005−1060. Available from https://doi.org/10.1152/physrev.00037.2011.

Zawieja, D. C. (2009). Contractile physiology of lymphatics. *Lymphatic Research and Biology*, *7*(2), 87−96. Available from https://doi.org/10.1089/lrb.2009.0007.

Zawieja, D. C., Davis, K. L., Schuster, R., Hinds, W. M., & Granger, H. J. (1993). Distribution, propagation, and coordination of contractile activity in lymphatics. *The American Journal of Physiology*, *264*(4 Pt 2), H1283−H1291. Available from https://doi.org/10.1152/ajpheart.1993.264.4.H1283.

Zeng, Y., Zhang, X. F., Fu, B. M., & Tarbell, J. M. (2018). The role of endothelial surface glycocalyx in mechanosensing and transduction. *Advances in Experimental Medicine and Biology*, *1097*, 1−27. Available from https://doi.org/10.1007/978-3-319-96445-4_1.

Zhang, W. B., Aukland, K., Lund, T., & Wiig, H. (2000). Distribution of interstitial fluid pressure and fluid volumes in hind-limb skin of rats: Relation to meridians? *Clinical Physiology (Oxford, England)*, *20*(3), 242−249. Available from https://doi.org/10.1046/j.1365-2281.2000.00254.x.

CHAPTER 3

Mechanisms of lymphatic spread in colon cancer: insights from molecular and genetic studies

Mary Smithson[1] and Karin Hardiman[2]

[1]Department of Surgery, University of Alabama Birmingham, Birmingham, AL, United States
[2]Department of Surgery, Birmingham Veterans Affairs Medical Center, Birmingham, AL, United States

3.1 Introduction

The presence of lymph node metastases (LNMs) is one of the most important features of colorectal cancer (CRC) staging and helps guide treatment. The lymphatic system is critical for maintaining tissue fluid balance, transporting cells (including antigen and antigen-presenting cells to generate immune responses), and carrying lipids. The lymph nodes are the most common site of metastasis and are generally the first known area of spread in CRC. Currently, the 5-year survival with node-negative disease is 70%–80%, node-positive disease is anywhere from 30% to 60%, and with spread to distant metastasis survival is around 20% (Ong & Schofield, 2016). While identifying and understanding the spread to lymph nodes helps guide treatment, current mechanisms of spread and evolution of metastasis are under investigation.

3.2 Mechanisms of the lymphatic system

3.2.1 Lymph formation and movement

The lymphatic system is an interconnected system of conduits including capillaries, collecting vessels, lymph nodes, trunks, and ducts throughout the body. This system is a one-way transport system for fluids and proteins that collects them from the interstitial space and returns them to the blood (Swartz, 2001). Lymph is created from extracellular fluid but contains unique components that change in relation to physiological or pathological processes. Lymph is formed when tissue fluid pressure is greater than the pressure in the lymphatics which causes lymphatic endothelial cell valves to open (Padera et al., 2016). When the pressure inside the lymphatic system is higher, the valves close and trap the fluid inside. The endothelial cell valves also allow dendritic cells and antigen-presenting cells to enter. These cells are initially attracted by cytokines as well as the flow gradient into the lymphatic system. After the fluid and cells flow into the lymph system at the cell level, they then travel toward lymph nodes and eventually into the blood circulation. Initially, metastatic spread to the lymph system was thought to be a passive process but now is thought to be regulated at multiple steps. Solid tumors cause expansion of surrounding lymphatic channels, but these channels are restricted to the margin of the tumor, resulting in elevation of interstitial fluid pressure which then results in entry of cancer cells into the lymphatic network. Tumor-derived vascular endothelial growth factor-C (VEGF-C) and VEGF-D proteins increase the contraction of proximal collecting lymphatic vessels, potentially increasing lymph flow and thus tumor cell dissemination (Gogineni et al., 2013). For instance, Gogineni et al. found evidence of tumor-induced lymphatic vessel remodeling. They showed that when VEGF-C was inhibited, the spread of cancer cells to lymph nodes was reduced, indicating that remodeling is important for spread. Tumors are also constantly growing and as they grow, resistance to lymphatic flow in the tumor increases promoting diversion to other collateral channels. Kwon et al. (2014) utilized a murine model of melanoma in the limbs of mice and found that if the draining lymph nodes were removed, the animal showed different patterns of metastasis indicating that the lymph system is changeable and can adapt.

3.3 Tumor microenvironment

Other factors leading to spread to the lymph system are found in the microenvironment. In order to move from the primary tumor to the lymph nodes, tumor cells need to have lymphatics to travel through and evade the immune system while doing so. Harrell et al. (2007) showed lymphangiogenesis is associated with cancer metastasis and happens prior to cancer found at these distant sites. They did this by assessing draining lymph nodes in a mouse model with melanoma. They found that footpad tumors showed no lymphatic growth or blood vessel growth, yet the tumor-draining popliteal lymph node showed greatly increased lymphatic sinuses indicating that the primary tumors induce these distant alterations. In cancer models, lymph nodes also undergo recruitment of myeloid cells and reduction of effector lymphocyte numbers and function (Ogawa et al., 2014). The theory is that the lymph-transported molecules that originate from the primary tumors orchestrate these events as they arrive in the sentinel lymph node. There are also cytokines at work in these microenvironments such as interlukin-10, transforming growth factor-β (TGF-β), and granulocyte-macrophage colony stimulating factor, which assist in immunosuppression of the microenvironment (Leong et al., 2002). Another important step in the spread of cancer involves evading the immune system. Lymphatic endothelial cells lack the molecules necessary to active CD8 + T cells as well as express the immune checkpoint inhibitory ligand programmed death ligand 1 (PD-L1) (Tewalt et al., 2012). Due to this lack of a checkpoint, these endothelial cells can scavenge and cross-present self-antigens from the lymph on major histocompatibility complex class 1 molecules, leading to deletion of autoreactive CD8 + T cells (Lund et al., 2012). This process may promote survival of tumor cells in lymph nodes. Endothelial cells can also alter maturation of dendritic cells which can prevent an immune response. Kohrt et al. (2005) performed immunohistochemical analysis on sentinel axillary nodes in breast cancer patients and found that there were fewer effector T cells. Watanabe et al. (2008) found more myeloid-derived suppressor cells in tumor-draining lymph nodes compared to nontumor draining lymph nodes in a sarcoma mouse model, which results in inhibition of regulatory T cells and suppressed T-cell proliferation. Tumors evade the immune system as they metastasize via multiple mechanisms.

3.4 Colorectal cancer staging and treatment

3.4.1 Colorectal cancer staging

Staging in the United States utilizes the American Joint Committee on Cancer (AJCC) system (Amin et al., 2017). Tumor staging is estimated through a combination of imaging and pathologic data. TNM staging, which consists of characteristics of the primary tumor (T), lymph node (N) involvement, and distant metastasis (M), is used to predict disease recurrence and survival. T reflects the layer of the bowel wall invaded by the tumor with higher T stage indicating deeper invasion into the wall of the bowel or beyond, N refers to whether there are lymph nodes identified containing metastatic tumor cells, and M represents distinct metastasis. The standard is to assess 12 lymph nodes with the specimen. In order to achieve this, surgeons remove a swath of mesentery adjacent to the tumor dividing the feeding major vessel and then pathologists use a variety of methods to assess this tissue to identify lymph nodes which they assess further for metastatic tumor. The clearing of fat by pathology after surgical removal of the tumor and its associated mesentery can also increase nodes removed (Jin & Frankel, 2018). Studies have shown assessment of 12 lymph nodes is often hard to achieve and only 37% of sample meet this criteria (Baxter et al., 2005). Low node count can be influenced by depth of invasion, T-classification, AJCC stage, and location of tumor (Altintas & Bayrak, 2019). Interestingly, in the INT-0089 trial, improved survival was seen in patients where an increased number of nodes were assessed regardless if they are positive or negative (Haller et al., 2005). This suggests that node count may be affected by a biologic or tumor—host association or that an increased number of nodes may be necessary for proper stage categorization. This staging currently guides both our understanding of disease as well as treatment.

3.4.2 Colorectal cancer treatment

Treatment of CRC varies based on tumor staging and is often multimodal. When cancer has spread to the lymph nodes, this is labeled N1 disease if 1—3 nodes are seen and N2 disease if greater than 4 positive nodes are identified. This is stage 3 CRC. According to the National Comprehensive Cancer Network guidelines for colon cancer, any patient with positive lymph node identified on pathology of the surgical specimen should receive chemotherapy (National Comprehensive Cancer Network: Rectal Cancer). Following surgery, adjuvant chemotherapy has been shown to reduce the relative risk of cancer recurrence and cancer-related death by one-third in stage 3 patients but not stage 2 patients (negative lymph nodes) (Czaykowski et al., 2011). For T1–3, N1 patients undergo treatment with the combination of infusional 5-fluorouracil (5-FU), leucovorin, and oxaliplatin (FOLFOX) for 3–6 months (6 months preferred) after surgery or capecitabine and oxaliplatin (CAPEOX) for 3 months (National Comprehensive Cancer Networ: Colon Cancer). T4, N1–N2 patients undergo CAPEOX for 3–6 months or FOLFOX for 6 months. For rectal cancer, any patient with T3 or above or N1 or worse undergoes chemoradiation with 5-FU to address the primary tumor prior to an operation and systemic chemotherapy before or after their operation (National Comprehensive Cancer Networ: Colon Cancer). Systemic treatment with chemotherapy is thought to benefit patients with local metastasis to the lymph nodes because it is toxic to cancer cells not addressed by the surgery such as micrometastasis or circulating tumor cells.

3.5 Evolutionary mechanisms of lymph node metastasis

The spreading of cancer cells to the lymph nodes is one of the most important prognostic factors in CRC. The first sign of advanced disease is traditionally thought to be the presence of malignant cells in nearby lymph nodes. The mechanisms of how and timing of when cancer spreads to the lymph nodes is debated. The mechanism was initially thought to be passive, but now is thought to be regulated by cytokines and immune-modulating cells. Initially, the lymphatics attract cancer cells in order to expose them to immune cells so they can be destroyed, but this process becomes unregulated leading to cancer cell spread. In addition, it is not clear which tumor cells spread to the lymph nodes and whether they are genetically different than the cells that do not metastasize. There are three models of tumor evolution and progression that can be categorized into linear and parallel.

3.5.1 Linear models

3.5.1.1 Model 1

Models of tumor spread can be divided into linear and parallel systems and within these two systems are three models that represent possibilities of lymph node metastasis. The first system is linear progression and is the oldest and most well-known. The first model (Fig. 3.1) suggests that the primary tumor is genetically homogeneous. Eventually, the primary tumor acquires a genetic alteration (mutation or copy number change) that takes over the tumor in a process called a clonal sweep and then the tumor cells spread to the lymph nodes via the lymphatic vessels in order of proximity to the primary site (Fearon & Vogelstein, 1990). This linear model of tumor progression currently dictates staging and in turn guides decision-making and treatment across most types of adult cancer. Depending upon the type of cancer, determining which lymph nodes harbor cancer then dictates correct extent of lymphadenopathy which can often increase morbidity to a

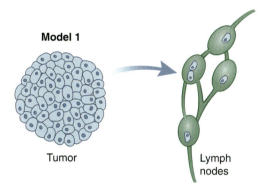
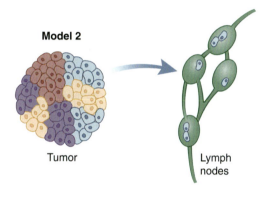
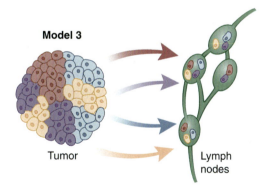

FIGURE 3.1 Three models of lymph node metastasis. Source: *From Kandagatla, P., Maguire, L. H., & Hardiman, K. M. (2018). Biology of nodal spread in colon cancer: Insights from molecular and genetic studies. European Surgical Research, 59 (5–6), 361–370. https://doi.org/10.1159/000494832.*

patient. This model supports the notion that lymph nodes closest to the tumor would have metastasis, but this is a linear progression, so if there was a lymph node without cancer, the nodes beyond would be negative and the dissection could stop which is the theory behind the treatment paradigm in breast cancer and melanoma where sentinel lymph node testing guides intraoperative dissection (Currie, 2019). In this model, the primary and metastatic lesions are genetically very similar.

3.5.1.2 Intratumor heterogeneity

An important recent finding over the last several years is the discovery that all adult tumors tested thus far exhibit genetic intratumor heterogeneity (Cros et al., 2018; Hardiman et al., 2016; Marusyk & Polyak, 2010; Sagaert et al., 2018). This means that when different areas of a tumor are assessed for mutations and for copy number changes, they are not the same. This heterogeneity is thought to be due to the presence and uneven distribution of genetically related subclones. When a genetic alteration (either a mutation or a copy number change) increases a cell's fitness, it will divide more and more of the subsequent cells will contain that alteration. Tumor cells with the same set of genetic alterations are called subclones. These subclones are genetically related through a phylogeny. For the most part, these subclones are defined mathematically but some researchers have been able to grow individual subclones and compare their characteristics (Frydrych et al., 2019; Hardiman et al., 2016; Roerinnk et al., 2018). Some mutations and copy number changes are identified throughout the tumor because they are early events in tumor evolution and others are private (not shared) which is because they occur later. These genetically related parts of a tumor can be arranged into a phylogeny. A phylogenetic classification is a system that arranges samples into groups based on their evolutionary origins and relationships. A hierarchical phylogeny is a phylogenetic tree with a recent common ancestor. This knowledge has been improved by advances in sequencing techniques and development of informatics tools. Intratumor heterogeneity occurs because when neoplastic cells divide, due to poor fidelity during DNA replication, they acquire further mutations which are not fixed, and then these altered cells separate spatially (Gerlinger et al., 2012). This results in spatially separated cells in a tumor that are more different from each other than close neighbors (Yachida et al., 2010). As a result, mutations that decrease DNA repair mechanisms (*MSH2*, *MSH6*, *MLH1*, and *PMS2*) as well as mutations that allow the tumor itself to better tolerate genomic instability (*TP53*, *ATM*) increase the mutational burden and in turn tumor heterogeneity. A study by Baldus et al. (2010) assessed colon cancer intratumor heterogeneity in the driver genes *KRAS*, *BRAF*, and *PIK3CA* and revealed that the primary tumor and metastatic lymph nodes often had discordance in the presence of mutations. Other similar studies have now been performed including one by Kosmidou et al. (2014), which identified discordance between the tumor center and the periphery 44% of the time when they assessed *KRAS* mutations. Jeantet et al. (2016) examined *KRAS* and *NRAS* mutations in patients with colon cancer and found intratumor heterogeneity 33% of the time among the samples from any one tumor.

3.5.1.3 Model 2

The discovery of intratumor heterogeneity and tumor subclones required generation of a new model of tumor spread. This second model suggests a heterogeneous tumor which then undergoes selective pressure until one subclone gains the ability to metastasize (Fig. 3.1). Genetically, in this model, as the tumor cells divide, selective pressures favor the development of multiple subclones which succeeds as the tumor undergoes molecular evolution. However, a single subclone then gains the ability to colonize regional lymph nodes and then from there goes on to create distant metastasis. In this model, some divergence is expected between primary and metastatic lesions, but genetic analysis should produce a hierarchical phylogeny where the primary tumor is parental and progresses in a linear way to the metastasis. Both of these models of linear progression have recently fallen out of favor due to more extensive genetic modeling of lymph node and distant metastasis revealing far more complexity.

3.5.1.4 Parallel model

The second system of tumor spread is a parallel one and has just one model, called a parallel progression model. This proposes that tumor subclones can metastasize early in tumorigenesis and evolve in parallel at each site with the primary tumor (Ulintz et al., 2018). Model 3 suggests that there are multiple distinct waves of early and late monoclonal and polyclonal events where cells are exiting the heterogeneous tumor in parallel (at the same time as) with primary tumor growth and evolution (Fig. 3.1). Subclones colonize lymph nodes and evolve temporarily in parallel to the primary tumor, but under a different set of environmental pressures. Mutational divergence in this theory would be expected to be high and the phylogeny should be complex but interrelated.

3.5.1.5 Model 3

This last model is supported by multiple recent publications assessing tumor evolution using sophisticated sequencing techniques alongside complex mathematical models. One publication from our group supporting this model was performed by assessing seven different primary CRCs and one to four matched lymph nodes per tumor (Ulintz et al., 2018). Each sample underwent assessment via next-generation whole-exome DNA sequencing, OncoScan arrays, and deep confirmatory sequencing. Mutations and copy number events were mapped between each primary tumor and their LNMs, and a computational program was used to define subclones within each tumor. Ulintz et al. (2018) found heterogeneity in mutations and copy number changes among all samples. For every patient, the primary tumor and matched LNMs were polyclonal, and for some these subclones differed from one lymph node to another even within the same patient. Supporting this finding in those patients, when mutations that were private to particular locations in the primary tumor were assessed for in individual lymph nodes, we found that there were mutations in individual lymph nodes that were private to multiple different regions of the primary tumor indicating that each lymph node contained subclones originated from multiple distinct regions with the primary tumor. When a similar analysis was performed assessing copy number alterations, the findings were the same. The authors proposed a theory where multiple waves of metastatic cells from the primary tumor seed lymph nodes over time. An additional study by Ryser et al. (2020) further supports the parallel progression model. They assessed ancestral, topographic, and phenotypic information in 12 colorectal tumors. They found that 60% of subclones share superficial and invasive phenotypes and 9 of 11 tumors assessed show evidence of multiclonal invasion and that these invasive subclones originate early in the phylogeny. Early multiclonal invasion indicates coevolving subclones with similar malignant potential as opposed to bottlenecks in tumor progression and suggests minimal barriers to invasion during CRC growth.

Naxerova et al. (2017) performed another study supporting parallel progression in which they assessed 213 CRC biopsy samples from 17 patients and used somatic variants in hypermutable DNA regions to make phylogenetic trees. They found that in 65% of cases, lymphatic and distant metastases arose from independent subclones from the primary tumor, supporting the multiple waves model. In only 35% of cases, the lymph nodes and distant metastases share a common subclonal origin. One theory to explain these similarities is attributed to a common origin in a subset of patients. Reiter et al. (2020) compared genetic differences in LNMs with more distant metastases utilizing 317 CRC samples from 20 patients. This study investigated the genetic makeup of lymph nodes versus distant metastases using a mathematical framework for quantifying phylogenetic diversity. They found that distant metastases are typically monophyletic and more genetically similar in contrast to lymph nodes metastases, which display high levels of intralesion diversity and heterogeneity. This support the theory that the two sites are subject to different levels of selection and that distant metastasis go through an evolutionary bottleneck that lymph node metastasis are not subject to. In order for a tumor to achieve metastasis, it has to overcome leaving the primary site, evading the immune system, and growth in a new environment. While there are clearly also barriers to metastasizing to local lymph nodes, these appear to be achievable for a greater variety of subclones from the primary tumor. There are different evolutionary pressures at different sites that can alter the genetics and spread of metastasis. Further research is still needed to better understand the role of tumor evolution in metastasis.

3.6 Molecular markers of colorectal cancer spread

There are multiple genetic alterations that commonly occur in CRC. Genetic alterations in the *APC*, *KRAS*, and *BRAF* genes are significant early events in the transition from normal colonic epithelium to adenoma (Huang et al., 2018). The Cancer Genome Atlas study of genetic alterations in CRC found that each CRC is unique and is characterized by either 100 or so mutations along with many copy number alterations (nonhypermutated tumors) or 1000 or more mutations (hypermutated tumors) (Cancer Genome Atlas, 2012). They established 32 recurrently mutated genes across both tumor subtypes. Pathway analysis revealed recurrent changes in the WNT (wingless/integrated), MAPK (mitogen-activated protein kinase), TGF-β, and p53 pathways. More recently, a study by Zaidi et al. (2020) reported mutation data for targeted sequencing of 205 genes and correlated survival data in 2105 patients. The study confirmed that patients with hypermutated tumors had improved survival which perhaps indicates that genomic heterogeneity only benefits a tumor to a certain extent and then after a certain point becomes detrimental. Additionally, after accounting for multiple testing and clinical factors, they did not identify any single gene whose alteration was correlated with survival. While multiple mutations and events have been established as important, which mutations are the key drivers involved in metastasis is still unknown and there are multiple contradictory studies. One study by Ko et al. (1998) showed no significant correlation between particular mutations with stage, tumor location, or metastasis. More recent studies have been performed revealing some evidence for a role of particular molecular markers which are discussed below.

3.7 Molecular markers of metastases

A study by Huang et al. (2018) identified six driver genes and through a meta-analysis showed that *KRAS* and *TP53* mutations were associated most often with CRC metastasis, including lymphatic and distant metastasis. Interestingly, a subgroup analysis stratified by patient ethnicity showed that the *BRAF* mutation was related to CRC metastasis. Another study investigated FOXM1 expression in 203 cases of primary colon cancer and found FOXM1 was an independent prognostic factor and experimentally, overexpression via transfection experiments significantly promoted growth and metastasis of colon cancer cells (Li et al., 2013). Li et al. found that eIF-3, an essential factor for initiation of protein synthesis in mammalian cells, has increased expression that is significantly correlated with both lymph node and distant metastasis (Li et al., 2013). Another potentially important biomarker is ZEB2 (zinc finger E box-binding homeobox 2) (Sreekumar et al., 2018). One study assessed ZEB2 expression using a validated scoring system. ZEB2 scoring was analyzed for 126 patients and ZEB2 positivity was found to predict reduced overall survival and disease-free survival after resection independent of stage in CRC patients. The authors also assessed the addition of ZEB2 to existing TNM staging and found that it improves the ability to stratify patients for risk of recurrence. Other markers that have been associated with metastasis include SMAD4, VEGF, insulin-like growth factor 2 mRNA-binding protein 3, and TRAF2- and NICK-interactive kinase (Peluso et al., 2017). Takahashi et al. (2015) studied expression profiles of cancer cells in 152 patients with stage 1–3 CRC patients using microarray analysis. They found relapse-free survival was significantly worse in the TNIK high expression group than in the TNIK low expression group in stage 2 and stage 3 patients. High TNIK expression was identified as an independent risk factor of distant recurrence in stage 3 patients.

Some alterations have been specifically associated with colorectal lymph nodes metastasis. For example, Kasem et al. (2014) studied *JK-1* gene DNA copy number changes in 211 CRCs. JK-1 is a gene in the p-arm region of chromosome 5, which plays a crucial role in the pathogenesis of cancer. This gene is found to be overexpressed in human esophageal squamous carcinoma cells lines and has shown to increase growth rate, colony formation, and foci formation in mouse and human cell lines. Copy number deletion of *JK-1* was found to be more frequent in colorectal adenocarcinomas than adenomas and conferred an increased risk of nodal disease. Other studies have looked at factors involving the tumor microenvironment as predictors of metastasis and poor outcomes. TGF-β is a cytokine that mediates many biological processes including cells homeostasis, angiogenesis, and immune regulation. This cytokine can act as both a tumor suppressor and a tumor promotor (Huang et al., 2018). TGF-β may suppress cancer initiation but in cancers that do form, there is increased expression of TGF-β ligands which are correlated with increasing tumor stage, depth of tumor invasion, and metastasis (Shim et al., 1999). Tauriello et al. (2018) studied four CRC mutations in mice (*KRAS, APC, p53,* and *TGFBR2*). They found inhibition of the PD-1/PD-L1 (programmed cell death protein 1/programmed death ligand 1) immune checkpoint provoked a limited response but by inhibiting TGF-β with a TGF-β receptor specific inhibitor, there was a potent and enduring cytotoxic T-cell response against tumor cells that prevented metastasis. When cytotoxic T lymphocytes were removed, the therapeutic effect was abolished. They concluded that increased TGF-β in the microenvironment represents a primary mechanism of immune evasion. A separate study by Rokavec et al. (2017) assessed a cellular model of colon cancer progression (DLD 1 colon cancer cells) and performed a genome-wide characterization of transcriptional and epigenetic profiles as well as a correlative analysis of expression and methylation. They found global hypomethylation of gene-regulatory regions was associated with metastases; upregulation of an axon-guidance-related gene signature was the most significant feature of metastatic tumor cells; and elevated expression of H19 lncRNA due to promoter demethylation was observed in cells from metastasis. The largest study reporting targeted sequencing of 1134 metastatic CRC s identified differences between right- and left-sided tumors with right-sided tumors showing enrichment for alterations in *KRAS, BRAF, PIK3CA, AKT1, RNF43,* and *SMAD4* (Yaeger et al., 2018). The mechanisms whereby these genetic defects are related to clinical differences seen in patients are yet to be determined. A diverse array of genetic modifications has been found to influence metastasis.

3.8 Conclusions

Metastasis to lymph nodes is a very important feature of colon cancer progression, but the mechanisms of spread are still being elucidated. The traditional view of linear progression of a clonal cancer from the primary tumor to the lymph nodes and distant sites has fallen out of favor in light of recent evidence. A parallel progression model whereby a heterogeneous tumor sends multiples waves of subclones to seed lymph nodes heterogeneously has been supported by more recent evidence. This discovery as well as multiple genetic and transcriptional alterations that have been identified that contribute to metastasis are vital to understanding the process of metastasis. The therapeutic implications of these new models of tumor spread are widespread. Individual subclones potentially harbor heterogeneous mechanisms of resistance pointing to the need for a more targeted and individualized approach.

Acknowledgments

We have no acknowledgments.

Conflict of interest

All authors declare that they have no conflict of interest.

References

Altintas, S., & Bayrak, M. (2019). Assessment of factors influencing lymph node count in colorectal cancer. *Journal of the College of Physicians and Surgeons—Pakistan: JCPSP, 29*(12), 1173–1178. Available from https://doi.org/10.29271/jcpsp.2019.12.1173.

Amin, M. B., Edge, S., Greene, F., Byrd, D., Brookland, R. K., Washington, M. K., Gershenwald, J. E., Compton, C. C., ... Hess, K. R. (2017). *AJCC Cancer Staging Manual* (8th ed.). Springer International Publishing: American Joint Commission on Cancer.

Available from https://www.nccn.org/professionals/physician_gls/pdf/colon.pdf. (2021), version 3.2021.

Available from https://www.nccn.org/professionals/physician_gls/pdf/rectal.pdf. (2021), version 2.2021.

Baldus, S. E., Schaefer, K. L., Engers, R., Hartleb, D., Stoecklein, N. H., & Gabbert, H. E. (2010). Prevalence and heterogeneity of KRAS, BRAF, and PIK3CA mutations in primary colorectal adenocarcinomas and their corresponding metastases. *Clinical Cancer Research: An Official Journal of the American Association for Cancer Research*, 16(3), 790−799. Available from https://doi.org/10.1158/1078-0432.CCR-09-2446.

Baxter, N. N., Virnig, D. J., Rothenberger, D. A., Morris, A. M., Jessurun, J., & Virnig, B. A. (2005). Lymph node evaluation in colorectal cancer patients: A population-based study. *Journal of the National Cancer Institute*, 97(3), 219−225. Available from https://doi.org/10.1093/jnci/dji020.

Cancer Genome Atlas Network. (2012). Comprehensive molecular characterization of human colon and rectal cancer. *Nature*, 487(7407), 330−337. Available from https://doi.org/10.1038/nature11252.

Cros, J., Raffenne, J., Couvelard, A., & Pote, N. (2018). Tumor heterogeneity in pancreatic adenocarcinoma. *Pathobiology: Journal of Immunopathology, Molecular and Cellular Biology*, 85(1−2), 64−71. Available from https://doi.org/10.1159/000477773.

Currie, A. C. (2019). Intraoperative sentinel node mapping in the colon: Potential and pitfalls. *European Surgical Research*, 60(1−2), 45−52. Available from https://doi.org/10.1159/000494833.

Czaykowski, P. M., Gill, S., Kennecke, H. F., Gordon, V. L., & Turner, D. (2011). Adjuvant chemotherapy for stage III colon cancer: Does timing matter? *Diseases of the Colon and Rectum*, 54(9), 1082−1089. Available from https://doi.org/10.1097/DCR.0b013e318223c3d6.

Fearon, E. R., & Vogelstein, B. (1990). A genetic model for colorectal tumorigenesis. *Cell*, 61, 759−767.

Frydrych, L. M., Ulintz, P., Bankhead, A., Sifuentes, C., Greenson, J., Maguire, L., Irwin, R., Fearon, E. R., & Hardiman, K. M. (2019). Rectal cancer sub-clones respond differentially to neoadjuvant therapy. *Neoplasia (New York, N.Y.)*, 21(10), 1051−1062. Available from https://doi.org/10.1016/j.neo.2019.08.004.

Gerlinger, M., Rowan, A. J., Horswell, S., Math, M., Larkin, J., Endesfelder, D., Gronroos, E., Martinez, P., Matthews, N., Stewart, A., Tarpey, P., Varela, I., Phillimore, B., Begum, S., McDonald, N. Q., Butler, A., Jones, D., Raine, K., Latimer, C., ... Swanton, C. (2012). Intratumor heterogeneity and branched evolution revealed by multiregion sequencing. *The New England Journal of Medicine*, 366(10), 883−892. Available from https://doi.org/10.1056/NEJMoa1113205.

Gogineni, A., Caunt, M., Crow, A., Lee, C. V., Fuh, G., van Bruggen, N., Ye, W., & Weimer, R. M. (2013). Inhibition of VEGF-C modulates distal lymphatic remodeling and secondary metastasis. *PLoS One*, 8(7), e68755. Available from https://doi.org/10.1371/journal.pone.0068755.

Haller, D. G., Catalano, P. J., Macdonald, J. S., O'Rourke, M. A., Frontiera, M. S., Jackson, D. V., & Mayer, R. J. (2005). Phase III study of fluorouracil, leucovorin, and levamisole in high-risk stage II and III colon cancer: Final report of Intergroup 0089. *Journal of Clinical Oncology: Official Journal of the American Society of Clinical Oncology*, 23(34), 8671−8678. Available from https://doi.org/10.1200/JCO.2004.00.5686.

Hardiman, K. M., Ulintz, P. J., Kuick, R. D., Hovelson, D. H., Gates, C. M., Bhasi, A., Rodrigues Grant, A., Liu, J., Cani, A. K., Greenson, J. K., Tomlins, S. A., & Fearon, E. R. (2016). Intra-tumor genetic heterogeneity in rectal cancer. *Laboratory Investigation; A Journal of Technical Methods and Pathology*, 96(1), 4−15. Available from https://doi.org/10.1038/labinvest.2015.131.

Harrell, M. I., Iritani, B. M., & Ruddell, A. (2007). Tumor-induced sentinel lymph node lymphangiogenesis and increased lymph flow precede melanoma metastasis. *The American Journal of Pathology*, 170(2), 774−786. Available from https://doi.org/10.2353/ajpath.2007.060761.

Huang, D., Sun, W., Zhou, Y., Li, P., Chen, F., Chen, H., Xia, D., Xu, E., Lai, M., Wu, Y., & Zhang, H. (2018). Mutations of key driver genes in colorectal cancer progression and metastasis. *Cancer Metastasis Reviews*, 37(1), 173−187. Available from https://doi.org/10.1007/s10555-017-9726-5.

Jeantet, M., Tougeron, D., Tachon, G., Cortes, U., Archambaut, C., Fromont, G., & Karayan-Tapon, L. (2016). High intra- and inter-tumoral heterogeneity of RAS mutations in colorectal cancer. *International Journal of Molecular Science*, 17(12). Available from https://doi.org/10.3390/ijms17122015.

Jin, M., & Frankel, W. L. (2018). Lymph node metastasis in colorectal cancer. *Surgical Oncology Clinics of North America*, 27(2), 401−412. Available from https://doi.org/10.1016/j.soc.2017.11.011.

Kasem, K., Gopalan, V., Salajegheh, A., Lu, C. T., Smith, R. A., & Lam, A. K. (2014). JK1 (FAM134B) gene and colorectal cancer: a pilot study on the gene copy number alterations and correlations with clinicopathological parameters. *Experimental and Molecular Pathology*, 97(1), 31−36. Available from https://doi.org/10.1016/j.yexmp.2014.05.001.

Ko, J., Cheung, M., Wong, C., Lau, K., Tang, C., Kwan, M., & Lung, M. (1998). Ki-ras codon 1 point mutational activation in Hong Kong colorectal carcinoma patients. *Cancer Letters*, 134, 169−176.

Kohrt, H. E., Nouri, N., Nowels, K., Johnson, D., Holmes, S., & Lee, P. P. (2005). Profile of immune cells in axillary lymph nodes predicts disease-free survival in breast cancer. *PLoS Medicine*, 2(9), e284. Available from https://doi.org/10.1371/journal.pmed.0020284.

Kosmidou, V., Oikonomou, E., Vlassi, M., Avlonitis, S., Katseli, A., Tsipras, I., Mourtzoukou, D., Kontogeorgos, G., Zografos, G., & Pintzas, A. (2014). Tumor heterogeneity revealed by KRAS, BRAF, and PIK3CA pyrosequencing: KRAS and PIK3CA intratumor mutation profile differences and their therapeutic implications. *Human Mutation*, *35*(3), 329–340. Available from https://doi.org/10.1002/humu.22496.

Kwon, S., Agollah, G. D., Wu, G., & Sevick-Muraca, E. M. (2014). Spatio-temporal changes of lymphatic contractility and drainage patterns following lymphadenectomy in mice. *PLoS One*, *9*(8), e106034. Available from https://doi.org/10.1371/journal.pone.0106034.

Leong, S., Peng, M., Zhou, Y., Vaquerano, J., & Chang, J. (2002). Cytokine profiles of sentinel lymph nodes draining the primary melanoma. *Annals of Surgical Oncology*, *9*(1), 82–87.

Li, D., Wei, P., Peng, Z., Huang, C., Tang, H., Jia, Z., Cui, J., Le, X., Huang, S., & Xie, K. (2013). The critical role of dysregulated FOXM1-PLAUR signaling in human colon cancer progression and metastasis. *Clinical Cancer Research: An Official Journal of the American Association for Cancer Research*, *19*(1), 62–72. Available from https://doi.org/10.1158/1078-0432.CCR-12-1588.

Lund, A. W., Duraes, F. V., Hirosue, S., Raghavan, V. R., Nembrini, C., Thomas, S. N., Issa, A., Hugues, S., & Swartz, M. A. (2012). VEGF-C promotes immune tolerance in B16 melanomas and cross-presentation of tumor antigen by lymph node lymphatics. *Cell Reports*, *1*(3), 191–199. Available from https://doi.org/10.1016/j.celrep.2012.01.005.

Marusyk, A., & Polyak, K. (2010). Tumor heterogeneity: Causes and consequences. *Biochimica et Biophysica Acta*, *1805*(1), 105–117. Available from https://doi.org/10.1016/j.bbcan.2009.11.002.

Naxerova, K., Reiter, J. G., Brachtel, E., Lennerz, J. K., van de Wetering, M., Rowan, A., Cai, T., Clevers, H., Swanton, C., Nowak, M. A., Elledge, S. J., & Jain, R. K. (2017). Origins of lymphatic and distant metastases in human colorectal cancer. *Science (New York, N.Y.)*, *357*(6346), 55–60. Available from https://doi.org/10.1126/science.aai8515.

Ogawa, F., Amano, H., Eshima, K., Ito, Y., Matsui, Y., Hosono, K., Kitasato, H., Iyoda, A., Iwabuchi, K., Kumagai, Y., Satoh, Y., Narumiya, S., & Majima, M. (2014). Prostanoid induces premetastatic niche in regional lymph nodes. *The Journal of Clinical Investigation*, *124*(11), 4882–4894. Available from https://doi.org/10.1172/JCI73530.

Ong, M. L., & Schofield, J. B. (2016). Assessment of lymph node involvement in colorectal cancer. *World Journal of Gastrointestinal Surgery*, *8*(3), 179–192. Available from https://doi.org/10.4240/wjgs.v8.i3.179.

Padera, T. P., Meijer, E. F., & Munn, L. L. (2016). The lymphatic system in disease processes and cancer progression. *Annual Review of Biomedical Engineering*, *18*, 125–158. Available from https://doi.org/10.1146/annurev-bioeng-112315-031200.

Peluso, G., Incollingo, P., Calogero, A., Tammaro, V., Rupealta, N., Chiaccio, G., Sandoval Sotelo, M. L., Minieri, G., Pisani, A., Riccio, E., Sabbatini, M., Bracale, U. M., Dodaro, C. A., & Carlomagno, N. (2017). Current tissue molecular markers in colorectal cancer: A literature review. *BioMed Research International*, *2017*, 2605628. Available from https://doi.org/10.1155/2017/2605628.

Reiter, J. G., Hung, W. T., Lee, I. H., Nagpal, S., Giunta, P., Degner, S., Liu, G., Wassenaar, E. C. E., Jeck, W. R., Taylor, M. S., Farahani, A. A., Marble, H. D., Knott, S., Kranenburg, O., Lennerz, J. K., & Naxerova, K. (2020). Lymph node metastases develop through a wider evolutionary bottleneck than distant metastases. *Nature Genetics*, *52*(7), 692–700. Available from https://doi.org/10.1038/s41588-020-0633-2.

Roerinnk, S. F., Sasaki, N., Lee-Six, H., Young, M., Alexandrov, L., Behjati, S., Mitchell, T., Grossman, S., Lightfoot, H., Egan, D., Pronk, A., Smakman, N., Van Gorp, J., Anderson, E., Gamble, S., Alder, C., Van de Wetering, M., Campbell, P., Stratton, M., ... Clevers, H. (2018). Intra-tumour diversification in colorectal cancer at the single-cell level. *Nature*, *556*(7702), 457–462.

Rokavec, M., Horst, D., & Hermeking, H. (2017). Cellular model of colon cancer progression reveals signatures of mRNAs, miRNA, lncRNAs, and epigenetic modifications associated with metastasis. *Cancer Research*, *77*(8), 1854–1867. Available from https://doi.org/10.1158/0008-5472.CAN-16-3236.

Ryser, M. D., Mallo, D., Hall, A., Hardman, T., King, L. M., Tatishchev, S., Sorribes, I. C., Maley, C. C., Marks, J. R., Hwang, E. S., & Shibata, D. (2020). Minimal barriers to invasion during human colorectal tumor growth. *Nature Communications*, *11*(1), 1280. Available from https://doi.org/10.1038/s41467-020-14908-7.

Sagaert, X., Vanstapel, A., & Verbeek, S. (2018). Tumor heterogeneity in colorectal cancer: What do we know so far? *Pathobiology: Journal of Immunopathology, Molecular and Cellular Biology*, *85*(1–2), 72–84. Available from https://doi.org/10.1159/000486721.

Shim, K., Kim, K., Han, W., & PArk, E. (1999). Elevated serum levels of transforming growth factor-B1 in patients with colorectal carcinoma. *Cancer, 85*(3), 554–561.

Sreekumar, R., Harris, S., Moutasim, K., DeMateos, R., Patel, A., Emo, K., White, S., Yagci, T., Tulchinsky, E., Thomas, G., Primrose, J. N., Sayan, A. E., & Mirnezami, A. H. (2018). Assessment of nuclear ZEB2 as a biomarker for colorectal cancer outcome and TNM risk stratification. *JAMA Network Open, 1*(6), e183115. Available from https://doi.org/10.1001/jamanetworkopen.2018.3115.

Swartz, M. A. (2001). The physiology of the lympahtic system. *Advanced Drug Delivery Review, 50*, 3–20.

Takahashi, H., Ishikawa, T., Ishiguro, M., O'kazaki, S., Mogushi, K., Kobayashi, H., Lida, S., Mizushima, H., Tanaka, H., Uetake, H., & Sugihara, K. (2015). Prognostic significance of Traf2- and Nck- interacting kinase (TNIK) in colorectal cancer. *BMC Cancer, 15*, 794.

Tauriello, D. V. F., Palomo-Ponce, S., Stork, D., Berenguer-Llergo, A., Badia-Ramentol, J., Iglesias, M., Sevillano, M., Ibiza, S., Canellas, A., Hernando-Momblona, X., Byrom, D., Matarin, J. A., Calon, A., Rivas, E. I., Nebreda, A. R., Riera, A., Attolini, C. S., & Batlle, E. (2018). TGFbeta drives immune evasion in genetically reconstituted colon cancer metastasis. *Nature, 554*(7693), 538–543. Available from https://doi.org/10.1038/nature25492.

Tewalt, E. F., Cohen, J. N., Rouhani, S. J., Guidi, C. J., Qiao, H., Fahl, S. P., Conaway, M. R., Bender, T. P., Tung, K. S., Vella, A. T., Adler, A. J., Chen, L., & Engelhard, V. H. (2012). Lymphatic endothelial cells induce tolerance via PD-L1 and lack of costimulation leading to high-level PD-1 expression on CD8 T cells. *Blood, 120*(24), 4772–4782. Available from https://doi.org/10.1182/blood-2012-04-427013.

Ulintz, P. J., Greenson, J. K., Wu, R., Fearon, E. R., & Hardiman, K. M. (2018). Lymph node metastases in colon cancer are polyclonal. *Clinical Cancer Research: An Official Journal of the American Association for Cancer Research, 24*(9), 2214–2224. Available from https://doi.org/10.1158/1078-0432.CCR-17-1425.

Watanabe, S., Deguchi, K., Zheng, R., Tamai, H., Wang, L. X., Cohen, P. A., & Shu, S. (2008). Tumor-induced CD11b + Gr-1 + myeloid cells suppress T cell sensitization in tumor-draining lymph nodes. *Journal of Immunology, 181*(5), 3291–3300. Available from https://doi.org/10.4049/jimmunol.181.5.3291.

Yachida, S., Jones, S., Bozic, I., Antal, T., Leary, R., Fu, B., Kamiyama, M., Hruban, R. H., Eshleman, J. R., Nowak, M. A., Velculescu, V. E., Kinzler, K. W., Vogelstein, B., & Iacobuzio-Donahue, C. A. (2010). Distant metastasis occurs late during the genetic evolution of pancreatic cancer. *Nature, 467*(7319), 1114–1117. Available from https://doi.org/10.1038/nature09515.

Yaeger, R., Chatila, W. K., Lipsyc, M. D., Hechtman, J. F., Cercek, A., Sanchez-Vega, F., Jayakumaran, G., Middha, S., Zehir, A., Donoghue, M. T. A., You, D., Viale, A., Kemeny, N., Segal, N. H., Stadler, Z. K., Varghese, A. M., Kundra, R., Gao, J., Syed, A., ... Schultz, N. (2018). Clinical sequencing defines the genomic landscape of metastatic colorectal cancer. *Cancer Cell, 33*(1), 125–136. Available from https://doi.org/10.1016/j.ccell.2017.12.004, e3.

Zaidi, S. H., Harrison, T. A., Phipps, A. I., Steinfelder, R., Trinh, Q. M., Qu, C., Banbury, B. L., Georgeson, P., Grasso, C. S., Giannakis, M., Adams, J. B., Alwers, E., Amitay, E. L., Barfield, R. T., Berndt, S. I., Borozan, I., Brenner, H., Brezina, S., Buchanan, D. D., ... Peters, U. (2020). Landscape of somatic single nucleotide variants and indels in colorectal cancer and impact on survival. *Nature Communications, 11*(1), 3644. Available from https://doi.org/10.1038/s41467-020-17386-z.

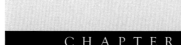

Anatomy and embryology of the lymphatic system of the colon and rectum

Wouter Willaert

Department of Human Structure and Repair, Ghent University, Ghent, Belgium

4.1 Introduction

Compared with blood vessels, the anatomy of the lymphatic system or lymphoid system is highly complex and varied. This system includes up to 1000 lymph nodes and lymphatic vessels that are predominantly very small. Together they form a large network of interlacing vessels with incorporated lymph nodes that spans nearly the entire human body. Unlike the vascular system, lymph vessels do not have a central pump but contain contractile compartments termed lymphangions, which propel fluid called lymph through a one-way low-pressure drainage system. This system starts blindly in the interstitial space and contains numerous independent compartments with different functions that progressively coalesce to empty into the venous circulation.

The main function of the lymphatic circulation is collecting and returning redundant interstitial fluid as transparent lymph from the soft tissues to the veins using lymphovenous connections. Other important functions are absorption and transportation of dietary fat as milky chyle by the intestinal lymphatic vessels and the delivery of hepatic proteins by the liver lymphatics into the systemic circulation. Lymph nodes, which are incorporated in the lymph vessel system, are responsible for immune defense (Goswami et al., 2020; Hsu & Itkin, 2016; Swartz, 2001). Emerging evidence reveals that lymphatic endothelium plays an active role in immune regulation both by expression of immunomodulatory genes and by antigen presentation (Ulvmar & Mäkinen, 2016).

4.2 Embryology of the lymphatic system

4.2.1 Molecular mechanisms

Research on the molecular mechanisms of lymphatic development has been initiated more than two decades ago. Here, a brief overview focuses on the most essential molecules involved in lymphatic development. Surely, the list of molecules with a lymphatic connection will further grow in the near future (Jha et al., 2018). In humans, lymphatic system development starts at about gestational weeks 6–7 and has a venous origin, as postulated already in 1902 by the American anatomist Florence Sabin (Tammela & Alitalo, 2010; Wigle & Oliver, 1999). Cardinal veins, which are responsible for the main venous drainage of the cranial and caudal parts of the embryo, express high levels of vascular endothelial growth factor receptor 3 (VEGFR-3, also known as FLT4), while a subpopulation of endothelial precursor cells in these veins upregulate lymphatic vessel hyaluronan receptor-1 (LYVE-1) (Kaipainen et al., 2019). The homeobox transcription factor sex-determining region Y box 18 is expressed in the LYVE-1-positive lymphatic endothelial cell (LEC) precursors. LECs also differ from most other venous cells by direct activation of prospero homeobox 1 (PROX1), the first marker for LEC determination (François et al., 2008; Wigle & Oliver, 1999). LEC specification occurs at the ventral side of the cardinal vein (Nicenboim et al., 2015). Subsequently, these cells migrate to the dorsal aspect of this vein. PROX1 regulates lymphatic identity and its continuance by directly expressing LEC-specific genes and suppressing blood endothelial cell-specific genes in cooperation with its binding partners, such as the nuclear receptor chicken ovalbumin upstream promoter-transcription factor II (Vaahtomeri et al., 2017). VEGFR-3 expression is downregulated in the blood vessels but remains elevated in the LEC precursors, which also start to express neuropilin-2, making them more responsive to VEGF-C signals derived from the lateral mesenchyme. These signals are required for sprouting of the LECs, which form primordial lumenized lymphatic vascular structures or lymph sacs on the lateral aspects of the cardinal veins. The formation of lymph sacs is considered as the first known event of lymph node development. The LECs initiate the expression of podoplanin that stimulates the C-type lectin-like receptor 2 (CLEC-2) in platelets, leading to the activation of platelet tyrosine kinase Syk and platelet aggregation. This process blocks lymphovenous connections and leads to the separation of the blood and lymphatic vascular system.

4.2.2 The lymph sacs

Next, from these primitive lymph sacs, LECs begin to sprout and connective tissue protrudes into the lymph sac, establishing the first formation of the lymph nodes. The lymph sacs give rise to the lymphatic vascular network by sprouting of lymph vessels. This peripheral lymphatic system is generated by centrifugal sprouting and proliferation of LECs belonging to the lymph sacs, driven by VEGF-C/VEGFR-3 and metalloproteinase with thrombospondin motifs 3 (ADAMTS3)/collagen- and calcium-binding EGF domains 1 (Ccbe1) signals (Jha, Rauniyar, & Jeltsch, 2018). The final steps in developmental lymphangiogenesis involve differentiation to lymphatic capillaries, precollectors, collecting vessels, and lymph nodes (Vaahtomeri et al., 2017). Of note, although the majority of the lymphatic vascular plexuses arise from PROX1-positive venous endothelia, nonvenous sources also contribute to lymphatic development in various tissues. For example, part of the mesenteric lymphatic vessels are formed from isolated groups of LECs, originating from blood-forming hemogenic endothelium (Stanczuk et al., 2015). Sabin, first, described the existence of the lymph sacs, which originate from endothelial budding from the veins (Wigle & Oliver, 1999). At the end of the embryonic period, six primitive lymph sacs are present. The paired enlarged jugular lymph sacs first arise at the confluence of the internal jugular and subclavian veins and produce a network of lymphatic capillaries that spreads to the upper limbs, upper trunk, head, and neck. Some plexuses enlarge and form lymphatic vessels in these regions. Both jugular lymph sacs retain a connection with their respective jugular vein. The left jugular lymph sac eventually develops into the cranial part of the thoracic duct. Next, during the 8th week, the singular retroperitoneal lymph sac, originating from the primitive vena cava and mesonephric veins, appears at the root of the mesentery. Lymphatic vessels spread from the retroperitoneal lymph sac to the abdominal viscera. The retroperitoneal lymph sac also connects with the cisterna chyli but loses its connection with its venous origin. After the appearance of the retroperitoneal sac, the cisterna chyli forms near the Wolffian bodies located dorsally to the retroperitoneal sac. It gives rise to the development of the caudal part of the final thoracic duct and loses its communication with its venous origin as well. Later in development, the paired posterior lymph sacs originate near the junctions of the primitive iliac veins and the posterior cardinal veins. These emit a lymphatic network of vessels in the abdominal wall, pelvis, and lower limbs. The posterior lymph sacs communicate with the cisterna chyli and no longer have venous connections. During gestational weeks 6 and 7, two major lymphatic vessels connect the cisterna chyli inferiorly with the paired jugular lymph sacs superiorly, leading to the formation of the embryonic right and left thoracic duct which have multiple interconnections. Portions of both ducts ultimately obliterate and the definitive thoracic duct arises from the caudal part of the right channel and the cranial part of the left channel (Loukas et al., 2007). Variations in the course of the thoracic duct are frequent and also duplicated thoracic ducts have been described (Chen et al., 2006).

4.3 Anatomy of the lymphatic system

There are five main categories of conduits in the lymphatic system: capillaries, collecting vessels, trunks, ducts—ranging from 10 μm to 2 mm in diameter—and lymph nodes, varying in size from 1 to 10 mm in diameter.

4.3.1 Lymphatic capillaries

The lymphatic system originates as dilated blind-ended tubes in the interstitial space where excess fluid moves into a network of lymphatic capillaries. These capillaries are also called initial or terminal lymphatics, measuring 10–60 μm in diameter, and are generally in a collapsed or partly collapsed state (Leak, 1976; Swartz, 2001). Unlike capillaries of the blood system, lymphatic capillaries are highly permeable. They have loose cell–cell contacts, so-called "button-like" junctions, and are characterized by a unique overlapping structure. These leaky flaps function as a one-way valvular structure and increase permeability, allowing passive passage of interstitial fluid and its contents into the capillary lumen. Moreover, these capillaries consist of one endothelial cell layer surrounded by a discontinuous or absent basement membrane, but they have no smooth muscle and are not contractile (Leak, 1976; Swartz, 2001). Anchoring filaments connect the basal lamina of the endothelium to the structural components of the extracellular matrix, preventing lymphatic capillaries from collapsing during interstitial pressure changes. This facilitates their drainage function, which needs only small transient pressure gradients from the interstitium into the initial lymphatics (Leak & Burke, 1968; Schmid-Schönbein, 1990; Swartz, 2001). Lymphatic capillaries are present in most tissues with the exception of cartilage, bone, bone marrow, epithelia, placenta, central nervous system, and placenta. Once the interstitial fluid has entered the lymphatic capillaries, it is termed lymph. The content of this clear fluid depends on the organ or tissue from which drainage occurs, but it is similar to plasma, consisting mainly of water alongside nutrients, plasma proteins, immunologic factors, etc. (Goswami et al., 2020). Starting from the capillary system, the lymphatic vessels constantly enlarge.

4.3.2 Collecting lymphatic vessels and lymph nodes

Lymphatic capillaries empty into precollecting lymphatic vessels, draining lymph into contractile collecting lymphatics, which are not tethered to the extracellular matrix but have tighter "zipper-like" junctions and smooth muscles that propel lymph. Subunits of collecting lymphatics between valves are called lymphangions. Each lymphangion sequentially contracts to propagate lymph through the lymphatics. The valves facilitate this peristalsis of lymph by allowing filling and emptying of every lymphangion, leading to stepwise pressure changes across subsequent lymphangions and a net pressure drop along the entire length of the collecting vessels (Ohhashi et al., 1980). Like the lymphatic capillaries, collecting vessels have macroscopic bicuspid valves formed from endothelial cells and elastin, preventing retrograde lymph flow (Schmid-Schönbein, 1990). Collecting lymphatics can be classified as prenodal or postnodal (or afferent and efferent), corresponding to whether they move lymph to or from the lymph nodes. Lymph nodes are mostly oval shaped and have an outer cortex and an inner medulla and are organized in clusters throughout the lymphatic system. They receive multiple afferent vessels at the convex surface. They have an indent at the concave side, termed the hilum, where a few hilar efferent lymphatic vessels leave and vessels enter and leave (Swartz, 2001). Afferent lymphatic vessels transport the lymph to the subcapsular space (also called the marginal sinus), which is located between the capsule and the cortex. The subcapsular space continues with the trabecular sinuses and joins the medullary sinus in the medulla of the lymph node. The medullary sinus drains the lymph into the efferent vessels. Lymph nodes mainly fill immunologic roles, serving as first responders against infectious and malignant agents but they also concentrate lymph. Therefore they have a vascular compartment that opposes each lymph compartment for cell transport and absorption of redundant water through osmosis (Moore & Bertram, 2018; Schmid-Schönbein, 1990).

4.3.3 Lymphatic trunks and ducts

Large lymphatic trunks are responsible for the drainage of lymph from a final set of lymph nodes into the lymphatic ducts (Swartz, 2001). Lymph is returned to the venous circulation via bilateral lymphovenous connections, typically found at the junction of the jugular and subclavian veins. Most of the body's lymph (80%–90%), including chyle, flows through the largest and longest (up to 45 cm) branch of the lymphatic system being the thoracic duct before draining into the junction of the left internal jugular and subclavian veins. Lymph from the right hemithorax, right head and neck, and right arm drains into the right lymphatic duct, which empties in the right jugulovenous angle. Besides these major lymphovenous drainage routes, there are many other connections at various locations in the body (Goswami et al., 2020; Swartz, 2001; Threefoot, 1968).

4.4 Anatomy of the lymphatic system of the colon and rectum

The intestinal lymph vessels include much of the lymphatic vessels and drain the small and large intestines, with the small intestine lymphatics being the largest network. Besides the general lymphatic functions such as fluid balance and immune surveillance, the intestinal lymphatics absorb dietary fats from the intestines that are eventually emptied into the venous circulation (Goswami et al., 2020). Lymph from the intestines and liver is called "chyle" and typically has higher concentrations of proteins and chylomicrons than lymph derived from other tissues. In contrast with lymph, chyle has a more milky white appearance.

4.4.1 Macroscopic anatomy of the colon and rectum

The abdominal lymphatic channels and lymph nodes typically accompany the blood vessels supplying or draining the organs and are located subperitoneally within the ligaments, mesentery, mesocolon, and mesorectum. To deeply understand the lymphatic system of the colon and rectum, knowledge of the anatomy of both hollow tubes and their vascularization, located in the mesentery, is essential. The colon has several anatomical divisions. A pouch-like structure, termed the cecum, is the beginning of the large bowel. Located in the right iliac fossa, the cecum extends caudally to the upper lip of the ileocecal valve. On its posteromedial side, the appendix vermiformis, a fingerlike structure, is connected with the cecum. Starting from the cecum, the large bowel continues cranially in the right flank as the ascending colon until it reaches the right (hepatic) colic flexure. From this curvature, the transverse colon crosses the peritoneal cavity from right to left until it reaches the left (splenic) colic flexure. The descending colon extends caudally in the left side of the peritoneal cavity to the root of the sigmoid colon, which is approximately at the level of the iliac crest. The sigmoid colon has an S-shaped appearance and runs downward to the level of the sacral promontory from where it is called the rectosigmoid. The latter ends at the inferior border of the second sacral vertebra and then continues as the upper rectum, which terminates at the peritoneal reflection. Most of the rectum is extraperitoneal, but anteriorly, the upper rectum is covered by the visceral peritoneum down to the peritoneal reflection in the floor of the pelvis. The lower rectum is the final segment, running from the peritoneal reflection to the superior border of the puborectal sling. Of note, opinions differ whether the rectosigmoid belongs to either the colon or the rectum (Kotake & Rectum, 2019).

4.4.2 Vascularization of the colon and rectum

The main arteries of the colon are the ileocolic, right colic, and middle colic (right and left branches), which are derived from the superior mesenteric artery, and the left colic and sigmoid arteries, branching off from the inferior mesenteric artery. The superior and inferior mesenteric arteries are connected by arcades, but their aforementioned branches are also interconnected. The inferior mesenteric artery continues downward in the mesocolon as the superior rectal artery, which splits into a branch on either side of the rectum. These branches break up into smaller branches that communicate with the inferior and middle rectal artery, derived from the internal iliac artery and internal pudendal artery, respectively.

4.4.3 Lymphatic vessel anatomy in the mesocolon

The lymphatic system draining the colon is located within the mesocolon. The latter is contiguous with the small intestinal mesentery and runs from the ileocecal to rectosigmoid level while attaching the large bowel to the posterior abdominal wall. Distally, it continues as the mesorectum. Throughout its length, the mesocolon is contained within a double fold of peritoneum, overlaying both upper and lower mesocolic surfaces (Coffey, 2013). The ascending and descending mesocolons are detached from the retroperitoneum by Toldt's fascia. This connective tissue layer is sandwiched between two peritoneal layers that cover the deep surface of the mesocolon and the anterior surface of the retroperitoneum (Culligan et al., 2012). Thus Toldt's fascia appears where mesocolic and retroperitoneal peritoneal surfaces come into permanent contact (Culligan et al., 2012). The mesocolon has compartments of adipocytes of varying numbers and sizes, firmly packed between fibrous septae, derived from submesothelial connective tissue layers. These compartments contain blood and lymphatic vessels together with nerves (Coffey, 2013). Culligan et al. described a complex connective tissue lattice within the small bowel mesentery and mesocolon, comprising subperitoneal monolayers and interlobular septations that compartmentalize the mesenteric adipocytes. Within the mesocolon, the histological structure and the distribution of lymphatic vessels are similar across each area. Specifically, there is a rich network of vessels measuring $1-20\,\mu m$ in diameter and located within 0.1 mm of both the anterior and posterior peritoneal surface. Within this dense network of lymphatic nodes and vessels, one can expect to find a lymphatic duct every 0.15−0.17 mm. Particularly within the mesocolon, lymphatic vessels run together with blood vessels, ranging in size from 1 to 40 μm in diameter (Culligan, Sehgal, et al., 2014; Culligan, Walsh, et al., 2014). Generally, lymphatic ducts run with the blood vessels and rarely in the avascular interval. Recent anatomical dissections of the right colon confirmed the existence of multiple long lymphatic vessels running a few millimeters from the middle and right colic arteries and ileocolic arteries as part of a lymphovascular bundle of mesenteric tissue but also detected in all 16 cadavers long lymph vessels in the intercolic space that cross both anteriorly and posteriorly to the superior mesenteric vessels and reach nodes at the superior mesenteric artery (Nesgaard et al., 2018). Despite a lower lymphatic vessel density, lymphatic channels were also found in approximately one-third of cadavers within Toldt's fascia interposed between the mesocolon and the retroperitoneum. The exact territory drained by the vessels in Toldt's fascia and their true clinical relevance are still unknown (Culligan, Sehgal, et al., 2014; Culligan, Walsh, et al., 2014).

4.4.4 Lymphatic vessel anatomy in the mesorectum

The lymphatic network draining the rectum is located in the mesorectal fatty tissue that also contains blood vessels and nerves. The rectum and its mesorectum are enclosed by the mesorectal fascia (also known as the fascia propria) (Heald et al., 1982). Recent research confirmed that lymphatic capillaries are predominant within the mesorectal lymphatic apparatus. According to their location, these capillaries are further subdivided into three types. First, lacteal capillaries originate as blind-ended vessels in the rectal submucosa and wrap around the rectal wall. Second, lymphatic capillaries located in the central part of the mesorectum pass around arterioles and venules along the mesorectal microsepta and communicate with collector channels that in turn connect with mesorectal lymph nodes. Third, the para-fascial capillaries run in parallel with the mesorectal fascia on its inner and outer surface and have intercommunicating branches, enabling passage of lymph to areas out the mesorectum. Of note, at light microscope level, no valves were found in the mesorectal lymphatic capillaries and collector channels (Reggiani-Bonetti et al., 2018).

4.4.5 Lymphatic drainage of the colon and rectum

In 1909 Jamieson and Dobson first described the macroscopic arrangement of lymph nodes and vessels draining the colon. The lymphatic glands that receive lymph from the colon are mainly dispersed along the blood vessels and vary in size and number. They classified colonic lymph nodes into four groups: epicolic, paracolic, intermediate, and main glands (Jamieson & Dobson, 1909). The Japanese Classification of Colorectal Carcinoma groups regional lymph similarly but considers the epicolic (on the bowel wall along the vasa recta) and paracolic (along the vascular arcades) glands as one group termed pericolic/perirectal lymph nodes (Jinnai, 1983). Below, the three-category classification and terminology for colon and rectum will be applied. There are between 100 and 150 lymph nodes in the mesocolon. The median number of lymph nodes in the sigmoid and ascending mesocolons is 71 (range: 24–116) and 61 (range: 33–71) (Ahmadi et al., 2015).

4.4.5.1 Lymphatic drainage of the colon

Lymphatics draining the large bowel are absent in the colonic mucosa between the crypts of Lieberkühn and appear to begin as a plexus wrapped around the muscularis mucosae. This plexus emits branches into the mucosa but not further than the base of the crypts (Kenney & Jain, 2008). At the level of the colon wall, escaping lymphatic ducts are interrupted almost immediately in pericolic lymph nodes that lie on the bowel wall immediately under the peritoneum. Although these glands are spread along the whole colon, they are particularly numerous on the sigmoid flexure. Because lymph nodes lying on the bowel wall are limited in number, much of the colon's lymph plausibly bypasses these nodes and flows through small lymphatic ducts to glands situated within the mesentery at the inner margin of the colon from the ileocolic transition to the rectum (Jamieson & Dobson, 1909). These glands, also termed pericolic lymph nodes, are mainly situated along the vascular arcades and along the terminal sigmoid artery and the superior rectal artery (Kotake & Rectum, 2019).

From there, lymph runs through ducts in parallel to surrounding vessels to intermediate nodes located between the first and the terminal branch of the main feeding artery. These nodes tend to group themselves midway between the root of the vessel and the gut (Jamieson & Dobson, 1909; Kotake & Rectum, 2019). Ileocolic, right colic, right and left middle colic, left colic, sigmoid colic, and inferior mesenteric trunk nodes belong to the intermediate nodes, lying along the corresponding arteries, and are related to the respective regions of the colon. Connecting glands are present between intermediate nodes and pericolic nodes and between intermediate nodes and the main group. Along the middle colic artery, intermediate glands are usually located midway between the base of the mesocolon anteriorly of the pancreas and the hepatic flexure, corresponding with the bifurcation of the middle colic artery in a left and right branch. Not infrequently, intermediate nodes are situated closer to the origin of the middle colic artery and continuous with the main group. Along the left colic artery, the chief mass of intermediate nodes can be detected in front of the point of crossing of the inferior mesenteric vein and the renal hilum. Intermediate nodes can be identified between the origin of the left colic artery and the terminal sigmoid artery and along the sigmoid arterial stems before these vessels divide. Intermediate glands receive most of the efferent ducts from the pericolic glands surrounding the vascular arcades together with some ducts that directly come from the gut. The presence of direct vessels between the gut and intermediate nodes varies in different colonic regions. Colonic dissections revealed that these ducts are constantly encountered in the upper right colon, hepatic flexure, right extremity of the transverse colon, and sigmoid colon but are infrequently found in the splenic flexure and descending colon and are absent in the central part of the transverse colon (Jamieson & Dobson, 1909; Kotake & Rectum, 2019).

The efferent ducts of the intermediate lymph nodes pass to the main glands, which are located at the origin of the main feeding artery being the ileocolic, right and middle colic artery, and along the root of the inferior mesenteric artery proximal to the origin of the left colic artery. The main middle colic nodes lie on the middle colic artery as it enters the transverse mesocolon. It is a distinct mass that, however, sometimes merges with the intermediate nodes or the superior mesenteric nodes. The inferior mesenteric root nodes lie on or under the origin of the inferior mesenteric artery and become continuous with the lumbar nodes (also termed lateral aortic nodes or para-aortic nodes). They have a complex relation with the lymphatic drainage of the colon because only the superior nodes receive intermediate and pericolic lymphatic ducts and infrequently direct vessels from the bowel wall. The lower lymph nodes, however, have a direct connection with the pericolic nodes located along the vascular arcade of the inferior portion of the sigmoid flexure and often receive direct vessels from the gut, whereas the lowest lymph nodes are roughly in contact with the terminal part of the sigmoid colon and the upper part of the rectum, receiving most of the direct vessels from the gut. Of note, starting from its lowest part, the inferior mesenteric root nodes emit some vessels to the upper nodes but also multiple efferents to the lumbar nodes located adjacent to the abdominal aorta. This means that the inferior mesenteric root nodes not only communicate at the upper part with the lumbar nodes but along the whole length (Jamieson & Dobson, 1909).

Remark that the lymphatic drainage of the splenic colic flexure occurs partly toward the superior mesenteric nodes and partly to the inferior mesenteric nodes (Vasey et al., 2018). In vivo laparoscopic assessment of the lymphatic drainage of the normal splenic flexure using Technetium-99m revealed that it is preferentially directed to the left colic pedicle and then to the inferior mesenteric lymph nodes.

4.4.5.2 Lymphatic drainage of the rectum

The average number of mesorectal lymph nodes reported varies considerably, ranging from a mean of 5.7 to 34.3. This variability in node yield is dependent on patient and surgical and pathologic factors (Ahmadi et al., 2015; Miscusi et al., 2010). Anatomical lymph node mapping in 12 human cadavers revealed that on average 34.3 (range: 31–37) nodes are present in the normal mesorectum. Almost two-thirds of nodes were detected in the upper third of the rectum, whereas the lowest number of nodes was harvested in the lower section. Almost half of the yielded nodes were found in the proximal posterior mesorectum (Miscusi et al., 2010). Prior cadaveric research confirmed this distribution and demonstrated that mesorectal lymph nodes are occasionally present within the lower third and are mainly located in the superior mesorectum above the peritoneal reflection and posteriorly (Canessa et al., 2001; Pirro et al., 2008; Topor et al., 2003). In 1925 Villemin et al. described that lymph drains from the mesorectum along two routes, involving three trunks with accompanying lymph glands, which the author termed superior, middle, and inferior trunk, referring to their course adjacent to the superior, middle, and inferior rectal vessels. The first route is responsible for drainage of the upper rectum located above the peritoneal reflection and occurs along the superior lymphatic trunk, which has short and long tributaries and is made up of perirectal lymph nodes. It adjoins the superior rectal and inferior mesenteric arterial axis. Lymphatic drainage of the rectum below the peritoneal reflection follows two pathways. One is upward lymphatic flow along the previously described first route; and the second route runs laterally through the middle and inferior lymphatic trunk along the middle and inferior rectal vessels and drains, respectively, into the lateral pelvic and inguinal lymph nodes. The pathway to lateral pelvic nodes involves lymph node stations along the middle rectal vessels and lateral sacral vessels, draining into nodes belonging to the distal and proximal internal iliac vessels. Drainage also passes lymph nodes along the external iliac vessels. The external iliac nodal group consists of three subgroups. The lateral chain positioned along the lateral aspect of the external iliac artery; the middle chain located between the external iliac artery and vein; and the medial chain (also termed obturator nodes) positioned medial and posterior to the external iliac vein. Next, lymph runs along the common iliac vessels. The common iliac nodal group involves three subgroups: lateral, middle, and medial. The lateral subgroup is an extension of the lateral chain of external iliac nodes situated lateral to the common iliac artery. The medial subgroup corresponds with the triangular area bordered by both common iliac arteries from the aortic bifurcation to the level of the bifurcation of the common iliac artery, comprising nodes along the median sacral vessels (also called promontorial nodes) and aortic bifurcation nodes. The middle subgroup consists of lymph nodes between the common iliac artery and vein and nodes located in a region bordered posteromedially by the lower lumbar or upper sacral vertebral bodies, anterolaterally by the psoas muscles and anteromedially by the common iliac vessels. This second route finishes at the lumbar nodes. Note that cadaveric examination of lateral pelvic lymph nodes in 16 human cadavers revealed a mean number of 28.6 nodes (range: 16—46) per pelvis and that the highest number of glands was harvested in the obturator fossa (mean: 7; range: 2—18) (Bell et al., 2009).

4.4.5.3 Preaortic and lumbar lymph nodes

Lymphatic drainage from most deep structures and regions of the body under the diaphragm converges chiefly to clustered lymph nodes and vessels related to the major blood vessels of the dorsal abdominal region. These clusters are subdivided into preaortic nodes, which are anterior to the abdominal aorta and left and right lumbar nodes, which are positioned on both sides of the abdominal aorta. The preaortic lymph nodes comprise the inferior and superior mesenteric lymph nodes and celiac nodes and aggregate around the three large anterior splanchnic branches of the abdominal aorta. These nodes receive lymph (i.e., chyle) from the organs supplied by the similarly named arteries. Nodes located at the root of the inferior mesenteric artery drain lymph from the splenic flexure, the descending colon, the sigmoid colon, and the rectum to the superior mesenteric lymph nodes but transport lymph to the lumbar nodes as well. Two to three lymph nodes at the root of the superior mesenteric artery collect lymph from the duodenum, pancreas, small intestines, cecum, appendix, ascending colon, transverse colon, splenic colic flexure, and inferior mesenteric lymph nodes and empty into the celiac nodes. Lymph from the stomach, duodenum, pancreas, liver, gallbladder, spleen, and inferior and superior mesenteric lymph nodes runs to one to three celiac lymph nodes located very close to the celiac ganglion. The lumbar nodes collect lymph from the iliac nodes and the inferior mesenteric root nodes, corresponding with drainage of the lower limb, the pelvic organs including the rectum, perineum, kidneys, suprarenal glands, anterior and posterior abdominal wall, and diaphragm. Table 4.1 summarizes the lymphatic drainage of the colon and rectum.

For each anatomical division of the large bowel, regional lymph nodes and further lymphatic drainage are tabulated. Remark that connecting channels that bypass one or more regional lymph node stations exist in different colonic regions.

4.4.5.4 Lumbar and intestinal lymph trunks and the cisterna chyli

The lumbar nodes form the right and left lumbar lymph trunks, while the preaortic nodes form the unpaired intestinal lymph trunk, which runs in the mesenteric root (Skandalakis et al., 2007). Fluid from the union of these three lymph trunks in turn drains into the cisterna chyli. This dilated structure can have a variable lymphatic origin, morphology, and location with respect to the vertebral level (AlShehri et al., 2020; Erden et al.,

TABLE 4.1 Lymphatic drainage of the colon and rectum.

Region drained	Regional nodes			Node groups proximal to the main nodes	Efferents to
	Pericolic/perirectal node	Intermediate nodes (along the vessels)	Main nodes (at the vessel root)		
Cecum	Pericolic nodes along the vascular arcades and vasa recta, and on the bowel wall	Ileocolic and right colic nodes	Ileocolic and right colic root nodes	Superior mesenteric arterial root nodes, celiac nodes around the trunk	Intestinal lymph trunk
Ascending colon	Pericolic nodes along the vascular arcades and vasa recta, and on the bowel wall	Ileocolic, right colic, and right middle colic nodes	Ileocolic, right colic, and middle colic root nodes	Superior mesenteric arterial root nodes, celiac nodes around the trunk	Intestinal lymph trunk
Transverse colon	Pericolic nodes along the vascular arcades and vasa recta, and on the bowel wall	Left and right middle colic nodes	Middle colic root nodes	Superior mesenteric arterial root nodes, celiac nodes around the trunk	Intestinal lymph trunk
Splenic colic flexure	Pericolic nodes along the vascular arcades and vasa recta, and on the bowel wall	Left middle colic nodes	Middle colic root nodes	Superior mesenteric arterial root nodes, celiac nodes around the trunk	Intestinal lymph trunk
		Left colic nodes	Inferior mesenteric root nodes	Lumbar nodes	Lumbar lymph trunks
				Superior mesenteric arterial root nodes, celiac nodes around the trunk	Intestinal lymph trunk
Descending colon	Pericolic nodes along the vascular arcades and vasa recta, and on the bowel wall	Left colic, sigmoid colic, and inferior mesenteric trunk nodes	Inferior mesenteric root nodes	Lumbar nodes	Lumbar lymph trunks
				Superior mesenteric arterial root nodes, celiac nodes around the trunk	Intestinal lymph trunk
Sigmoid colon	Pericolic and perirectal nodes along the vascular arcades and vasa recta, and on the bowel wall	Left colic, sigmoid colic, and inferior mesenteric trunk nodes	Inferior mesenteric root nodes	Lumbar nodes	Lumbar lymph trunks
				Superior mesenteric arterial root nodes, celiac nodes around the trunk	Intestinal lymph trunk
Upper and lower rectum	Perirectal nodes along the superior rectal vessels and along the terminal sigmoid artery, and on the rectum wall	Inferior mesenteric trunk nodes	Inferior mesenteric root nodes	Lumbar nodes	Lumbar lymph trunks
				Superior mesenteric arterial root nodes, celiac nodes around the trunk	Intestinal lymph trunk

(Continued)

TABLE 4.1 (Continued)

Region drained	Regional nodes			Node groups proximal to the main nodes	Efferents to
	Pericolic/perirectal node	Intermediate nodes (along the vessels)	Main nodes (at the vessel root)		
Lower rectum	Perirectal lymph nodes along the middle rectal vessels	Lateral pelvic lymph nodes along the lateral and median sacral vessels, obturator vessels and nerves, distal and proximal internal iliac vessels, external iliac vessels, aortic bifurcation nodes, common iliac nodes			Lumbar trunk
	Perirectal lymph nodes along the lower rectal vessels	Inguinal lymph nodes, external iliac lymph nodes, common iliac lymph nodes			

Table modified from the Japanese Society for Cancer of the Colon and Rectum and Kanehara & Co., Ltd.: Japanese Classification of Colorectal, Appendiceal, and Anal Carcinoma—the third English edition, 2019.

2005; Loukas et al., 2007). In most cases, it is a single structure, formed by the union of the intestinal trunk and the left lumbar lymph trunk. Within this tributary configuration, the right lumbar lymph trunk mostly joins the left lumbar lymph trunk and in slightly less than half of cases it drains into the intestinal trunk (Loukas et al., 2007). The cisterna chyli predominantly has a fusiform, saccular, or lobulated configuration, resembling a string-of-pearls configuration (AlShehri et al., 2020; Feuerlein et al., 2009; Loukas et al., 2007). The size of this confluence markedly varies with a mean length ranging on the order of a few centimeters and shorter anteroposterior and transverse diameters, measuring about 1 cm or less (Loukas et al., 2007; Rosenberger & Abrams, 1971; Smith & Grigoropoulos, 2002). The cisterna chyli has a diameter twice that of the thoracic duct, making it an accurate criterion for identification (Loukas et al., 2007). The cisterna chyli cannot always be distinctively identified as the trunks can combine without a focal dilatation (Johnson et al., 2016; Skandalakis et al., 2007). One study reported it in only 52% of cases. This lymphatic structure is usually situated at the midline and extends anteriorly of the body of the 11th thoracic vertebra down to the second lumbar vertebra with a predominant appearance at the lumbar vertebra 1–2 transition area (Erden et al., 2005; Loukas et al., 2007). It is mostly located on the right and behind the abdominal aorta in the retrocrural space, which is between the aortic hiatus and the right crus of the diaphragm, but it can be detected anteriorly and to the left of the aorta as well (Loukas et al., 2007).

4.4.5.5 The thoracic duct

From the superior aspect of the cisterna chyli, lymph drains cranially through the thoracic duct. This main lymphatic vessel of the body with a length of 38–45 cm and measuring up to 2–5 mm in diameter may vary in morphology along any aspect of its course (Johnson et al., 2016). Arising from the cisterna chyli, it usually courses cranially along the right anterior aspect of the vertebral column. Extending superiorly, it runs between the aorta and the azygos vein, entering the thorax through the aortic hiatus of the diaphragm. In the thorax, it courses in the posterior mediastinum to the right of the vertebral column between aorta (left), azygos vein (right), and esophagus (anterior). At the 7th thoracic vertebra, it ascends obliquely and dorsally of the esophagus. At approximately the 5th—6th thoracic vertebral levels, it crosses the midline to the left and passes behind the aorta and to the left of the esophagus as it ascends 2—3 cm above the clavicle. Once in the superior mediastinum, it runs behind the internal jugular vein, curving inferiorly to drain into the left venous angle between the subclavian vein and internal jugular vein (Skandalakis et al., 2007).

4.5 Conclusion

The lymphatic system has a unique and highly complex anatomical structure that extends over almost the entire human body. Originating in the interstitium as a highly permeable blind-ended network of lymphatic capillaries that is mainly responsible for collecting excess interstitial fluid, it transports lymph or chyle through a one-way low-pressure drainage system of interweaving collecting vessels and incorporated lymph nodes to large lymphatic trunks that in turn drain into lymphatic ducts, which communicate with the venous circulation. The venous origin of the lymphoid system arising from cardinal veins to form primitive lymph sacs was postulated more than 100 years ago. Recently, nonvenous contributions in developmental lymphangiogenesis in multiple tissues have been posited. During the last two to three decades, our knowledge of the extremely complex molecular mechanisms involved in the developmental differentiation of the lymphatic system from venous endothelium has grown dramatically. PROX1 is the major transcriptional regulator that maintains lymphatic identity and is a universal LEC marker. VEGFR-3 is the primary tyrosine kinase receptor that regulates LEC proliferation and migration. VEGF-C activates VEGFR-3 but before this can be established, it must be activated by CCBE1 and ADAMTS3.

The mesocolon, which is contiguous with the mesentery and the mesorectum, has compartments of adipocytes packed between fibrous septae and is rich in lymph nodes and is featured by a uniform and dense network of lymphatic vessels located just below the anterior and posterior peritoneal surface. These vessels typically run with the blood vessels supplying or coming from the colon and are infrequently found as long channels in the avascular interval. Regional lymph nodes draining the colon are classified as follows: pericolic nodes, corresponding to nodes positioned on the bowel wall along the vasa recta (also termed epicolic) and nodes adjacent to the vascular arcades (or paracolic); intermediate nodes, comprising ileocolic, right colic, right and left middle colic, left colic, sigmoid colic, and inferior mesenteric trunk nodes, located along the corresponding arteries and related to the respective colonic regions; and main glands positioned at the origin of the aforementioned arteries. Beyond the main glands, lymph particularly coming from the inferior mesenteric root nodes either drains to lumbar nodes positioned adjacent to the abdominal aorta or passes nodes grouped at the root of the superior mesenteric artery and celiac trunk (also called preaortic nodes), while lymph coming from the other main nodes (i.e., ileocolic, right colic, and middle colic root nodes) converges to clustered nodes at the superior mesenteric arterial root and nodes around the origin of the celiac trunk. The lumbar nodes form the right and left lumbar trunks, whereas the preaortic nodes form the singular intestinal lymph trunk. Lymph from the union of these three trunks drains into the cisterna chyli, which in turn communicates with the thoracic duct. The lymphatic network draining the rectum is located within the mesorectal fatty tissue surrounded by the mesorectal fascia. Mesorectal lymph nodes are predominantly located in the posterior superior mesorectum above the peritoneal reflection. Lymph draining from the rectum passes primarily through capillaries located within this mesorectum and follows an upward route along the superior rectal artery—inferior mesenteric artery axis but also runs laterally to the pelvic nodes and further to the lumbar nodes. The first route is responsible for drainage of the upper rectum located above the peritoneal reflection, while both routes collect lymph from the lower rectum.

References

Ahmadi, O., McCall, J. L., & Stringer, M. D. (2015). Mesocolic lymph node number, size, and density: an anatomical study. *Diseases of the Colon and Rectum*, 58(8), 726–735. Available from https://doi.org/10.1097/DCR.0000000000000413.

AlShehri, E., Lam, C. Z., Kamath, B. M., & Chavhan, G. B. (2020). Abdominal lymphatic system visibility, morphology, and abnormalities in children as seen on routine MCRP and its association with immune-mediated diseases. *European Radiology*. Available from https://doi.org/10.1007/s00330-020-07152-6.

Bell, S., Sasaki, J., Sinclair, G., Chapuis, P. H., & Bokey, E. L. (2009). Understanding the anatomy of lymphatic drainage and the use of blue-dye mapping to determine the extent of lymphadenectomy in rectal cancer surgery: unresolved issues. *Colorectal Disease: The Official Journal of the Association of Coloproctology of Great Britain and Ireland*, 11(5), 443–449. Available from https://doi.org/10.1111/j.1463-1318.2009.01769.x.

Canessa, C. E., Badía, F., Fierro, S., Fiol, V., & Háyek, G. (2001). Anatomic study of the lymph nodes of the mesorectum. *Diseases of the Colon and Rectum*, 44(9), 1333–1336.

Chen, H., Shoumura, S., & Emura, S. (2006). Bilateral thoracic ducts with coexistent persistent left superior vena cava. *Clinical Anatomy*, 19(4), 350–353.

Coffey, J. C. (2013). Surgical anatomy and anatomic surgery—Clinical and scientific mutualism. *The Surgeon: Journal of the Royal Colleges of Surgeons of Edinburgh and Ireland*, 11(4), 177–182. Available from https://doi.org/10.1016/j.surge.2013.03.002.

Culligan, K., Coffey, J. C., Kiran, R. P., Kalady, M., Lavery, I. C., & Remzi, F. H. (2012). The mesocolon: A prospective observational study. *Colorectal Disease: The Official Journal of the Association of Coloproctology of Great Britain and Ireland*, 14(4), 421–428. Available from https://doi.org/10.1111/j.1463-1318.2012.02935.x, discussion 428–430.

Culligan, K., Sehgal, R., Mulligan, D., et al. (2014). A detailed appraisal of mesocolic lymphangiology—An immunohistochemical and stereological analysis. *Journal of Anatomy*, 225(4), 463–472.

Culligan, K., Walsh, S., Dunne, C., et al. (2014). The mesocolon: A histological and electron microscopic characterization of the mesenteric attachment of the colon prior to and after surgical mobilization. *Annals of Surgery*, 260(6), 1048–1056.

Culligan, K., Sehgal, R., Mulligan, D., Dunne, C., Walsh, S., Quondamatteo, F., Dockery, P., & Coffey, J. C. (2014). A detailed appraisal of mesocolic lymphangiology—An immunohistochemical and stereological analysis. *Journal of Anatomy*, 225(4), 463–472. Available from https://doi.org/10.1111/joa.12219.

Culligan, K., Walsh, S., Dunne, C., Walsh, M., Ryan, S., Quondamatteo, F., Dockery, P., & Coffey, J. C. (2014). The mesocolon: A histological and electron microscopic characterization of the mesenteric attachment of the colon prior to and after surgical mobilization. *Annals of Surgery*, 260(6), 1048–1056. Available from https://doi.org/10.1097/SLA.0000000000000323.

Erden, A., Fitoz, S., Yagmurlu, B., & Erden, I. (2005). Abdominal confluence of lymph trunks: Detectability and morphology on heavily T2-weighted images. *American Journal of Roentgenology*, 184(1), 35–40.

Feuerlein, S., Kreuzer, G., Schmidt, S. A., Muche, R., Juchems, M. S., Aschoff, A. J., Brambs, H.-J., & Pauls, S. (2009). The cisterna chyli: Prevalence, characteristics and predisposing factors. *European Radiology*, 19(1), 73–78. Available from https://doi.org/10.1007/s00330-008-1116-5.

François, M., Caprini, A., Hosking, B., Orsenigo, F., Wilhelm, D., Browne, C., Paavonen, K., Karnezis, T., Shayan, R., Downes, M., Davidson, T., Tutt, D., Cheah, K. S. E., Stacker, S. A., Muscat, G. E. O., Achen, M. G., Dejana, E., & Koopman, P. (2008). Sox18 induces development of the lymphatic vasculature in mice. *Nature, 456*(7222), 643−647. Available from https://doi.org/10.1038/nature07391.

Jha, S. K., Rauniyar, K., & Jeltsch, M. (2018). Key molecules in lymphatic development, function, and identification. *Annals of Anatomy = Anatomischer Anzeiger: Official Organ of the Anatomische Gesellschaft, 219*, 25−34. Available from https://doi.org/10.1016/j.aanat.2018.05.003.

Johnson, O. W., Chick, J. F. B., Chauhan, N. R., Fairchild, A. H., Fan, C.-M., Stecker, M. S., Killoran, T. P., Suzuki-Han, A., et al. (2016). The thoracic duct: Clinical importance, anatomic variation, imaging, and embolization. *European Radiology, 26*(8), 2482−2493. Available from https://doi.org/10.1007/s00330-015-4112-6.

Jinnai, D. (1983). General rules for clinical and pathological-studies on cancer of the colon, rectum and anus. 1. Cinical classification. *The Japanese Journal of Surgery, 13*(6), 557−573.

Goswami, A. K., Khaja, M. S., Downing, T., Kokabi, N., Saad, W. E., & Majdalany, B. S. (2020). Lymphatic anatomy and physiology. *Seminars in Interventional Radiology, 37*(3), 227−236. Available from https://doi.org/10.1055/s-0040-1713440.

Heald, R. J., Husband, E. M., & Ryall, R. D. (1982). The mesorectum in rectal cancer surgery—The clue to pelvic recurrence? *The British Journal of Surgery, 69*(10), 613−616.

Hsu, M. C., & Itkin, M. (2016). Lymphatic anatomy. *Techniques in Vascular and Interventional Radiology, 19*(4), 247−254. Available from https://doi.org/10.1053/j.tvir.2016.10.003.

Jamieson, J. K., & Dobson, J. F. (1909). The lymphatics of the colon. *Proceedings of the Royal Society of Medicine, 2*, 149−174. (Surg Sect).

Kotake, K., & Rectum, J. S. C. C. (2019). Japanese Classification of Colorectal, Appendiceal, and Anal Carcinoma: The 3d English edition [secondary publication]. *Journal of Anus Rectum Colon, 3*(4), 175−195.

Kaipainen, A., Chen, E., Chang, L., Zhao, B., Shin, H., Stahl, A., Fishman, S. J., Mulliken, J. B., Folkman, J., Huang, S., & Fannon, M. (2019). Characterization of lymphatic malformations using primary cells and tissue transcriptomes. *Scandinavian Journal of Immunology, 90*(4), e12800. Available from https://doi.org/10.1111/sji.12800.

Kenney, B. C., & Jain, D. (2008). Identification of lymphatics within the colonic lamina propria in inflammation and neoplasia using the monoclonal antibody D2-40. *The Yale Journal of Biology and Medicine, 81*(3), 103−113.

Leak, L. V. (1976). The structure of lymphatic capillaries in lymph formation. *Federation Proceedings, 35*(8), 1863−1871.

Leak, L. V., & Burke, J. F. (1968). Ultrastructural studies on the lymphatic anchoring filaments. *The Journal of Cell Biology, 36*(1), 129−149.

Loukas, M., Wartmann, C. T., Louis, R. G., Tubbs, R. S., Salter, E. G., Gupta, A. A., & Curry, B. (2007). Cisterna chyli: A detailed anatomic investigation. *Clinical Anatomy, 20*(6), 683−688.

Miscusi, G., Di Gioia, C. R. T., Patrizi, G., Gravetz, A., Redler, A., & Petrozza, V. (2010). Anatomical lymph node mapping in normal mesorectal adipose tissue. *Diseases of the Colon and Rectum, 53*(12), 1640−1644. Available from https://doi.org/10.1007/DCR.0b013e3181f48f90.

Moore, J. E., & Bertram, C. D. (2018). Lymphatic system flows. *Annual Review of Fluid Mechanics, 50*, 459−482. Available from https://doi.org/10.1146/annurev-fluid-122316-045259.

Nesgaard, J. M., Stimec, B. V., Soulie, P., Edwin, B., Bakka, A., & Ignjatovic, D. (2018). Defining minimal clearances for adequate lymphatic resection relevant to right colectomy for cancer: a post-mortem study. *Surgical Endoscopy, 32*(9), 3806−3812. Available from https://doi.org/10.1007/s00464-018-6106-3.

Nicenboim, J., Malkinson, G., Lupo, T., Asaf, L., Sela, Y., Mayseless, O., Gibbs-Bar, L., Senderovich, N., Hashimshony, T., Shin, M., Jerafi-Vider, A., Avraham-Davidi, I., Krupalnik, V., Hofi, R., Almog, G., Astin, J. W., Golani, O., Ben-Dor, S., Crosier, P. S., … Yaniv, K. (2015). Lymphatic vessels arise from specialized angioblasts within a venous niche. *Nature, 522*(7554), 56−61.

Ohhashi, T., Azuma, T., & Sakaguchi, M. (1980). Active and passive mechanical characteristics of bovine mesenteric lymphatics. *The American Journal of Physiology, 239*(1), H88−H95.

Pirro, N., Pignodel, C., Cathala, P., Fabbro-Peray, P., Godlewski, G., & Prudhomme, M. (2008). The number of lymph nodes is correlated with mesorectal morphometry. *Surgical and Radiologic Anatomy, 30*(4), 297−302. Available from https://doi.org/10.1007/s00276-008-0322-9.

Reggiani-Bonetti, L., Barresi, V., Manenti, A., Domati, F., & Farinetti, A. (2018). Histology of the mesorectal lymphatics explains aspects of rectal cancer. *Clinics and Research in Hepatology and Gastroenterology, 42*(3), 285−287. Available from https://doi.org/10.1016/j.clinre.2017.12.009.

Rosenberger, A., & Abrams, H. L. (1971). Radiology of the thoracic duct. *The American Journal of Roentgenology, Radium Therapy, and Nuclear Medicine, 111*(4), 807−820.

Schmid-Schönbein, G. W. (1990). Microlymphatics and lymph flow. *Physiological Reviews, 70*(4), 987−1028.

Skandalakis, J. E., Skandalakis, L. J., & Skandalakis, P. N. (2007). Anatomy of the lymphatics. *Surgical Oncology Clinics of North America, 16*(1), 1−16.

Smith, T. R., & Grigoropoulos, J. (2002). The cisterna chyli—Incidence and characteristics on CT. *Clinical Imaging, 26*(1), 18−22.

Stanczuk, L., Martinez-Corral, I., Ulvmar, M. H., Zhang, Y., Laviña, B., Fruttiger, M., Adams, R. H., Saur, D., Betsholtz, C., Ortega, S., Alitalo, K., Graupera, M., & Mäkinen, T. (2015). cKit lineage hemogenic endothelium-derived cells contribute to mesenteric lymphatic vessels. *Cell Reports, 10*(10), 1708−1721. Available from https://doi.org/10.1016/j.celrep.2015.02.026.

Swartz, M. A. (2001). The physiology of the lymphatic system. *Advanced Drug Delivery Reviews, 50*(1−2), 3−20.

Tammela, T., & Alitalo, K. (2010). Lymphangiogenesis: Molecular mechanisms and future promise. *Cell, 140*(4), 460−476. Available from https://doi.org/10.1016/j.cell.2010.01.045.

Threefoot, S. A. (1968). Gross and microscopic anatomy of the lymphatic vessels and lymphaticovenous communications. *Cancer Chemotherapy Reports, 52*(1), 1−20.

Topor, B., Acland, R., Kolodko, V., & Galandiuk, S. (2003). Mesorectal lymph nodes: Their location and distribution within the mesorectum. *Diseases of the Colon and Rectum, 46*(6), 779−785.

Ulvmar, M. H., & Mäkinen, T. (2016). Heterogeneity in the lymphatic vascular system and its origin. *Cardiovascular Research, 111*(4), 310−321. Available from https://doi.org/10.1093/cvr/cvw175.

Vaahtomeri, K., Karaman, S., Mäkinen, T., & Alitalo, K. (2017). Lymphangiogenesis guidance by paracrine and pericellular factors. *Genes & Development, 31*(16), 1615−1634. Available from https://doi.org/10.1101/gad.303776.117.

Vasey, C. E., Rajaratnam, S., O'Grady, G., & Hulme-Moir, M. (2018). Lymphatic dfrainage of the splenic flexure defined by intraoperative scintigraphic mapping. *Diseases of the Colon and Rectum, 61*(4), 441−446. Available from https://doi.org/10.1097/DCR.0000000000000986.

Wigle, J. T., & Oliver, G. (1999). Prox1 function is required for the development of the murine lymphatic system. *Cell, 98*(6), 769−778.

SECTION 2

Pathology and Imaging

CHAPTER 5

Imaging of colorectal nodal disease

Lishan Cai[1,2], Zuhir Bodalal[1,2], Stefano Trebeschi[1], Selam Waktola[1], Tania C. Sluckin[1,3], Miranda Kusters[3], Monique Maas[1], Regina Beets-Tan[1,2] and Sean Benson[1]

[1]Department of Radiology, The Netherlands Cancer Institute, Amsterdam, The Netherlands
[2]GROW School for Oncology and Developmental Biology, Maastricht University, Maastricht, The Netherlands [3]Department of Surgery, Cancer Center Amsterdam, Amsterdam University Medical Centers, Vrije Universiteit Amsterdam, Amsterdam, The Netherlands

5.1 Introduction

The use of imaging for patients with colorectal cancer (CRC) has significantly evolved over the last decades (McKeown et al., 2014) and has a role in diagnosis, staging, treatment selection, and follow-up (Iyer et al., 2002). The imaging modalities available for the detection and staging of tumors and lymph node metastases (LNMs) can be broadly classed as anatomical or functional. Anatomical imaging techniques comprise computed tomographic (CT) imaging for colon tumor staging and magnetic resonance imaging (MRI) for rectal tumor staging. Functional techniques such as positron emission tomography (PET-CT) and functional MRI techniques also give insights into tumor perfusion, density, metabolic and molecular phenotypes and is therefore increasingly important to understand the cytological mechanisms of tumor response that provide benefit to patients in the clinic, with a focus beyond the reduction in tumor burden that is traditionally associated with chemotherapy response (Thoeny & Ross, 2010).

For nodal (N) staging, MRI is the most important tool due to its high resolution, whole field of view, and high contrast. One study in particular that included the morphological criteria of nodes was able to achieve a sensitivity of 85% and specificity of 98%. The use of MRI has been found to be superior to endorectal ultrasonography and digital rectal examination in terms of clinical benefit, lymph node involvement, and circumferential resection margin status (Beets-Tan et al., 2013). The presence of LNMs detected with MRI upstages a patient from stage II to stage III. Current pathological TNM staging, corresponding to the primary tumor size, nodal status, and presence of metastases, respectively, appears to be suboptimal as it fails to stratify patients adequately and contains differences in distinctions between tumor deposits (TDs) and

LNMs, where the former is defined as a nodule without evidence of underlying lymph node structure. The presence of TDs is associated with a poorer prognosis (Lord et al., 2019).

In addition to providing staging for a patient, imaging is also useful in determining the molecular pathway of CRC through pathological microscopy of biopsied and resected tissue. Treatment of CRC is of varying effectiveness, which has been shown to be linked to the individual pathophysiologic pathways of CRC, where CRCs arising via the serrated pathways are strongly associated with poor prognosis and therapy resistance (Müller et al., 2016). Another factor determining prognosis is the route of metastasis, that is, via the hematological or lymphatic system.

The maturity of imaging technology is providing increasingly larger amounts of clinical data. This has the consequence and opportunity that applied imaging research increasingly is focused on the development of artificial intelligence (AI) models. The use of AI for the treatment response and characterization can be split into two approaches. The first approach involves the use of the segmented tumor of an MRI or CT scan to define the region of interest (ROI) against the background parenchyma with a subsequent determination of quantitative features from the ROI (Lambin et al., 2017). These so-called radiomics features can subsequently be combined with genomics data and clinical information. This then has the potential to develop imaging biomarkers that incorporate the tumor characteristics to predict response and to correlate genetic mutations with image-specific phenotypes (Bodalal et al., 2019).

The second approach uses deep learning to bypass the need for the input segmentation to directly link the input image to the endpoint of interest. The so-called deep learning approach often uses a convolutional neural network (CNN) to perform relevant feature extraction and subsequent classification, with UNets being a prominent example of CNN showing promise for automatic medical image annotations (Ronneberger et al., 2015). With the use of transfer learning, models developed to classify a wide range of pictures are adapted to perform other classification tasks, often with a smaller dataset (Zhuang et al., 2020). The use of AI in the field of histopathology and indeed radiology also provides consistency and reproducibility, which are notable issues when comparing analyses (Rock et al., 2014).

The rest of the chapter is divided into subsections according to the targeted prediction of the models. Section 5.2 describes the studies attempting to stage, delineate and identify CRC in imaging. Section 5.3 describes those focused metastasis classification and prediction, and Section 5.4 details the models that focus on the prediction of recurrence-free survival and treatment response.

5.2 Staging, segmentation, and endoscopic detection

5.2.1 Staging

Staging, strictly speaking, can be performed via two, well-defined methods: the Dukes staging (Astler & Coller, 1954), which defines four stages, describing tumors confined to mucosa to tumors spread into distant metastases; and the TNM stage (Rami-Porta, 2019; Sobin, 2003), which categorizes tumors in terms of primary tumor (T), nodal status (N), and metastases

(M). The literature of AI algorithms for the automatic assessment of the cancer spread according to either one of these two methods is still limited. Raju et al. (2020) proposed an algorithm based on deep learning for the prediction of TNM in whole-slide imaging (WSI) of CRC patients from a population of 1500 individuals. Their method reached an overall accuracy of 81.1% (no breakdown of the results provided). A more frequent approach, alternative to the prediction of the precise TNM stage, is to predict specific stages of either the tumor, the nodal status, or the metastases. In these settings, the problem is broken down to the binary classification, which answers a much simpler question of, for example, is T0 versus T1 + ? Following this mantra, we find the largest body of work in the early diagnostic stages, distinguishing between colorectal tumors and benign polyps (T0 vs T1 +). Among them, we can find examples of deep learning algorithms applied to endoscopy images (white-light imaging), with accuracy ranging between 82.4% and 97.8% (Mori et al., 2016; Takemura et al., 2012; Zhou et al., 2020), with the study of Zhou et al. (2020) showing that the algorithm performed comparably to the human operator. Similar research questions have been addressed with CT colonography (radiological imaging), where the authors looked at textures of polyps with machine-learning algorithms, also reaching significant results, but with much smaller datasets (Cao et al., 2019). Radiological imaging becomes more prevalent in the diagnosis of later stages. This is the case for segmentation algorithms, mostly based on MRI and/or CT data. These algorithms are tasked to identify T2−3 tumors (Huang et al., 2018; Panic et al., 2020; Soomro et al., 2018), nodal involvement (N +) in WSIs (Chuang et al., 2021; Takamatsu et al., 2019) as well as radiological imaging (Li et al., 2020, 2021), and metastatic involvement in distal organs, most notably the liver (Romero et al., 2019; Vorontsov et al., 2019; Xu & Zhang, 2018) and peritoneum (Huang et al., 2020; Yuan et al., 2020). Worth noticing is that the scope of these algorithms is to segment or, in other words, to detect which pixels of the image are tumor and which pixels are not. Segmentation algorithms are almost never compared against a healthy, tumor-free control for diagnostic performance.

5.2.2 Segmentation and endoscopic detection

AI, especially deep learning, has shown great promise in image segmentation and image classification (Jiang et al., 2021). The application of automatic segmentation with AI in colorectal nodal disease is quite popular. The focus is mainly on two aspects, one is the segmentation of colorectal polyps to prevent the development of tumors and nodal metastases and the other is the segmentation of colorectal tumors.

5.2.2.1 Colorectal polyps segmentation

Colonoscopy is regarded as the first option for detection and removal of colorectal lesions, which is closely related to the reduction of CRC mortality. Therefore marking the exact polyp area in endoscopy accurately and precisely is of vital importance in the diagnosis and prevention of colorectal nodal diseases in clinical practices.

Firstly, Li et al. (2017) proposed an end-to-end CNN in the task of the segmentation of colorectal polyps. Akbari et al. (2018) introduced a CNN-based method with two stages. The first stage is a novel image patch selection with CNN in the training process and the second stage is a postprocessing over the probability map obtained from the first stage.

Jha et al. (2021) propose ColonSegNet and the algorithm achieved a better trade-off between the dice score and the efficiency. To improve the generalization of the AI model, Yang et al. (2021) introduced a mutual-prototype adaptation network, which can eliminate domain shifts among colonoscopy images from different centers and different devices. Safarov and Whangbo (2021) developed a DenseUNet which combines multiscale semantic information to generate a global feature and encode such features to the decoder side alongside skip-connection features. Meanwhile, an attention mechanism is added to take inputs from features of different levels and filters noisy and ambiguous regions and captures long-range dependencies in the image. The modern methods described are able to achieve accuracies as high as 97.7% (sensitivity: 74.8%, specificity: 99.3%), which was observed by Akbari et al. (2018).

5.2.2.2 Colorectal tumors segmentation

Segmentation of colorectal tumors can be divided into three categories according to modalities. One is based on MRI; the second one is to segment colorectal tumor cells from normal cells in pathological images; and the third one is based on CT imaging. As can be seen from Fig. 5.1, encoder-decoder-based U shape network is one of the most popular networks based on a fully convolutional network (FCN) structure and is regarded as the baseline standard architecture in medical image segmentation. Most rectal tumor segmentation networks are based on the standard network but with different structure modifications.

MRI is the most preferable modality in initial CRC diagnosis and also for precise radiotherapy treatment planning (Soomro et al., 2017). The segmentation or delineation of MRI is usually accomplished manually, which is s time-consuming, laborious, and has high

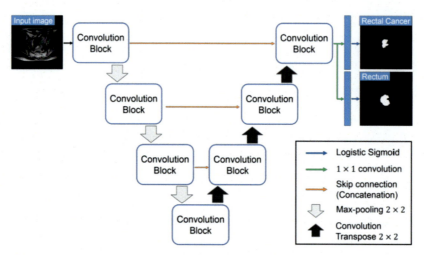

FIGURE 5.1 Rectal tumor segmentation network with an additional task of rectum segmentation (Lee et al., 2019). Most rectal tumor segmentation networks are based on UNet (Encoder-decoder networks) but with different structural modifications. Source: *From Lee, J., Oh, J. E., Kim, M. J., Hur, B. Y., & Sohn, D. K. (2019). Reducing the model variance of a rectal cancer segmentation network. IEEE Access, 7, 182725–182733. https://doi.org/10.1109/access.2019.2960371.*

inter- and intra-variability. Therefore efficient and accurate automatic colorectal tumor segmentation methods are in great demand.

Trebeschi et al. (2017) evaluate deep learning methods for automatic segmentation of rectal cancers on multiparametric MRI with patch extraction for the first time and demonstrate that deep learning can perform accurate localization and segmentation of rectal cancer in MRI in the majority of patients. Soomro et al. (2018) proposed a novel method to automatically segment CRC from 3D MR images based on the combination of 3D FCNNs and 3D level-set. The advantage of incorporating 3D level-set is to fine-tune the training stages and to refine the prediction in the test stage by merging smoothing function and prior information in a postprocessing step. The proposed method showed higher dice similarity coefficient (DSC) than 3D-FCNNs alone (0.94 vs 0.86). Later on, Soomro et al. (2018) presented a CNN framework, based on densely connected algorithm. The network provided a dense interconnectivity among the depth and scaled layers to couple finer and coarser features, which maintains features of all resolutions throughout the whole network. Meanwhile, a 3D level set is added. It has been shown that adding a 3D level set increases the performance of all deep learning-based approaches. Panic et al. (2020) propose a CNN-based system, which consists of a preprocessing, the classification with three CNNs, and a postprocessing (DSC: 0.6). The final segmentation mask is obtained by a majority voting. Huang et al. (2018) develop HL-FCN, which is a volume-to-volume FCNNs trained with hybrid loss. The incorporation of dice-based hybrid loss function resolves class-imbalance and low-contrast appearance issues and improves the performance by a significant margin (0.72 vs 0.70). Besides, a multiresolution model ensemble is used to decrease false positives and preserve boundary details. Afterward, Huang et al. (2020) proposed 3D RUNet, an encoder-decoder-based framework for 3D whole-volume segmentation. The model consists of a global image encoder for global understanding-based ROI localization, and a local region decoder that operates on pyramid-shaped in-region global features. Also, a dice-based multitask hybrid loss function was applied to facilitate global-to-local learning process and preserve contour details. A resulting DSC of 0.76 was achieved.

CT is another essential part of the diagnosis of rectal cancer metastases. As for the case of MRI, the segmentation is a time-consuming process with high variability. Therefore automatic CT segmentation has become more popular.

Men et al. (2017) introduce DDCNN, novel deep dilated CNN-based method for consistent automatic segmentation of CRC (DSC of 0.88 was observed for the clinical target volume). Liu et al. (2019) propose LAGAN, a postprocessing model to refine the segmentation of deep works. The LAGAN consists of a generating model, imitating the distribution of the ground truths and a discriminating model, distinguishing generations and ground truth (DSC of 0.91 observed). Barr et al. (2020) show a 3D implementation of U-Net and the method is evaluated with K-fold cross-validation (accuracy of 0.99 was observed, no DSC quoted).

5.3 Metastasis classification and prediction

Alongside the work done in automated segmentation, researchers have also analyzed the potential of imaging markers to predict the presence of metastases. Given that

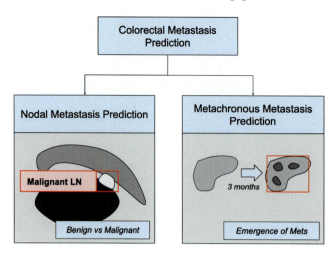

FIGURE 5.2 Schematic diagram illustrating the two main research questions covered in colorectal metastasis prediction literature.

metastasis prediction needs a complete overview of the body, work on this research question has been almost exclusively performed on radiological images. Pathological slides provide detailed insight to only a minuscule portion of the whole tumor burden, namely what is biopsied or resected during surgery. On the other hand, radiological images offer a (lower resolution) view of a much larger field of vision.

The involvement of regional and distant lymph nodes in CRC indicates advanced disease. In general, as the number of involved nodes and the distance of the nodes from the primary increases, patients experience worse prognostic outcomes (Cserni, 2003). Pretreatment assessment of lymph node status is therefore crucial for treatment planning, especially in determining the extent of needed lymph node dissection and the type and planning of neoadjuvant treatment.

Within the domain of colorectal metastasis prediction, two broad categories of research questions have been addressed (Fig. 5.2).

Prediction of nodal metastases: A lymph node would be delineated on an image. Based on pathology ground truth, imaging markers would predict whether it is benign or malignant.

Prediction of metachronous metastases: The term "metachronous" refers to the fact that the metastasis emerges more than 3 to 6 or even 12 months after the index (primary) cancer (Laubert et al., 2012).

Lymphatic and distant metastases are closely associated with a poorer prognosis in CRC patients (Bray et al., 2018), though the seeding of metastases is a debated topic. In the prevalent sequential progression model, tumor cells spread to lymph nodes and then subsequently to distant organs via the lymphatic system. Current CRC management guidelines are based on this model (Gunderson et al., 2010). However, liver metastases are believed primarily to be hematogenic. Lymph nodes higher up can be transferred cells through the lymphatic vessels. Therefore a part of liver and lung metastases can be lymphatic but most are regarded hematogenic. A recent study by Zhang et al. mapped the tumor genetic profile of 10 treatment-naive microsatellite-stable patients with CRC (Zhang et al., 2020). Their results show that while primary-to-metastasis was the major form of metastatic seeding (38/61, 62.3%) in CRC, metastasis to-metastasis seeding was identified in 9 of 10 patients, accounting for 37.8% (23/61) of all seeding events, where metastases from all sites were considered. Their results also showed that not all LNMs have the same metastatic potential. For the case of analyses using imaging as input for predictions, it is in general not known which metastases are seeded by LNM, therefore predictions and classifications of all metastases will be considered relevant.

5.3.1 Pathological classification

The diagnosis of colon cancer mainly depends on histopathology by reading pathological images from puncture biopsies and section staining but the procedure is time-consuming and also has high interobserver variability. Therefore computer-aided diagnosis of histopathological images of CRC has gained increasing attention. Tang et al. (2018) present SegNet, a gland segmentation method based on CNNs, which shows better performance compared with conventional methods for the first time. Qaiser et al. (2019) introduce a segmentation algorithm combining the continuous coherence method and deep learning for CRC histology whole-slide images (WSI) based on persistent homology profiles (PHP). They believe that PHPs can discriminate tumor regions from the morphological characteristics of tumor cell nuclei. Feng et al. (2021) introduced a framework containing a VGG net as backbone, and two schemes for training and inference. The framework aimed to analyze colonoscopy-derived histopathological WSIs including lesion segmentation and tissue diagnosis. To balance the numbers of benign and malignant samples, a class-wise dice loss function is applied. A DSC of 0.78 was observed.

Sitnik et al. (2020) have created an automated CAD system that can automatically analyze unstained metastasis specimens to aid intraoperative diagnosis during surgery. The analysis uses the CAD model in association with features from the images as a replacement for staining. An autoencoder and a U-net were compared and the latter was found to achieve the best results, with sensitivity of 94.5% and specificity of 72.9%. Due to the very large size of the WSIs, patch-based methods in which a large image is divided into small patches are common. However, it is difficult to determine which magnification level is most appropriate since the patches extracted at different magnification levels tend to provide different visual features. Xu and Zhang (2018) created multiscale context-aware networks to address this issue in the automatic identification of liver metastases. The multiscale network involved the concatenation of extracted features from images at a different scale based on a dataset of 30 ROIs from 18 WSIs. Better results were found for the custom multiscale network when compared against a transfer learning approach on individual images, with an F1 score of 95% for each output class attempted.

Detection of nodal micrometastasis with tumor size between 0.2 and 2.0 mm is challenging for pathologists due to the small size of metastatic foci. Chuang et al. (2021) were able to use transfer learning in association with a Resnet-50 pretrained network that was able to identify micrometastases with an AUC of 0.9956, based on a training dataset consisting of 1963 slides, including 973 positive and 990 negative ones, that were annotated by two expert pathologists. Ke et al. (2020) were able to go further than detecting the presence of metastases in WSIs by predicting microsatellite instability, which is the result of a defective DNA mismatch repair system.

5.3.2 Radiological LNM classification

The ground truth of such studies is very often pathology. The underlying hypothesis is that cell-level biological changes result in global changes to the morphological phenotype, quantified by radiomics and used for predictive purposes by AI methods. Radiomic features from preoperative CT scans in a cohort of 390 CRC patients were found to be predictive of LNM (accuracy = 64.87%) (Eresen et al., 2020). Li et al. (2021) explored the potential of a clinicalradiomics nomogram to predict LNM. The dataset consisted of clinical data (e.g., tumor markers and risk factors) and radiomic features derived from the primary lesion and LNs in preoperative CT scans. Combining radiomic and clinical characteristics, a logistic regression model achieved an AUC = 0.76 (Li et al., 2020). In the MRI domain, radiomic features baseline scans, specifically from the T2 sequence, were predictive of pretherapeutic lymph node status (AUCs > 0.70) (Song & Yin, 2020; Yang et al., 2019), though the performance is still inferior to visual evaluation. Current trends favor the use of deep learning over hand-crafted radiomic features. Ding et al. (2020) developed Faster R-CNN nomograms to predict metastatic lymph nodes in rectal cancer MR images.

A study from Kumamoto et al. (2019), while not building a radiomic model, constructed hand-crafted features from 173 lymph nodes to identify LNMs. The study found 2D short-axis diameter of the lymph node to be most predictive (AUC = 0.75), although more than 10 features were identified with significant performance, meaning that a potential multivariate classifier would be relevant for reliable classification. Li et al. (2021) were able to build a custom CNN model from MRI, in a dataset that contained 1646 positive and 1718 negative lymph nodes. The model was then able to extract and classify features from regions of interest with AUC values of 0.84 and 0.86, for the metastasis and nonmetastasis classes, respectively.

5.3.2.1 Metachronous metastasis prediction

According to this research line, the primary objective is to predict the emergence of currently unknown metastasis in otherwise morphologically healthy tissue. The biological basis of this type of metastasis is that parts of the healthy tissue can form a premetastatic niche, that is, microenvironment favorable to the growth of secondary lesions should circulating tumor cells become seeded or established secondary tumors become visible (Peinado et al., 2017). Conventional predictive markers for metachronous metastases have traditionally relied on biological ground truth, such as pathological staging and serum tumor marker levels (Segelman et al., 2014; Tsai et al., 2021). Taghavi et al. have among the first publications focused on radiomic features from macroscopically healthy parenchymal liver tissue. A schematic diagram of the radiomics workflow is shown in Fig. 5.3. Their multicenter liver CT cohort was divided into two groups; those that had developed colorectal liver metastases at ≥ 24 months after staging and those patients that didn't. Both groups had a metastasis-free liver at baseline. A predictive model trained only on radiomic features (AUC = 0.86) achieved a predictive performance similar to a model

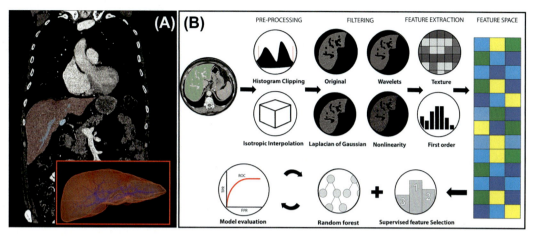

FIGURE 5.3 Radiomics machine learning workflow. *Reprinted with the permission of Springer Nature Limited.*

combining it with clinical parameters (AUC = 0.86) (Taghavi et al., 2021). These findings suggest that an element of the premetastatic niche, or even micrometastasis, can be encoded in the morphological phenotype of the healthy liver tissue.

5.4 Treatment response, recurrence, and survival

Due to the continued development of AI and its wide application in the medical field, it shows great promise in the prognosis and treatment of CRC patients. In addition, recent research has demonstrated that AI can play a crucial role in improving the survival rate of CRC patients as well (Wang et al., 2020).

Ding et al. (2019) designed an AI model by using faster R-CNN that exceeds radiologists in the evaluation of MRI images of pelvic metastatic LNs of rectal cancer. The study collected 414 patients with rectal cancer from six clinical centers, and the AI model was compared with radiologist-based diagnoses and pathologist-based diagnoses for methodological verification. For clinical verification, they followed up patients for 36 months with postoperative evaluations of recurrence of rectal cancer. The results demonstrate that the performance of the faster R-CNN model was much closer to the pathological criteria in N staging evaluation. Moreover, the faster R-CNN model was highly consistent with radiologists in N staging assessment with the kappa coefficient of 0.926. This study also shows the accumulative recurrence-free survival rate of patients at stage N2 evaluated by pathologists is significantly lower than for those separately evaluated the R-CNN model and radiologists.

Deep transfer learning models are used to classify LNM related to recurrence and design of treatment plans (Wang et al., 2020). The study collected 3364 samples with lymph nodes of a diameter of at least 3 mm. All patients underwent 3.0T MRI before surgery and experienced radiologists manually segmented all images patches and classified them as negative or positive. Their experimental results showed that compared to machine learning methods, transfer deep learning models achieved better with an accuracy of 0.75 and an AUC of 0.79. Since medical images are usually limited especially in CRC patients' data, using transfer learning could be a better alternative technique to train AI models with small imaging datasets. Worth noting is that it is common at the time of writing not to receive direct histology proof as most patients with nodal metastases receive chemoradiation that sterilizes nodes. For the case of a small locally excised rectal cancer, accurate staging of nodes is important to prevent local recurrence and distant metastases.

Sometimes it is technically challenging to determine an appropriate treatment strategy after endoscopic resection such as for local recurrence and LNM. Hence, early prediction of CRC is critical for the choice of treatment methods (Ito et al., 2019). Recently, machine learning on WSI was used to predict LNM for submucosal invasive (cT1) CRC (Takamatsu et al., 2019). The study was conducted by training a random forest model with a dataset of 397 cT1 CRCs and the result showed that the machine learning predictive model is better than the conventional method on the same datasets (AUC: 0.94). Similarly, Kudo et al. (2021) have developed a machine-learning artificial neural network (ANN) using a multi-center dataset of 3134 patients. Their ANN model was able to identify patients with LNM and an initial endoscopic resection with an AUC of 0.84.

To predict survival in patients with CRC, recent studies used supervised (Kather et al., 2019) and unsupervised (Muhammad et al., 2019) CNNs learning methods. Supervised learning is a method of training an AI model by using part annotation information from the dataset. However, collecting annotations is time consuming, challenging, and difficult to scale up. To resolve this issue, some researchers proposed to find localized parts in an unsupervised manner. Thus it is not necessary to manually label the dataset by qualified experts (Gu et al., 2018). By using the whole-slide images DeepConvSurv (Zhu et al., 2017) was one of the first unsupervised learning methods and demonstrated a training technique based on graphs of global topological representations. Another study (Abbet et al., 2020) also proposed a self-supervised learning method related to survival prediction based on WSI. The study collected 374 patients from a clinicopathological dataset, including their survival time and treatment information. The paper presents a new technique named divide-and-rule to learn pathological patterns in the regions of cancer tissue. Their technique allows training a convolutional autoencoder model by using colorization learning as a proxy task for extracting generalizable features from the unlabeled images itself and achieved a better result as compared to the state-of-the-art.

Although recent AI models have shown remarkable performance on most clinical prediction tasks in treatment response, recurrence, and survival of CRC patients, the lack of interpretability makes them difficult to be practically adopted in real clinical settings. In fact, typical AI models like deep learning are black-boxes, which are not transparent, and the prediction outcomes are difficult to interpret (Montavon et al., 2018). However, explainable and interpretable AI models are a crucial element for implementation into clinical practice and treatment decisions (Chen et al., 2020). Weakly supervised deep learning techniques are used to predict disease-specific survival from WSIs and generate human-interpretable histological features (Wulczyn et al., 2021). The study included stage II and III CRC using 3652 cases and their results from the deep-learning-based image-similarity interpretable model showed that it explained around 80% of the variance in the survival scores, which could be understood, described, and reproducibly identified by pathologists. Approximately 20% of the variance remains unexplained and needs to be addressed in future works.

5.5 Summary and future directions

Recent decades have seen significant advances in imaging technologies that have resulted in extensive utilization in the clinic, be it in the screening, diagnosis, staging, management, and follow-up of nodal CRC. This has resulted in large patient datasets that facilitate the retrospective development of AI models, which in turn has resulted in significant applied research into deriving pretreatment annotation and treatment predictions.

Retrospective studies of imaging data have yielded important insights into the progression of nodal disease and automated ways to identify lesions using imaging. With growing use of personalized cancer care, imaging plays a critical role in the management and treatment choice for patients with nodal CRC. While retrospective analyses are far from fully explored, future prospective analyses are required to confirm and validate results and increase predictive power of models. Developed AI models are potentially capable of providing early identification of treatment response to prevent side effects, reduce the costs of ineffective treatment, and prevent delays of a second, potentially more effective, therapy. However, the real-world use of such models requires extensive reporting, validation, and certification, of which prospective studies are an essential ingredient.

The complete exploitation of the underlying imaging technology will play an essential role in reducing the mortality of this preventable and manageable cancer. However, challenges remain in the form of limited coherent cohorts, which necessitates multicenter research.

References

Abbet, C., et al. (2020). Divide-and-rule: Self-supervised learning for survival analysis in colorectal cancer. *Medical image computing and computer assisted intervention — MICCAI 20 20* (pp. 480–489). Springer International Publishing.
Akbari, M., et al. (2018). Polyp segmentation in colonoscopy images using fully convolutional network. IEEE.
Ding, L., et al. (2019). Artificial intelligence system of faster region-based convolutional neural network surpassing senior radiologists in evaluation of metastatic lymph nodes of rectal cancer. *Chinese Medical Journal, 132*(4), 379–387.
Astler, V. B., & Coller, F. A. (1954). The prognostic significance of direct extension of carcinoma of the colon and rectum. *Annals of Surgery, 139*(6), 846–852.
Barr, K., et al. (2020). Automated segmentation of computed tomography colonography images using a 3D U-Net. SPIE.
Beets-Tan, R. G. H., et al. (2013). Magnetic resonance imaging for the clinical management of rectal cancer patients: Recommendations from the 2012 European Society of Gastrointestinal and Abdominal Radiology (ESGAR) consensus meeting. *European Radiology, 23*(9), 2522–2531.
Bodalal, Z., et al. (2019). Radiogenomics: Bridging imaging and genomics. *Abdominal Radiology, 44*(6), 1960–1984.
Bray, F., et al. (2018). Global cancer statistics 2018: GLOBOCAN estimates of incidence and mortality worldwide for 36 cancers in 185 countries. *CA: A Cancer Journal for Clinicians, 68*(6), 394–424.
Cao, W., et al. (2019). Multilayer feature selection method for polyp classification via computed tomographic colonography. *Journal of Medical Imaging, 6*(4), 1.
Chen, P., et al. (2020). Interpretable clinical prediction via attention-based neural network. *BMC Medical Informatics and Decision Making, 20*(3), 22–24.
Chuang, W.-Y., et al. (2021). Identification of nodal micrometastasis in colorectal cancer using deep learning on annotation-free whole-slide images. *Modern Pathology*.
Cserni, G. (2003). Nodal staging of colorectal carcinomas and sentinel nodes. *Journal of Clinical Pathology, 56*(5), 327–335.
Ding, L., et al. (2020). A deep learning nomogram kit for predicting metastatic lymph nodes in rectal cancer. *Cancer Medicine, 9*(23), 8809–8820.

Eresen, A., et al. (2020). Preoperative assessment of lymph node metastasis in colon cancer patients using machine learning: A pilot study. *Cancer Imaging: The Official Publication of the International Cancer Imaging Society, 20*(1).

Feng, R., et al. (2021). A deep learning approach for colonoscopy pathology WSI analysis: Accurate segmentation and classification. *IEEE Journal of Biomedical and Health Informatics*, 1, -1.

Gu, J., et al., (2018) Recent advances in convolutional neural networks.

Gunderson, L. L., et al. (2010). Revised TN categorization for colon cancer based on national survival outcomes data. *Journal of Clinical Oncology, 28*(2), 264−271.

Huang, Y.-J., et al. (2018). HL-FCN: Hybrid loss guided FCN for colorectal cancer segmentation. IEEE.

Huang, Y.-J., et al. (2020). 3-D RoI-aware U-Net for accurate and efficient colorectal tumor segmentation. *IEEE Transactions on Cybernetics*, 1−12.

Ito, S., et al. (2019). Treatment strategy for local recurrences after endoscopic resection of a colorectal neoplasm. *Surgical Endoscopy, 33*(4), 1140−1146.

Iyer, R. B., et al. (2002). Imaging in the diagnosis, staging, and follow-up of colorectal cancer. *American Journal of Roentgenology, 179*(1), 3−13.

Jha, D., et al. (2021). Real-time polyp detection, localization and segmentation in colonoscopy using deep learning. *IEEE Access, 9*, 40496−40510.

Jiang, H., Diao, Z., & Yao, Y.-D. (2021). Deep learning techniques for tumor segmentation: A review. *The Journal of Supercomputing*.

Kather, J. N., et al. (2019). Predicting survival from colorectal cancer histology slides using deep learning: A retrospective multicenter study. *PLoS Medicine, 16*(1).

Ke, J., et al. (2020). Identifying patch-level MSI from histological images of colorectal cancer by a knowledge distillation model. IEEE.

Kudo, S., et al. (2021). artificial intelligence system to determine risk of T1 colorectal cancer metastasis to lymph node. *Gastroenterology, 160*(4), 1075−1084, e2.

Kumamoto, T., et al. (2019). Optimal diagnostic method using multidetector-row computed tomography for predicting lymph node metastasis in colorectal cancer. *World Journal of Surgical Oncology, 17*(1).

Lambin, P., et al. (2017). Radiomics: The bridge between medical imaging and personalized medicine. *Nature Reviews Clinical Oncology, 14*(12), 749−762.

Laubert, T., et al. (2012). Metachronous metastasis- and survival-analysis show prognostic importa nce of lymphadenectomy for colon carcinomas. *BMC Gastroenterology, 12*(1).

Lee, J., et al. (2019). Reducing the model variance of a rectal cancer segmentation network. *IEEE Access, 7*, 182725−182733.

Li, J., et al. (2021). A novel classification method of lymph node metastasis in colorectal cancer. *Bioengineered., 12*(1), 2007−2021.

Li, J., et al. (2021). Different machine learning and deep learning methods for the classification of colorectal cancer lymph node metastasis images. *Frontiers in Bioengineering and Biotechnology, 8*.

Li, M., et al. (2020). A clinical-radiomics nomogram for the preoperative prediction of lymph node metastasis in colorectal cancer. *Journal of Translational Medicine, 18*(1).

Li, Q., et al. (2017). Colorectal polyp segmentation using a fully convolutional neural network. IEEE.

Liu, X., et al. (2019). Accurate colorectal tumor segmentation for CT scans based on the label assignment generative adversarial network. *Medical Physics, 46*(8), 3532−3542.

Lord, A., et al. (2019). The current status of nodal staging in rectal cancer. *Current Colorectal Cancer Reports, 15*(5), 143−148.

McKeown, E., et al. (2014). Current approaches and challenges for monitoring treatment response in colon and rectal cancer. *Journal of Cancer., 5*(1), 31−43.

Men, K., Dai, J., & Li, Y. (2017). Automatic segmentation of the clinical target volume and organs at risk in the planning CT for rectal cancer using deep dilated convolutional neural networks. *Medical Physics, 44*(12), 6377−6389.

Montavon, G., Samek, W., & Müller, K. R. (2018). Methods for interpreting and understanding deep neural networks. *Digital Signal Processing: A Review Journal, 73*, 1−15.

Mori, Y., et al. (2016). Impact of an automated system for endocytoscopic diagnosis of small colorectal lesions: An international web-based study. *Endoscopy, 48*(12), 1110−1118.

Muhammad, H., et al., (2019). Towards unsupervised cancer subtyping: Predicting prognosis using a histologic visual dictionary.

Müller, M. F., Ibrahim, A. E. K., & Arends, M. J. (2016). Molecular pathological classification of colorectal cancer. *Virchows Archiv: An International Journal of Pathology, 469*(2), 125–134.

Panic, J., et al. (2020). A convolutional neural network based system for colorectal cancer segmentation on MRI images. IEEE.

Peinado, H., et al. (2017). Pre-metastatic niches: Organ-specific homes for metastases. *Nature Reviews. Cancer, 17*(5), 302–317.

Qaiser, T., et al. (2019). Fast and accurate tumor segmentation of histology images using persistent homology and deep convolutional features. *Medical Image Analysis, 55*, 1–14.

Raju, A., et al. (2020). Graph attention multi-instance learning for accurate colorectal cancer staging. *Medical image computing and computer assisted intervention − MICCAI 20 20* (pp. 529–539). Springer International Publishing.

Rami-Porta, R., (2019). Towards the 9th edition of the tumour, node and metastasis classification of lung cancer. A historical appraisal and future perspectives. ATD.

Rock, J. B., et al. (2014). Debating deposits: An interobserver variability study of lymph nodes a nd pericolonic tumor deposits in colonic adenocarcinoma. *Archives of Pathology & Laboratory Medicine, 138*(5), 636–642.

Romero, F. P., et al. (2019). End-to-end discriminative deep network for liver lesion classification. IEEE.

Ronneberger, O., Fischer, P., & Brox, T. (2015). U-Net: Convolutional networks for biomedical image segmentation. *Lecture notes in computer science* (pp. 234–241). Springer International Publishing.

Safarov, S., & Whangbo, T. K. (2021). A-DenseUNet: Adaptive densely connected UNet for polyp segmentation in colonoscopy images with atrous convolution. *Sensors, 21*(4), 1441.

Segelman, J., et al. (2014). Individualized prediction of risk of metachronous peritoneal carcinomatosis from colorectal cancer. *Colorectal Disease, 16*(5), 359–367.

Sitnik, D., et al. (2020). Deep learning approaches for intraoperative pixel-based diagnosis of colon cancer metastasis in a liver from phase-contrast images of unstai ned specimens. SPIE.

Sobin, L. H. (2003). TNM: Evolution and relation to other prognostic factors. *Seminars in Surgical Oncology, 21*(1), 3–7.

Song, L., & Yin, J. (2020). Application of texture analysis based on sagittal fat-suppression and oblique axial T2-weighted magnetic resonance imaging to identify lymph node invasion status of rectal cancer. *Frontiers in Oncology, 10*.

Soomro, M. H., et al. (2018) Automatic segmentation of colorectal cancer in 3D MRI by combining deep learning and 3D level-set algorithm-a preliminary study. IEEE.

Soomro, M. H., et al. (2017). Haralick's texture analysis applied to colorectal T2-weighted MRI: A preliminary study of significance for cancer evolution. ACTAPRESS.

Taghavi, M., et al. (2021). Machine learning-based analysis of CT radiomics model for prediction of colorectal metachronous liver metastases. *Abdominal Radiology, 46*(1), 249–256.

Takamatsu, M., et al. (2019). Prediction of early colorectal cancer metastasis by machine learning using digital slide images. *Computer Methods and Programs in Biomedicine, 178*, 155–161.

Takemura, Y., et al. (2012). Computer-aided system for predicting the histology of colorectal tumors by using narrow-band imaging magnifying colonoscopy (with video). *Gastrointestinal Endoscopy, 75*(1), 179–185.

Tang, J., Li J., & Xu X. (2018). Segnet-based gland segmentation from colon cancer histology images. IEEE.

Thoeny, H. C., & Ross, B. D. (2010). Predicting and monitoring cancer treatment response with diffusion-weighted MRI. *Journal of Magnetic Resonance Imaging, 32*(1), 2–16.

Trebeschi, S., et al. (2017). Deep learning for fully-automated localization and segmentation of rectal cancer on multiparametric MR. *Scientific Reports, 7*(1).

Tsai, T.-Y., et al. (2021). A prediction model for metachronous peritoneal carcinomatosis in patients with stage T4 colon cancer after curative resection. *Cancers, 13*(11), 2808.

Vorontsov, E., et al. (2019). Deep learning for automated segmentation of liver lesions at CT in patients with colorectal cancer liver metastases. *Radiology: Artificial Intelligence, 1*(2)180014.

Wang, Y., et al. (2020). Application of artificial intelligence to the diagnosis and therapy of colorectal cancer. *American Journal of Cancer Research, 10*(11), 3575–3598.

Wulczyn, E., et al. (2021). Interpretable survival prediction for colorectal cancer using deep learning. *npj Digital Medicine., 4*(1), 1–13.

Xu, Z. & Zhang Q. (2018). Multi-scale context-aware networks for quantitative assessment of colo rectal liver metastases. IEEE.

Yang, C., et al., (2021). Mutual-prototype adaptation for cross-domain polyp segmentation. *IEEE Journal of Biomedical and Health Informatics*: p. 1-1.

Yang, L., et al. (2019). Rectal cancer: Can T2WI histogram of the primary tumor help predict the existence of lymph node metastasis? *European Radiology, 29*(12), 6469–6476.

Yuan, Z., et al., (2020). Development and validation of an image-based deep learning algorithm f or detection of synchronous peritoneal carcinomatosis in colorectal cancer. Annals of Surgery. Publish Ahead of Print.

Zhang, C., et al. (2020). Mapping the spreading routes of lymphatic metastases in human colorect al cancer. *Nature Communications, 11*(1).

Zhou, D., et al. (2020). Diagnostic evaluation of a deep learning model for optical diagnosis of colorectal cancer. *Nature Communications, 11*(1).

Zhu, X., et al. (2017) WSISA: Making survival prediction from whole slide histopathological i mages. Institute of Electrical.

Zhuang, F., et al., A comprehensive survey on transfer learning. Proc. IEEE. 2020 109(1): p. 43–76.

CHAPTER 6

Tumor deposits in colorectal cancer

Nelleke Pietronella Maria Brouwer, Kai Francke and Iris D. Nagtegaal

Department of Pathology, Radboud University Medical Center, Nijmegen, The Netherlands

6.1 Introduction

Cancer staging plays an essential role in the considerations for cancer treatment. The TNM staging system is set to determine the prognostic outcomes for a patient and to make different cohorts more comparable. The nodal status is especially important in the staging process: the presence of lymph node (LN) metastases in the absence of distant metastases is the main indication for (neo)adjuvant systemic therapy in colorectal cancer (CRC) patients (Benson et al., 2014). In recent years, the prominent role of LN metastases in the TNM staging system has been challenged by the identification of other types of locoregional spread with prognostic significance such as perineural invasion (PNI), extramural vascular invasion (EMVI), and tumor deposits (TDs) of which the latter have the most prognostic impact (Chand et al., 2016; Knijn et al., 2016; Lord et al., 2017; Nagtegaal et al., 2017). When studying LN staging in CRC, it is essential to be aware of the existence of TD because of their close relation to LN metastases. TD, aggregates of tumor cells in the fatty tissue surrounding the bowel wall, can be identified as metastatic LNs on imaging and macroscopic examination. However, microscopically, TD lacks the typical characteristics of LNs metastases. In this chapter, we will describe the current insights regarding the definition of TD, the innovations and difficulties in the radiological and histopathological diagnosis of TD, and their prognostic value, as well as the ongoing discussion about the incorporation of TD in the TNM staging system.

6.2 Definition of tumor deposits

TDs are tumor cell aggregates in the LN draining area of cancers, that is in the mesorectal or mesocolonic fat column (Fig. 6.1). Several different terms have been used for TD, including tumor satellites, tumor nodules, microfoci, neoplastic foci, tumor aggregates,

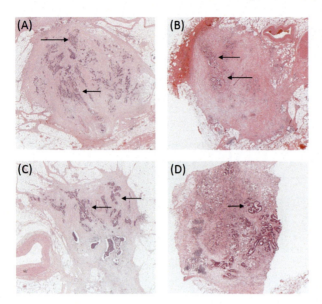

FIGURE 6.1 Histology of tumor deposits A−D Examples of TDs (H&E, 20×). Arrows indicate examples of adenocarcinoma cells. Source: *Images obtained from Brouwer, Lord et al. (14).*

discontinuous carcinoma, nonnodal deposits, and extranodal deposits. They vary in size but are usually between 1 and 5 mm in diameter and occur in approximately 20%−25% or all CRC cases. While TDs have first been described in 1935 (Gabriel et al., 1935), they were not formally acknowledged in the staging system until the 5th edition of the TNM [1997, Sobin & Wittekind (1997)]. In the 5th edition, TDs were classified as LN metastasis if larger than 3 mm in size. Smaller TDs were considered discontinuous extensions in the T category, that is, the presence of these small TDs would automatically classify these tumors as at least pT3. Although the size criterion was arbitrarily set and—in hindsight—rejected by the TNM committee based on the lack of evidence for this criterion, the obvious benefit was good reproducibility (Sobin, 2003). The 6th edition [2002, Greene et al. (2002)] moved away from the size criterion and defined TD based on contour. A smooth contour, similar to an uninvolved LN, was a reason to classify TD as LN metastases. In case of an irregular contour, TD should be classified as a discontinuous extension. After considerable criticism (Nagtegaal & Quirke, 2007; Quirke et al., 2007), this specific contour criterium was more or less removed in the 7th edition of TNM [2010, Sobin et al. (2010)]. Guidelines in that edition were vague and the decision was left to the reporting pathologist: "If a nodule is considered to be a totally replaced lymph node (generally having a smooth contour), it should be recorded as a positive lymph node and not as a TD." Furthermore, a new nodal subcategory was created to register TD in the absence of LN metastases: N1c. This new subcategory remained in the 8th edition [2017, Brierley et al. (2017)]. The definition of TD is currently as follows: "TD are discrete macroscopic or microscopic nodules of cancer in the pericolorectal adipose tissue's lymph drainage area of a primary carcinoma that are discontinuous from the primary, without histological evidence of residual lymph node or identifiable vascular or neural structures" (Brierley et al., 2017). Despite this definition, the discussion is continuing.

6.3 The origins and biology of tumor deposits

6.3.1 Where do tumor deposits come from?

The most recent definition of TD excludes those cases with clues about the origin (evidence of residual LNs, vascular or neural structures). However, these structures form, most likely, the origin of TD. As the first ones to describe TD, Gabriel et al. (1935) viewed vascular invasion to be the origin of TDs. Since then more studies have suggested that this is a potential route of origin, albeit not the only one. There are several strategies that have been applied to determine the origin of TD.

The first is based on the *association* of TD with other forms of locoregional spread, such as LN metastases, EMVI, lymphatic invasion, or PNI. In a recent meta-analysis and systematic review, evidence from multiple cohorts was summarized (Nagtegaal et al., 2017). TDs are more common in patients with LN metastases ($n = 7583$, 41.6% vs 8.7%) and patients with extramural vascular invasion ($n = 2805$, 31.6% vs 20.9%). This suggests that some TDs originate from LN metastases or vascular invasion. However, they are also associated with PNI.

The second method is the in vivo visualization of the location of TD by high-resolution magnetic resonance imaging (MRI). This method can distinguish between LN metastases and EMVI. Along the veins, TD can be observed (Balyasnikova & Brown, 2016), which, by some observers, is considered as evidence for the vascular origin of TD. However, the lymphovascular and neurovascular bundles also run along veins, locating these structures also in close relation to these TDs (Breslin et al., 2018).

The most conclusive identification of the origins of TD comes from studies using step-sectioning to examine the origins. The first was by Goldstein and Turner (2000), which was limited to the nonsmooth TD. From the 30 TDs that were investigated in 10–18 different levels, "83% had at least one location of intravascular growth, 77% had at least one focus of PNI, and 73% had at least one focus of peri-large vessel growth." As is clear from Fig. 6.2, TDs are associated with multiple origins, or at least have access to different anatomical highways. Ten years later, Wünsch et al. (2010), using on average 10 levels per TD, showed venous invasion in 26% of the cases, lymphatic invasion in 4%, nerve sheath invasion in 9%, and continuous growth in 12%, while in 49% of the cases no distinct morphological structures were found. In this study, all different contours were taken into account, in contrast to the Goldstein paper. This might explain the relatively high number of cases

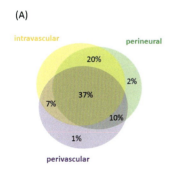
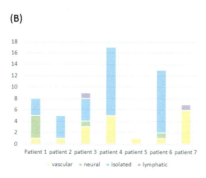

FIGURE 6.2 On the origins of tumor deposit (TDs): (A) origins of TDs based on the data by Goldstein and Turner (16); and (B) the number of TDs and their origins of each patient. *Source: From Wunsch et al. (17).*

without morphological structures. On a per patient basis, different origins were present in at least six patients. All patients ($n = 7$) showed at least one TD with vascular invasion (Fig. 6.2). It can be concluded that the origin of TD remains unclear due to their association with multiple histological structures, and it is currently unknown whether the origin of TD is even prognostically relevant.

6.3.2 The biological mechanisms underlying tumor deposit development

While histological studies are important to partially explain the origins of TD, the molecular mechanisms causing their formation are largely unknown. The potential role of epithelial–mesenchymal transition (EMT) was studied by Fan et al. (2013). EMT is a process by which epithelial cells gain invasive properties, which enables tumor cells from epithelial origin to migrate and invade lymphatic and blood vessels. This way they can spreadtoward LNs or distant tissues, hereby initiating the metastatic process (Pećina-Slaus, 2003). The investigators focused on the expression of Twist and Snail in the primary tumor, transcription factors that induce EMT by suppressing E-cadherin. They found differential expression, with increased Snail expression in patients with LN metastases but no TD, while Twist was upregulated in tumors of patients with TD without LN metastases. These results reaffirm the notion of the biological difference between TD and LN metastases, and suggest that there might not be a common origin for TD and LN metastases. Unfortunately, this is the only paper that has investigated the molecular background of TD up to now and it has not focused specifically on the biology of the tumor cells that constitute the TD or LN metastases.

6.4 Staging of tumor deposits

The changing definitions of TD have had profound impact on staging: with every change in definition, stage migration for large cohorts of patients occurred (Fig. 6.3).

Changes in staging systems can have major impact on the treatment of patients in two ways. First, the direct effect: patients with stage III CRC are eligible for adjuvant therapy. If more patients are considered stage III due to the inclusion in this stage, more patients will receive adjuvant chemotherapy, although it is unsure whether they would benefit from it. This would add cost to the health care systems. Second, the indirect effect: treatment choices are often made based on clinical trials. Inclusion in trials is often (partly) based on tumor stage. If there is stage migration during the running of a trial, different cohorts are included, which might influence the results. If stage migration occurs after the results of trials have been published, the current patients differ from those included and might not benefit from these treatment as much as could be expected.

The changes in T classification are less profound and less likely to alter treatment decisions. Compared to the 4th edition of TNM, when TDs were not described specifically, 4% of pT classification might have changed (Nagtegaal, Quirke, et al., 2011). For the subsequent changes to the 6th and the 7th editions, these numbers were 4.3% and 5.5%, respectively. More extensive changes were observed in the N category with 64% change in

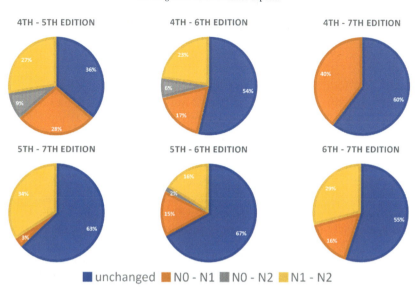

FIGURE 6.3 Stage migration due to changes in definitions. Changes in nodal stage due to changes in TNM definitions. Source: *Based on Nagtegaal et al. (39).*

patient with TD due to the introduction of the 5th edition, 33.2% after the 6th edition, and again 44.8% due to the introduction of the N1c category in the 7th edition. Since in the 7th edition, it was left to the discretion of the pathologist to determine whether a deposit is actually an LN, and no definitions are given. The percentage thus varied according to the pathologist reporting the case.

The current staging guidelines only incorporate TD in the absence of LN metastases, they are not considered when LN metastasesare present. The poor outcome associated with the presence of TD in combination with LN metastases does not support this view. Based on the literature, several suggestions have been made. The most simple proposal is to add the number of TDs to the number of LN metastasesto derive a final N stage because of the prognostic properties of TD, although TDs are not the same as LN metastases (Nagtegaal et al., 2017). Support comes from a recent study (Liu et al., 2019) that shows that the prognosis of patients with a limited number of LN metastases (pN1) and TD have similar outcomes as patients with pN2. In contrast, patients with pN2 and TD have a worse outcome. The same has been observed in several cohorts (Lino-Silva et al., 2019; Puppa et al., 2009), who suggest that the outcome of patients with TD is comparable to stage IV patients. They view TD as "in transit metastases" and feel that these should be classified as metastatic disease. This statement is still under discussion.

6.5 Prognostic value of tumor deposits

TDs have been identified as an adverse prognostic factor in multiple studies. A meta-analysis from 2017 presented overwhelming evidence that TD^+ patients have a higher hazard ratio for disease-free survival, disease-specific survival, and overall survival (Nagtegaal et al., 2017). Since then, more studies have been performed that corroborate the evidence (Lino-Silva et al., 2019; Liu et al., 2019; Wong-Chong et al., 2018; Zheng et al., 2020).

Several features of TD are associated with variation in outcome. The size of TDs is important: large TDs (defined as over 12 mm in diameter) are associated with a poor outcome in comparison with small TDs (less than 3 mm in diameter) (Ueno et al., 2011). This was confirmed in study by Shimada. The prognostic effect of the contour of TD is still a point of discussion, although there are suggestions that the irregular contour is associated with a poor outcome. The number of TDs is relevant: with increasing number of TD, there is an increasing risk of poor outcome (Goldstein & Turner, 2000; Jin et al., 2015).

One explanation of the poor prognosis associated with TD is the strong association with other features representing locoregional spread (like LN metastases, EMVI, and PNI). The features in themselves also carry an increased risk for the development of metastatic disease. In multivariate analyses, TD shows independent prognostic value in addition to nodal status (Goldstein & Turner, 2000; Lin et al., 2014; Nagayoshi et al., 2014; Shimada & Takii, 2010; Song et al., 2012), lymphovascular invasion (Al Sahaf et al., 2011; Goldstein & Turner, 2000; Nagayoshi et al., 2014; Shimada & Takii, 2010; Song et al., 2012; Yabata et al., 2014), venous invasion (Goldstein & Turner, 2000; Lin et al., 2014; Nagayoshi et al., 2014; Yabata et al., 2014), and PNI (Lin et al., 2014).

By combining multiple of these high-risk features, even better risk profiling can be performed. The combination of TD with LN metastasesresults in a high risk of metastatic disease. In several studies (Goldstein & Turner, 2000; Wünsch et al., 2010), the presence of TD is mainly associated with intraabdominal metastases; however, in an individual case meta-analysis, it was shown that there is a powerful correlation with the development of liver and lung metastases as well (Nagtegaal et al., 2017).

6.6 Tumor deposits and neoadjuvant therapies

A confusing problem occurs after neoadjuvant therapy when considering TD. Tumor fragmentation and regression due to treatment of a larger primary tumor can result in remnants in the form of tumor cell islands in the mesentery. These TDs would thus be present as a result of the treatment instead of being a primary phenomenon (Wünsch et al., 2010). As a consequence, this would be associated with a good outcome rather than the poor outcome of TD in the treatment-naïve setting. Currently, there is no guideline to accurately classify TD in this situation.

Most of the studies investigating the prognostic value of TDs exclude patients that have undergone neoadjuvant therapy. In a recent meta-analysis of eight studies that included patients that had undergone neoadjuvant chemoradiotherapy (NCRT), Lord, Graham Martínez, et al. (2019) examined the prognostic implications of TD after NCRT. They showed that TDs are a poor prognostic marker in patients that have undergone NCRT, similarly to the prognostic implications in treatment-naïve patients. The presence of TD is associated with higher TNM state, especially advanced T stage, LN metastases, and distant metastases, again comparable to untreated patients. Notably, this has also been described in an analysis of a large database of surveillance, epidemiology, and end results (SEER) data

(Wei et al., 2016). However, this study was not included in the meta-analysis by Lord, Graham Martínez, et al. (2019) because of the lack of histological review and limited registration of TD which was due to the inclusion of TDs only in the absence of LN metastases. This resulted in a lower prevalence of TD (10.7%) in the SEER data set compared to the prevalence found by the meta-analysis (21.6%). The prevalence of 21.6% is similar to the one reported for untreated patients (Nagtegaal et al., 2017). Lord et al. argue that the great variation in the reported prevalence represents differences in pathology techniques and classification, high interobserver variation, and differences in case-mix between institutions.

Because of the comparable prevalence in both treated and untreated settings, it can be speculated that the effect of NCRT on TD^+ cases is minimal. One study that compared cases within the same institute showed indeed no differences (Gopal et al., 2014). This study is one of only two that provide a direct comparison between treated and untreated patients to investigate the prognostic importance of TD (Gopal et al., 2014; Ratto et al., 2002). However, both studies were nonrandomized retrospective studies which leave the possibility of several confounding factors, a critical one being that patients with advanced cancers were more likely to be selected for NCRT.

Unlike in untreated patients (Goldstein & Turner, 2000; Wünsch et al., 2010), no correlation could be found in patients with NCRT for the possible origins of TDs: lymphatic invasion and vascular invasion were not associated with the presence of TD, only PNI. Due to the lack of correlations, other mechanisms might be involved in TD formation in patients treated with NCRT. Differences in radiation sensitivity might have caused the disappearance of these histological structures (Wünsch et al., 2010) or the TD associated with these structures. New TDs might have appeared due to tumor fragmentation. This specific type of tumor regression is also associated with a relatively poor prognosis. Ratto et al. (2007) show that at least half of the patients with TD after neoadjuvant therapy have multiple origins for their TD, albeit that lymphatic association never appears on its own (Fig. 6.4). However, 11.8% of the patients did have TD without evidence of an associated structure.

Since TDs have been noted to be visible on MRI, Lord et al. propose that the relationship of the origin of TD and tumor fragmentation in the neoadjuvant setting could be elucidated

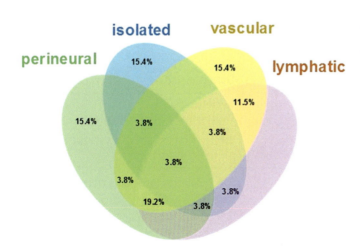

FIGURE 6.4 Origins of tumor deposits found in patients treated with neoadjuvant therapy. Source: *Based on Ratto et al. (38).*

by pre- and posttreatment analysis by MRI. This way, a clear 3D map of the primary tumor and presence of TD from before the treatment could be compared to a 3D map after treatment. In combination with pathologists, pre- and posttreatment findings could be correlated to establish which TDs were a present in the first place and which established over time.

6.7 Pathological assessment of tumor deposits: interobserver variation

Due to their prognostic importance, TDs have been incorporated in TNM staging in a changing manner over the years as described previously. Currently, there is still no clear definition of TDs and how to differentiate them from other types of locoregional spread which causes concern regarding interobserver variation.

6.7.1 Interobserver agreement regarding tumor deposits in previous TNM editions

With changing definitions for TDs in different TNM editions, interobserver variability has also varied (Table 6.1). A low interobserver variation (0.84) was found when identifying all nodules larger than 3 mm as TDs (TNM5) (Nagtegaal, Quirke, et al., 2011), whereas the definition based on contour (TNM6) led to high interobserver variation (0.21) (Howarth et al., 2004; Nagtegaal, Tot, et al., 2011). Since previous definitions were based on weak and unsubstantial data, TNM7 incorporated TDs into a new N category (N1c), based on the presence of histological structures with the following definition "Any cancerous nodule, either microscopic or macroscopic, located in the lymph drainage area of the peritumoral fatty tissue, irrespective of size or shape, as long as there is histologically proven absence of both residual lymphatic tissue in the nodule and regional lymph node metastasis (N0)" (Sobin et al., 2010). This definition led to moderate interobserver variation (0.48) when using challenging nodules (Rock et al., 2014).

6.7.2 Interobserver agreement regarding tumor deposits in the TNM 8th edition

The definition for TDs has increased even further in complexity in the TNM8 in an attempt to refine the classification, adding that "histological evidence of residual lymph

TABLE 6.1 Overview of interobserver variation using different TNM editions.

TNM edition	Definition tumor deposits	Interobserver variation (κ)	Publication
TNM 5th	Size > 3 mm = tumor deposit	0.84	Nagtegaal et al. (2011)
TNM 6th	Contour *Irregular* = tumor deposit	0.21	Howarth et al. (2004)
TNM 7th	Histological features *Absence of lymphoid, venous, neural structures = tumor deposit*	0.48	Rock et al. (2014)
TNM 8th	Histological features as TNM 7 *Addition of EMVI/PNI as separate classifications*	0.49	Lord/Brouwer et al. (2020)

node or identifiable vascular or neural structures should be absent" and that "if a vessel wall is identifiable on H&E, elastic or other stains, it should be classified as venous invasion (V1/2) or lymphatic invasion (L1). Similarly, if neural structures are identifiable, the lesion should be classified as PNI (Pn1)" (Brierley et al., 2017). However, this edition does not include guidelines about how to differentiate between TDs and LN metastases, extramural venous invasion, or PNI. Previous studies have shown that interobserver variation is also high for the diagnosis of EMVI and lymphovascular invasion in CRC, as well as for PNI in other types of cancer (Chi et al., 2016; Harris et al., 2008; Littleford et al., 2009). Therefore it is not surprising that the interobserver variation remained moderate when assessing nodules using the TNM8 (0.49) (Lord et al., 2020).

6.7.3 How to improve interobserver variation regarding tumor deposits?

The definition of TD in the TNM8 is based on the presence of histological structures, and it has been shown that certain histological features are used when assessing tumor nodules (e.g., lymphoid follicles, peripheral lymphocyte rim, and capsule) (Rock et al., 2014). However, there are currently no guidelines as to what or how many specific features need to be present and whether some carry more weight than others, leaving this decision to the discretion of the pathologist which leads to high interobserver variation. Although histological characteristics certainly play a part in diagnostic decision-making, it seems that pathologists also use other variables when deciding on the classification of nodule, and it is currently unclear what variables we are missing with only relying on histological structures.

It is clear that the classification system for locoregional spread in CRC needs to be improved. Current problems with interobserver variation will inevitably lead to heterogeneity in patients grouped within one stage. This will significantly affect research findings and have implications on clinical practice which may result in stage migration and inadequate treatment. When taking into account the poor interobserver agreement using definitions based on histological characteristics deposits, one could wonder whether including origin of a tumor nodule in the classification is practical or justified. There is still discussion about what TDs really are when it comes to their biological origin. It has been suggested by several authors that they represent a stage of the invasion process and should be defined as distinct groups, classifying them as either the lymphovascular, perineural, or nodular type (Belt et al., 2010; Goldstein & Turner, 2000; Ueno et al., 2007). Grouping tumor nodules based on their supposed origin would ultimately provide the data to determine whether biological origin of tumor nodules is even prognostically relevant. However, earlier TNM definitions of TDs yielded a lower interobserver variation as they were based on more objective criteria, such as the 3 mm rule in the TNM5, which, importantly, can also be related to findings on MRI and computed tomography (CT) (Balyasnikova & Brown, 2016; D'Souza et al., 2019). Developing more understanding of the impact of the origin of locoregional spread on prognosis will ultimately lead to optimized classification. However, until there is more evidence regarding the importance of origin, we should try to find a balance between a prognostically accurate, precise, and easily reproducible definition, which incorporates more objective criteria such as size, shape, and contour together with histological characteristics.

With varying TNM editions, the definition of TDs changed, leading to different scores for interobserver variation. Interobserver variation scores are shown as the kappa score and corresponding publications are provided.

6.8 Radiological assessment of tumor deposits: advances and challenges

Optimal treatment planning in CRC is achieved through staging of the disease, highlighting the importance of accurate imaging techniques that play a key role in the process of preoperative staging. In rectal cancer, nodal staging using good-quality, high-resolution MRI is very important, as it identifies high-risk, node positive patients who are usually treated with neoadjuvant radiation-based therapy (Glynne-Jones et al., 2018; Poston et al., 2011; Van De Velde et al., 2014). There is increasing awareness that preoperative identification of these high-risk patients on imaging has improved rectal cancer outcomes (Glynne-Jones et al., 2013; MERCURY Study Group, 2006).

In colon cancer, preoperative staging is performed by CT, but nodal stage has not been relevant to preoperative decision-making as neoadjuvant therapy is traditionally not part of treatment options in colon cancer (Rollvén et al., 2017). Most patients will be treated with resection of the colon, after which postoperative pathological assessment will identify those patients who are candidates for adjuvant systemic therapy (Bender et al., 2019). However, in recent years, novel treatment options such as complete mesocolic excision, neoadjuvant chemotherapy, and hyperthermic intraperitoneal therapy have been developed for locally advanced colon cancer (Arjona-Sánchez et al., 2018; Foxtrot Collaborative, 2012; Wang et al., 2017). These potentially beneficial treatment regimens have renewed the interest in preoperative staging and its diagnostic accuracy in colon cancer.

The preoperative staging patients using MRI or CT is based on the TNM system, with the main focus on LN status. However, recent evidence challenges the focus on LN metastases as the gateway to distant metastatic disease by showing that 65% of distant metastases were not seeded by LN metastases (Naxerova et al., 2017). Furthermore, other types of locoregional spread which have been proven to influence patient outcome and can be accurately identified by imaging, for example, EMVI and TDs (Chand et al., 2016; Lord et al., 2017; Nagtegaal et al., 2017). With the new insight that the process of CRC spread is far more complex than previously thought, these other forms of locoregional spread are increasingly incorporated into the staging of CRC patients. Based on their prognostic value, TDs are added to the TNM staging system in the form of N1c, which means they are also increasingly taken into consideration when assessing preoperative scans to determine treatment options (Patel et al., 2018).

6.8.1 Tumor deposits in colon cancer, identification by CT

For colon cancer, nodal disease is no longer recommended for clinical staging because it cannot be reliably identified by CT. Meta-analyses found a specificity of 55%–67% for the detection of LN metastases and population-based data representing daily practice showed

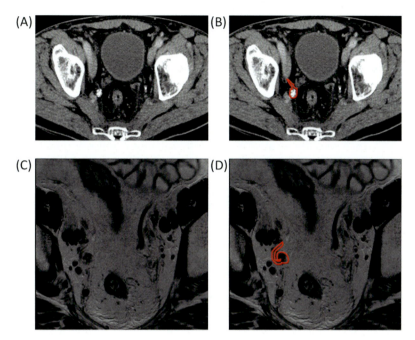

FIGURE 6.5 Radiology of tumor deposits: Examples of tumor deposits identified on CT and MRI. (A) Unmarked CT scan with a tumor deposit along the course of a vein. (B) Tumor deposit is traced in red in the CT scan. (C) Unmarked MRI scan with a tumor deposit with a tail-like shape. (D) Tumor deposit is traced in red in the MRI scan. *CT*, Computed tomography; *MRI*, magnetic resonance imaging. Source: *Images provided by Prof. G. Brown.*

a positive predictive value of only 59% (Brouwer et al., 2018; Nerad et al., 2016; Rollvén et al., 2017).

Recently, Brown et al. developed a staging system for colon cancer based on CT findings, which shows that TDs are detected on CT in 20% of patients. TDs could be differentiated from LNs on CT (Fig. 6.5) and were associated with a significantly poorer prognosis on multivariate analysis (D'Souza et al., 2019). Currently, the impact of radiologically detected TDs is limited, but it is likely that neoadjuvant therapy becomes more important in the treatment of locally advanced colon cancer in the future (Arredondo et al., 2020), and when this happens the presence of TDs should play a role in identifying high-risk patients.

6.8.2 Tumor deposits in rectal cancer, identification by MRI

Morphological features, such as heterogeneous signal intensity and LN irregular capsule border, are accurate predictors or metastatic spread within the nodes (Brown et al., 2003; Kim et al., 2004). The assessment of LN status on MRI had a specificity of 71% in a meta-analysis and population-based date yielded a specificity that varied from 67% for patients that received neoadjuvant treatment to 87% for patients that did not (Al-Sukhni et al., 2012; Brouwer et al., 2018). Importantly, if good-quality surgery is performed, nodal status seems to be of no prognostic importance for local recurrence (Chand et al., 2013, 2014). Therefore it is essential to accurately identify biomarkers with more prognostic impact. It has been found that some nodular structures on MRI could represent TDs. On histopathology, it can be difficult to distinguish TDs from an LN metastases with affected capsule. On high-resolution MRI, TDs have a very different appearance to LN metastases (Fig. 6.5C and D) and are quite common with a prevalence of around 40% (Chand et al., 2016; Knijn et al., 2016; Lord et al., 2017; Nagtegaal et al., 2017). They are defined as a nodule of tumor within the mesorectum which appear to directly interrupt the course of a vein, as opposed to LNs which may be located adjacent to veins but will not interrupt their course when seen on two orthogonal views. As on pathology, TDs (or venous deposits as they are often called in the literature) are likely to have a prognostic effect which is worse than that of LN metastases, and they are classified according to the TNM 8th edition as N1c (Taylor et al., 2011).

6.8.3 A radiological concept regarding the origin of tumor deposits

Even though the prognostic evidence on TDs is overwhelming, the biological questions, especially regarding the origin of TDs, are the subject of ongoing discussion. Due to remnants of different histological structures found in TDs (e.g., vessels, lymphoid structures, or nerves), it has been suggested that TDs can originate from different forms of invasion, such as EMVI, LN metastases, and PNI, and that the final TDs should thus be classified such as lymphovascular, perineural, or nodular (Belt et al., 2010; Goldstein & Turner, 2000; Ueno et al., 2007).

In contrast, radiological studies show that the presence of TDs is closely related to that of vascular invasion and that TDs themselves are also in close proximity of, or directly related to, vessels. Therefore it has been suggested that all TDs seen on CT or MRI represent metastases in transit after EMVI (Balyasnikova & Brown, 2016; Lord, D'Souza, et al., 2019). Although discussion about the concept of TD development is interesting, future studies are needed to determine whether all TDs seen on imaging indeed originate from vascular invasion and whether this is even clinically relevant.

6.8.4 The future of tumor deposits in radiology

Currently, studies are being performed to compare the features detected on MRI or CT with histopathological findings, which could have several benefits. First, imaging diagnostics find TDs in 20% of colon cancer patients and 40% of rectal cancer patients, whereas this is currently 22% for all CRC patients on histopathology (D'Souza et al., 2019; Lord, Graham Martínez, et al., 2019; Nagtegaal et al., 2017). If it is established that the findings on MRI or CT are accurate, they could then be used to guide pathologists and improve detection rates. Second, correlation of radiological findings with histopathology could provide evidence for a better and more universal definition of TDs. This, in combination with the prognostic evidence from both radiological and histological studies, could form a solid scientific basis for a more prominent role of TDs in both pre- and postoperative staging, then is currently being done and thereby improve the identification of high-risk tumors.

References

Al Sahaf, O., Myers, E., Jawad, M., Browne, T. J., Winter, D. C., & Redmond, H. P. (2011). The prognostic significance of extramural deposits and extracapsular lymph node invasion in colon cancer. *Diseases of the Colon and Rectum, 54*(8), 982−988. Available from https://doi.org/10.1097/DCR.0b013e31821c4944.

Al-Sukhni, E., Milot, L., Fruitman, M., Beyene, J., Victor, J. C., Schmocker, S., Brown, G., McLeod, R., & Kennedy, E. (2012). Diagnostic accuracy of MRI for assessment of T category, lymph node metastases, and circumferential resection margin involvement in patients with rectal cancer: A systematic review and meta-analysis. *Annals of Surgical Oncology, 19*(7), 2212−2223. Available from https://doi.org/10.1245/s10434-011-2210-5.

Arjona-Sánchez, A., Barrios, P., Boldo-Roda, E., Camps, B., Carrasco-Campos, J., Concepción Martín, V., García-Fadrique, A., Gutiérrez-Calvo, A., Morales, R., Ortega-Pérez, G., Pérez-Viejo, E., Prada-Villaverde, A., Torres-Melero, J., Vicente, E., Villarejo-Campos, P., Sánchez-Hidalgo, J. M., Casado-Adam, A., García-Martin, R., Medina, M., ... Rufián-Peña, S. (2018). HIPECT4: Multicentre, randomized clinical trial to evaluate safety and efficacy of hyperthermic intraperitoneal chemotherapy (HIPEC) with Mitomycin C used during surgery for treatment of locally advanced colorectal carcinoma. *BMC Cancer, 18*(1), 183. Available from https://doi.org/10.1186/s12885-018-4096-0.

Arredondo, J., Pastor, E., Simó, V., Beltrán, M., Castañón, C., Magdaleno, M. C., Matanza, I., Notarnicola, M., & Ielpo, B. (2020). Neoadjuvant chemotherapy in locally advanced colon cancer: A systematic review. *Techniques in Coloproctology, 24*(10), 1001−1015. Available from https://doi.org/10.1007/s10151-020-02289-4.

Balyasnikova, S., & Brown, G. (2016). Imaging advances in colorectal cancer. *Current Colorectal Cancer Reports, 12*, 162−169. Available from https://doi.org/10.1007/s11888-016-0321-x.

Belt, E. J., Van Stijn, M. F., Bril, H., De Lange-de Klerk, E. S., Meijer, G. A., Meijer, S., & Stockmann, H. B. (2010). Lymph node negative colorectal cancers with isolated tumor deposits should be classified and treated as stage III. *Annals of Surgical Oncology, 17*(12), 3203−3211. Available from https://doi.org/10.1245/s10434-010-1152-7.

Bender, U., Rho, Y. S., Barrera, I., Aghajanyan, S., Acoba, J., & Kavan, P. (2019). Adjuvant therapy for stages II and III colon cancer: Risk stratification, treatment duration, and future directions. *Current Oncology, 26*(1), S43−s52. Available from https://doi.org/10.3747/co.26.5605.

Benson, A. B., 3rd, Venook, A. P., Bekaii-Saab, T., Chan, E., Chen, Y. J., Cooper, H. S., Engstrom, P. F., Enzinger, P. C., Fenton, M. J., Fuchs, C. S., Grem, J. L., Hunt, S., Kamel, A., Leong, L. A., Lin, E., Messersmith, W., Mulcahy, M. F., Murphy, J. D., Nurkin, S., ... Freedman-Cass, D. A. (2014). Colon cancer, version 3.2014. *Journal of the National Comprehensive Cancer Network, 12*(7), 1028−1059. Available from https://doi.org/10.6004/jnccn.2014.0099.

Breslin, J. W., Yang, Y., Scallan, J. P., Sweat, R. S., Adderley, S. P., & Murfee, W. L. (2018). Lymphatic vessel network structure and physiology. *Comprehensive Physiology, 9*(1), 207−299. Available from https://doi.org/10.1002/cphy.c180015.

Brierley, J. D., Gospodarowicz, M. K., & Wittekind, C. H. (2017). *International Union Against Cancer TNM Classification of Malignant Tumours (8th ed)*. Wiley-Blackwell.

Brouwer, N. P. M., Stijns, R. C. H., Lemmens, V., Nagtegaal, I. D., Beets-Tan, R. G. H., Fütterer, J. J., Tanis, P. J., Verhoeven, R. H. A., & De Wilt, J. H. W. (2018). Clinical lymph node staging in colorectal cancer: A flip of the coin? *European Journal of Surgical Oncology: The Journal of the European Society of Surgical Oncology and the British Association of Surgical Oncology, 44*(8), 1241−1246. Available from https://doi.org/10.1016/j.ejso.2018.04.008.

Brown, G., Richards, C. J., Bourne, M. W., Newcombe, R. G., Radcliffe, A. G., Dallimore, N. S., & Williams, G. T. (2003). Morphologic predictors of lymph node status in rectal cancer with use of high-spatial-resolution MR imaging with histopathologic comparison. *Radiology, 227*(2), 371−377. Available from https://doi.org/10.1148/radiol.2272011747.

Chand, M., Bhangu, A., Wotherspoon, A., Stamp, G. W. H., Swift, R. I., Chau, I., Tekkis, P. P., & Brown, G. (2014). EMVI-positive stage II rectal cancer has similar clinical outcomes as stage III disease following pre-operative chemoradiotherapy. *Annals of Oncology: Official Journal of the European Society for Medical Oncology, 25*(4), 858−863. Available from https://doi.org/10.1093/annonc/mdu029.

Chand, M., Heald, R. J., & Brown, G. (2013). The importance of not overstaging mesorectal lymph nodes seen on MRI. *Colorectal Disease: The Official Journal of the Association of Coloproctology of Great Britain and Ireland, 15*(10), 1201−1204. Available from https://doi.org/10.1111/codi.12435.

Chand, M., Siddiqui, M. R., Swift, I., & Brown, G. (2016). Systematic review of prognostic importance of extramural venous invasion in rectal cancer. *World Journal of Gastroenterology*, *22*(4), 1721−1726. Available from https://doi.org/10.3748/wjg.v22.i4.1721.

Chi, A. C., Katabi, N., Chen, H. S., & Cheng, Y. L. (2016). Interobserver variation among pathologists in evaluating perineural invasion for oral squamous cell carcinoma. *Head and Neck Pathology*, *10*(4), 451−464. Available from https://doi.org/10.1007/s12105-016-0722-9.

D'Souza, N., Shaw, A., Lord, A., Balyasnikova, S., Abulafi, M., Tekkis, P., & Brown, G. (2019). Assessment of a staging system for sigmoid colon cancer based on tumor deposits and extramural venous invasion on computed tomography. *JAMA Network Open*, *2*(12), e1916987. Available from https://doi.org/10.1001/jamanetworkopen.2019.16987.

Fan, X. J., Wan, X. B., Yang, Z. L., Fu, X. H., Huang, Y., Chen, D. K., Song, S. X., Liu, Q., Xiao, H. Y., Wang, L., & Wang, J. P. (2013). Snail promotes lymph node metastasis and twist enhances tumor deposit formation through epithelial-mesenchymal transition in colorectal cancer. *Human Pathology*, *44*(2), 173−180. Available from https://doi.org/10.1016/j.humpath.2012.03.029.

Foxtrot Collaborative, G. (2012). Feasibility of preoperative chemotherapy for locally advanced, operable colon cancer: The pilot phase of a randomised controlled trial. *The Lancet Oncology*, *13*(11), 1152−1160. Available from https://doi.org/10.1016/s1470-2045(12)70348-0.

Gabriel, W. B., Dukes, C., & Bussey, H. J. R. (1935). Lymphatic spread in cancer of the rectum. *British Journal of Surgery*, *23*(90), 395−413. Available from https://doi.org/10.1002/bjs.1800239017.

Glynne-Jones, R., Harrison, M., & Hughes, R. (2013). Challenges in the neoadjuvant treatment of rectal cancer: Balancing the risk of recurrence and quality of life. *Cancer Radiotherapie: Journal de la Societe Francaise de Radiotherapie Oncologique*, *17*(7), 675−685. Available from https://doi.org/10.1016/j.canrad.2013.06.043.

Glynne-Jones, R., Wyrwicz, L., Tiret, E., Brown, G., Rödel, C., Cervantes, A., & Arnold, D. (2018). Rectal cancer: ESMO Clinical Practice Guidelines for diagnosis, treatment and follow-up. *Annals of Oncology: Official Journal of the European Society for Medical Oncology*, *29*(4), iv263. Available from https://doi.org/10.1093/annonc/mdy161.

Goldstein, N. S., & Turner, J. R. (2000). Pericolonic tumor deposits in patients with T3N + M0 colon adenocarcinomas: Markers of reduced disease free survival and intra-abdominal metastases and their implications for TNM classification. *Cancer*, *88*(10), 2228−2238.

Gopal, P., Lu, P., Ayers, G. D., Herline, A. J., & Washington, M. K. (2014). Tumor deposits in rectal adenocarcinoma after neoadjuvant chemoradiation are associated with poor prognosis. *Modern Pathology: An Official Journal of the United States and Canadian Academy of Pathology, Inc*, *27*(9), 1281−1287. Available from https://doi.org/10.1038/modpathol.2013.239.

Greene, F. L., Page, D. L., Fleming, I. D., Fritz, A. G., Balch, C. M., Haller, D. G., & Morrow, M. (2002). *International Union Against Cancer TNM Classification of Malignant Tumours (6th ed)*. John Wiley & Sons.

Harris, E. I., Lewin, D. N., Wang, H. L., Lauwers, G. Y., Srivastava, A., Shyr, Y., Shakhtour, B., Revetta, F., & Washington, M. K. (2008). Lymphovascular invasion in colorectal cancer: An interobserver variability study. *The American Journal of Surgical Pathology*, *32*(12), 1816−1821. Available from https://doi.org/10.1097/PAS.0b013e3181816083.

Howarth, S. M., Morgan, J. M., & Williams, G. T. (2004). The new (6th edition) TNM classification of colorectal cancer---A stage too far. *Gut*, *53*, A21.

Jin, M., Roth, R., Rock, J. B., Washington, M. K., Lehman, A., & Frankel, W. L. (2015). The impact of tumor deposits on colonic adenocarcinoma AJCC TNM staging and outcome. *The American Journal of Surgical Pathology*, *39*(1), 109−115. Available from https://doi.org/10.1097/pas.0000000000000320.

Kim, J. H., Beets, G. L., Kim, M. J., Kessels, A. G., & Beets-Tan, R. G. (2004). High-resolution MR imaging for nodal staging in rectal cancer: Are there any criteria in addition to the size? *European Journal of Radiology*, *52*(1), 78−83. Available from https://doi.org/10.1016/j.ejrad.2003.12.005.

Knijn, N., Mogk, S. C., Teerenstra, S., Simmer, F., & Nagtegaal, I. D. (2016). Perineural invasion is a strong prognostic factor in colorectal cancer: A systematic review. *The American Journal of Surgical Pathology*, *40*(1), 103−112. Available from https://doi.org/10.1097/pas.0000000000000518.

Lin, Q., Ye, Q., Zhu, D., Wei, Y., Ren, L., Ye, L., Feng, Q., Xu, P., Zheng, P., Lv, M., Fan, J., & Xu, J. (2014). Determinants of long-term outcome in patients undergoing simultaneous resection of synchronous colorectal liver metastases. *PLoS One*, *9*(8), e105747. Available from https://doi.org/10.1371/journal.pone.0105747.

Lino-Silva, L. S., Anchondo-Núñez, P., Chit-Huerta, A., Aguilar-Romero, E., Morales-Soto, J., Salazar-García, J. A., Guzmán-López, C. J., Maldonado-Martínez, H. A., Meneses-García, A., & Salcedo-Hernández, R. A.

(2019). Stage I--III colon cancer patients with tumor deposits behave similarly to stage IV patients. Cross-section analysis of 392 patients. *Journal of Surgical Oncology*, 120(2), 300−307. Available from https://doi.org/10.1002/jso.25482.

Littleford, S. E., Baird, A., Rotimi, O., Verbeke, C. S., & Scott, N. (2009). Interobserver variation in the reporting of local peritoneal involvement and extramural venous invasion in colonic cancer. *Histopathology*, 55(4), 407−413. Available from https://doi.org/10.1111/j.1365-2559.2009.03397.x.

Liu, F., Zhao, J., Li, C., Wu, Y., Song, W., Guo, T., Chen, S., Cai, S., Huang, D., & Xu, Y. (2019). The unique prognostic characteristics of tumor deposits in colorectal cancer patients. *Annals of Translational Medicine*, 7(23), 769. Available from https://doi.org/10.21037/atm.2019.11.69.

Lord, A. C., Brouwer, N. P. M., Terlizzo, M., Bateman, A. C., West, N. P., Goldin, R., Martinez, A., Wong, N. A. C. S., Novelli, M., Nagtegaal, I. D., & Brown, G. (2020). Interobserver variation in the histopathological diagnosis of tumour deposits in rectal cancer. Manuscript Submitted for Publication.

Lord, A. C., D'Souza, N., Pucher, P. H., Moran, B. J., Abulafi, A. M., Wotherspoon, A., Rasheed, S., & Brown, G. (2017). Significance of extranodal tumour deposits in colorectal cancer: A systematic review and meta-analysis. *European Journal of Cancer*, 82, 92−102. Available from https://doi.org/10.1016/j.ejca.2017.05.027.

Lord, A. C., Graham Martínez, C., D'Souza, N., Pucher, P. H., Brown, G., & Nagtegaal, I. D. (2019). The significance of tumour deposits in rectal cancer after neoadjuvant therapy: A systematic review and meta-analysis. *European Journal of Cancer*, 122, 1−8. Available from https://doi.org/10.1016/j.ejca.2019.08.020.

Lord, A., D'Souza, N., Shaw, A., Day, N., & Brown, G. (2019). The current status of nodal staging in rectal cancer. *Current Colorectal Cancer Reports*, 15(5), 143−148. Available from https://doi.org/10.1007/s11888-019-00441-3.

MERCURY Study Group. (2006). Diagnostic accuracy of preoperative magnetic resonance imaging in predicting curative resection of rectal cancer: Prospective observational study. *British Medical Journal*, 333(7572), 779. Available from https://doi.org/10.1136/bmj.38937.646400.55.

Nagayoshi, K., Ueki, T., Nishioka, Y., Manabe, T., Mizuuchi, Y., Hirahashi, M., Oda, Y., & Tanaka, M. (2014). Tumor deposit is a poor prognostic indicator for patients who have stage II and III colorectal cancer with fewer than 4 lymph node metastases but not for those with 4 or more. *Diseases of the Colon and Rectum*, 57(4), 467−474. Available from https://doi.org/10.1097/dcr.0000000000000059.

Nagtegaal, I. D., & Quirke, P. (2007). Colorectal tumour deposits in the mesorectum and pericolon: A critical review. *Histopathology*, 51(2), 141−149. Available from https://doi.org/10.1111/j.1365-2559.2007.02720.x.

Nagtegaal, I. D., Knijn, N., Hugen, N., Marshall, H. C., Sugihara, K., Tot, T., Ueno, H., & Quirke, P. (2017). Tumor deposits in colorectal cancer: Improving the value of modern staging—A systematic review and meta-analysis. *Journal of Clinical Oncology: Official Journal of the American Society of Clinical Oncology*, 35(10), 1119−1127. Available from https://doi.org/10.1200/jco.2016.68.9091.

Nagtegaal, I. D., Quirke, P., & Schmoll, H. J. (2011). Has the new TNM classification for colorectal cancer improved care? *Nature Reviews Clinical Oncology*, 9(2), 119−123. Available from https://doi.org/10.1038/nrclinonc.2011.157.

Nagtegaal, I. D., Tot, T., Jayne, D. G., McShane, P., Nihlberg, A., Marshall, H. C., Påhlman, L., Brown, J. M., Guillou, P. J., & Quirke, P. (2011). Lymph nodes, tumor deposits, and TNM: Are we getting better? *Journal of Clinical Oncology: Official Journal of the American Society of Clinical Oncology*, 29(18), 2487−2492. Available from https://doi.org/10.1200/jco.2011.34.6429.

Naxerova, K., Reiter, J. G., Brachtel, E., Lennerz, J. K., van de Wetering, M., Rowan, A., Cai, T., Clevers, H., Swanton, C., Nowak, M. A., Elledge, S. J., & Jain, R. K. (2017). Origins of lymphatic and distant metastases in human colorectal cancer. *Science*, 357(6346), 55−60. Available from https://doi.org/10.1126/science.aai8515.

Nerad, E., Lahaye, M. J., Maas, M., Nelemans, P., Bakers, F. C., Beets, G. L., & Beets-Tan, R. G. (2016). Diagnostic accuracy of CT for local staging of colon cancer: A systematic review and meta-analysis. *American Journal of Roentgenology*, 207(5), 984−995. Available from https://doi.org/10.2214/ajr.15.15785.

Patel, A., Rockall, A., Guthrie, A., Gleeson, F., Worthy, S., Grubnic, S., Burling, D., Allen, C., Padhani, A., Carey, B., Cavanagh, P., Peake, M. D., & Brown, G. (2018). Can the completeness of radiological cancer staging reports be improved using proforma reporting? A prospective multicentre non-blinded interventional study across 21 centres in the UK. *British Medical Journal Open*, 8(10), e018499. Available from https://doi.org/10.1136/bmjopen-2017-018499.

Pećina-Slaus, N. (2003). Tumor suppressor gene E-cadherin and its role in normal and malignant cells. *Cancer Cell International*, 3(1), 17. Available from https://doi.org/10.1186/1475-2867-3-17.

Poston, G. J., Tait, D., O'Connell, S., Bennett, A., & Berendse, S. (2011). Diagnosis and management of colorectal cancer: Summary of NICE guidance. *British Medical Journal, 343*, d6751. Available from https://doi.org/10.1136/bmj.d6751.

Puppa, G., Ueno, H., Kayahara, M., Capelli, P., Canzonieri, V., Colombari, R., Maisonneuve, P., & Pelosi, G. (2009). Tumor deposits are encountered in advanced colorectal cancer and other adenocarcinomas: An expanded classification with implications for colorectal cancer staging system including a unifying concept of in-transit metastases. *Modern Pathology: An Official Journal of the United States and Canadian Academy of Pathology, Inc, 22*(3), 410–415. Available from https://doi.org/10.1038/modpathol.2008.198.

Quirke, P., Williams, G. T., Ectors, N., Ensari, A., Piard, F., & Nagtegaal, I. (2007). The future of the TNM staging system in colorectal cancer: Time for a debate? *The Lancet Oncology, 8*(7), 651–657. Available from https://doi.org/10.1016/s1470-2045(07)70205-x.

Ratto, C., Ricci, R., Rossi, C., Morelli, U., Vecchio, F. M., & Doglietto, G. B. (2002). Mesorectal microfoci adversely affect the prognosis of patients with rectal cancer. *Diseases of the Colon and Rectum, 45*(6), 733–742. Available from https://doi.org/10.1007/s10350-004-6288-8, discussion 742-3.

Ratto, C., Ricci, R., Valentini, V., Castri, F., Parello, A., Gambacorta, M. A., Cellini, N., Vecchio, F. M., & Doglietto, G. B. (2007). Neoplastic mesorectal microfoci (MMF) following neoadjuvant chemoradiotherapy: Clinical and prognostic implications. *Annals of Surgical Oncology, 14*(2), 853–861. Available from https://doi.org/10.1245/s10434-006-9163-0.

Rock, J. B., Washington, M. K., Adsay, N. V., Greenson, J. K., Montgomery, E. A., Robert, M. E., Yantiss, R. K., Lehman, A. M., & Frankel, W. L. (2014). Debating deposits: An interobserver variability study of lymph nodes and pericolonic tumor deposits in colonic adenocarcinoma. *Archives of Pathology & Laboratory Medicine, 138*(5), 636–642. Available from https://doi.org/10.5858/arpa.2013-0166-OA.

Rollvén, E., Abraham-Nordling, M., Holm, T., & Blomqvist, L. (2017). Assessment and diagnostic accuracy of lymph node status to predict stage III colon cancer using computed tomography. *Cancer Imaging: The Official Publication of the International Cancer Imaging Society, 17*(1), 3. Available from https://doi.org/10.1186/s40644-016-0104-2.

Shimada, Y., & Takii, Y. (2010). Clinical impact of mesorectal extranodal cancer tissue in rectal cancer: Detailed pathological assessment using whole-mount sections. *Diseases of the Colon and Rectum, 53*(5), 771–778. Available from https://doi.org/10.1007/DCR.0b013e3181cf7fd8.

Sobin, L. H. (2003). TNM, sixth edition: New developments in general concepts and rules. *Seminars in Surgical Oncology, 21*(1), 19–22. Available from https://doi.org/10.1002/ssu.10017.

Sobin, L. H., & Wittekind, C. H. (1997). *International Union Against Cancer TNM Classification of Malignant Tumours (5th ed)*. John Wiley & Sons.

Sobin, L. H., Gospodarowicz, M. K., & Wittekind, C. H. (2010). *International Union Against Cancer TNM Classification of Malignant Tumours (7th ed)*. Wiley-Blackwell.

Song, Y. X., Gao, P., Wang, Z. N., Liang, J. W., Sun, Z., Wang, M. X., Dong, Y. L., Wang, X. F., & Xu, H. M. (2012). Can the tumor deposits be counted as metastatic lymph nodes in the UICC TNM staging system for colorectal cancer? *PLoS One, 7*(3), e34087. Available from https://doi.org/10.1371/journal.pone.0034087.

Taylor, F. G., Quirke, P., Heald, R. J., Moran, B., Blomqvist, L., Swift, I., Sebag-Montefiore, D. J., Tekkis, P., & Brown, G. (2011). Preoperative high-resolution magnetic resonance imaging can identify good prognosis stage I, II, and III rectal cancer best managed by surgery alone: A prospective, multicenter, European study. *Annals of Surgery, 253*(4), 711–719. Available from https://doi.org/10.1097/SLA.0b013e31820b8d52.

Ueno, H., Mochizuki, H., Hashiguchi, Y., Ishiguro, M., Miyoshi, M., Kajiwara, Y., Sato, T., Shimazaki, H., & Hase, K. (2007). Extramural cancer deposits without nodal structure in colorectal cancer: Optimal categorization for prognostic staging. *American Journal of Clinical Pathology, 127*(2), 287–294. Available from https://doi.org/10.1309/903ut10vq3lc7b8l.

Ueno, H., Mochizuki, H., Shirouzu, K., Kusumi, T., Yamada, K., Ikegami, M., Kawachi, H., Kameoka, S., Ohkura, Y., Masaki, T., Kushima, R., Takahashi, K., Ajioka, Y., Hase, K., Ochiai, A., Wada, R., Iwaya, K., Nakamura, T., & Sugihara, K. (2011). Actual status of distribution and prognostic impact of extramural discontinuous cancer spread in colorectal cancer. *Journal of Clinical Oncology: Official Journal of the American Society of Clinical Oncology, 29*(18), 2550–2556. Available from https://doi.org/10.1200/jco.2010.33.7725.

Van De Velde, C. J., Boelens, P. G., Borras, J. M., Coebergh, J. W., Cervantes, A., Blomqvist, L., Beets-Tan, R. G., van den Broek, C. B., Brown, G., Van Cutsem, E., Espin, E., Haustermans, K., Glimelius, B., Iversen, L. H., van Krieken, J. H., Marijnen, C. A., Henning, G., Gore-Booth, J., Meldolesi, E., ... Valentini, V. (2014). EURECCA colorectal: Multidisciplinary management: European consensus conference colon & rectum. *European Journal of Cancer*, *50*(1), 1. Available from https://doi.org/10.1016/j.ejca.2013.06.048, e1-1.e34.

Wang, C., Gao, Z., Shen, K., Shen, Z., Jiang, K., Liang, B., Yin, M., Yang, X., Wang, S., & Ye, Y. (2017). Safety, quality and effect of complete mesocolic excision vs non-complete mesocolic excision in patients with colon cancer: A systemic review and meta-analysis. *Colorectal Disease: The Official Journal of the Association of Coloproctology of Great Britain and Ireland*, *19*(11), 962–972. Available from https://doi.org/10.1111/codi.13900.

Wei, X. L., Qiu, M. Z., Zhou, Y. X., He, M. M., Luo, H. Y., Wang, F. H., Zhang, D. S., Li, Y. H., & Xu, R. H. (2016). The clinicopathologic relevance and prognostic value of tumor deposits and the applicability of N1c category in rectal cancer with preoperative radiotherapy. *Oncotarget*, *7*(46), 75094–75103. Available from https://doi.org/10.18632/oncotarget.12058.

Wong-Chong, N., Motl, J., Hwang, G., Nassif, G. J., Jr., Albert, M. R., Monson, J. R. T., & Lee, L. (2018). Impact of tumor deposits on oncologic outcomes in stage III colon cancer. *Diseases of the Colon and Rectum*, *61*(9), 1043–1052. Available from https://doi.org/10.1097/dcr.0000000000001152.

Wünsch, K., Müller, J., Jähnig, H., Herrmann, R. A., Arnholdt, H. M., & Märkl, B. (2010). Shape is not associated with the origin of pericolonic tumor deposits. *American Journal of Clinical Pathology*, *133*(3), 388–394. Available from https://doi.org/10.1309/ajcpawolx7adzq2k.

Yabata, E., Udagawa, M., & Okamoto, H. (2014). Effect of tumor deposits on overall survival in colorectal cancer patients with regional lymph node metastases. *Journal of Rural Medicine*, *9*(1), 20–26. Available from https://doi.org/10.2185/jrm.2880.

Zheng, P., Chen, Q., Li, J., Jin, C., Kang, L., & Chen, D. (2020). Prognostic significance of tumor deposits in patients with stage III colon cancer: A nomogram study. *The Journal of Surgical Research*, *245*, 475–482. Available from https://doi.org/10.1016/j.jss.2019.07.099.

CHAPTER 7

Lymph node classification in colorectal cancer: tumor node metastasis versus the Japanese system

Kozo Kataoka[1], Yukihide Kanemitsu[2], Manabu Shiozawa[3] and Masataka Ikeda[1]

[1]Hyogo College of Medicine, Japan [2]National Cancer Center, Japan [3]Kanagawa Cancer Center, Japan

7.1 Japanese D3 lymphadenectomy

7.1.1 Japanese lymph node classification

According to the Japanese Classification of Colorectal, Appendiceal, and Anal Carcinoma (JCCRC) system, the lymph node station of the colon and rectum is labeled using three-digit numbers in the 2000's. Lymph nodes are classified into three groups, according to anatomical location; paracolic, intermediate, and main lymph nodes. Paracolic lymph nodes are denoted as "2X1," intermediate lymph nodes as "2X2," and main lymph nodes as "2X3".

The second digit represents the location of feeding artery of tumor; the lymph nodes along the ileocolic artery are denoted as "20X," right colic artery as "21X," middle colic artery as "22X," left colic artery as "23X," sigmoid artery as "24X," and inferior mesenteric artery as "25X."

The frequency of metastatic lymph node involvement is highest in the pericolic lymph nodes and second highest in the intermediate lymph nodes, followed by the main lymph nodes, which is reported to be 1% to 8% (Paquette et al., 2018).

Regarding the lateral lymph nodes, lymph nodes along the internal iliac arteries are denoted as 263P in proximal side and 263D in distal side, divided by umbilical artery. Lymph nodes around the obturator foramen are labeled as 283 (obturator nodes).

Lymph nodes along the common iliac artery are denoted as 273 and external iliac artery as 293. The location of lateral lymph nodes is represented using "rt" for the right side and "lt" for the left side followed by three digits. Lymph nodes along the median sacral artery are denoted as 260 (lateral sacral nodes), 270 (median sacral nodes), and 280 (aortic bifurcation nodes). Other lymph nodes include inguinal lymph nodes that are represented by 292, superior mesenteric arterial root nodes by 214, and paraaortic nodes by 216. These three lymph node stations are numbered according to the gastric cancer classification. A chart of the different lymph node stations is shown in Fig. 7.1.

7.1.2 Concept of Japanese lymphadenectomy

According to the Japanese Classification of Colorectal Carcinoma, the anatomical extent of lymphadenectomy is expressed with the D number (Fig. 7.2) (Hashiguchi et al., 2020).

D3 lymphadenectomy requires the removal of lymph nodes along all three regional lymph node stations or levels (pericolic, intermediate, and main lymph nodes). For right-sided colonic cancers, the vascular pedicles are ligated at their origin together with removal of the main lymph nodes (203, 213, 223) along the superior mesenteric vein. For left-sided tumors, the inferior mesenteric artery is ligated at its root or the superior rectal

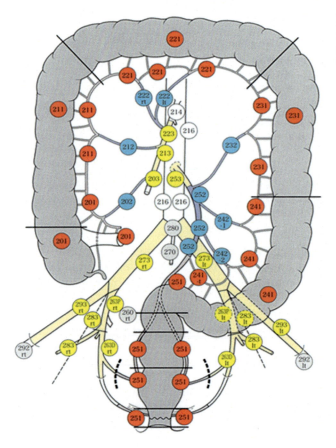

FIGURE 7.1 Lymph Node (LN) classification of JSCCR (quoted from JSCCR guidelines). *JSCCR*, Japanese Society for Cancer of the Colon and Rectum.

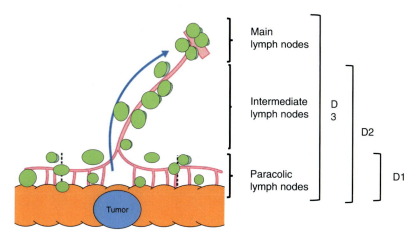

FIGURE 7.2 Extent of lymphadenectomy in Japan.

artery is ligated at its origin preserving the left colic artery, with removal of the main lymph nodes.

Complete dissection of the pericolic and intermediate lymph nodes is labeled as D2 lymphadenectomy. The anatomical boundary between D2 and D3 is clear in left-sided colorectal cancer (i.e., the left colic artery), whereas it is not clear in right-sided colon cancer, although many Japanese surgeons agree that the boundary is at the right edge of superior mesenteric vein.

D1 lymphadenectomy is defined as a surgical procedure in which only the pericolic lymph nodes are completely dissected. Incomplete dissection of the pericolic lymph nodes is labeled as D0.

Regarding the optimal extent of bowel resection, there are no standardized international criteria for regional lymph nodes in the pericolic region. In the tumor node metastasis (TNM) classification, all pericolic lymph nodes are resected with no definition of the extent (Brierley et al., 2017).

In Japan, it is assumed that no metastatic pericolic lymph nodes are present further than 10 cm from either edge of the tumor. As a consequence, the "10 cm rule" was adopted in Japanese daily surgical practice (Hida et al., 1997). Since 2006, this 10 cm rule has been modified by the Japanese Classification of Colorectal Carcinoma. According to this classification, the distribution of the feeding artery is classified into four types: A, there is a feeding artery in close proximity to the tumor; B, there is only one feeding artery within 10 cm from the tumor; C, there are two feeding arteries within 10 cm from the tumor; and D, when there is no feeding artery within 10 cm from the tumor, the artery closest to the tumor is regarded as its feeding artery (Fig. 7.3). To evaluate the optimal extent of central lymph node dissection and length of bowel resection in colon cancer surgery, the International Prospective Observational Cohort Study for Optimal Bowel Resection Extent and Central Radicality for Colon Cancer (T-REX) has completed accrual and the results are awaited (Shiozawa et al., 2021) (ClinicalTrials.gov NCT02938481).

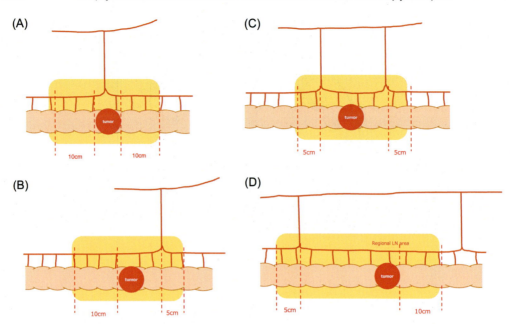

FIGURE 7.3 Regional pericolic (Lymph Nodes) LNs in colon cancer according to the modified 10-cm rule by the Japanese Classification of Colorectal Carcinoma, 8th edition (2012).

7.1.3 Lateral lymph node dissection

In Japan, lateral lymph node dissection is performed for clinical stage II or III lower rectal cancer because when the lower margin of the tumor is located at or below the peritoneal reflection, metastases to lateral pelvic lymph nodes (common iliac or internal iliac) may develop. The incidence of pelvic lymph node metastasis in patients with T3 or T4 lower rectal cancer was reported to be 18.1% according to a retrospective multicenter study in Japan (Sugihara et al., 2006). Based on this background, lateral lymph node dissection combined with total mesorectal excision (TME) has been the standard treatment for clinical stage II and III lower third rectal cancer. To evaluate the efficacy of lateral lymph node dissection, JCOG0212 compared total mesorectal excision alone with total mesorectal excision in combination with lateral lymph node dissection. Since the primary analysis failed to demonstrate the noninferiority of total mesorectal excision alone, the combination of total mesorectal excision with lateral lymph node dissection is still considered to a standard treatment in Japan, especially for Stage III lower third rectal cancer.

The extent of lateral lymph node dissection is classified in the same way as Japanese D3 lymphadenectomy. Label LD3 is used when all lateral lymph nodes (263D, 263P, 283, 273, 293, 260, 270, and 280) are removed. LD2 is labeled when the dissection of 263D, 263P, and 283 is performed. LD1 is defined as lateral lymph node dissection which does not satisfy LD2. LD0 is used when lateral lymph node dissection is not performed. LDX is labeled when the extent of LD cannot be assessed. In clinical practice, when lateral lymph node dissection is performed, the common iliac nodes (273), internal iliac nodes (263), and obturator nodes (283) are resected. Dissection of external iliac nodes (293) and median sacral nodes (270) is not mandatory as metastasis to these nodes is rare (Kobayashi et al., 2009). The autonomic nerves are usually preserved.

7.1.4 Japanese D3 versus complete mesocolic excision

Japanese D3 lymphadenectomy is often considered to be similar to complete mesocolic excision (CME). Although both surgical techniques have similar concepts, there are some differences to be noted. First of all, the concept of vascular ligation is different between the two. The CME procedure is defined as sharp anatomic dissection along the embryologic planes with the preservation of an intact visceral fascia of the mesocolon, the concept of which is derived from the that of TME, which involves sharp dissection along the fascia propria of the mesorectum (Heald & Ryall, 1986; Heald et al., 1982). Proximal vascular ligation is required in CME; however, dissection at the origin of the feeding vessels is not specified. Actually, the term "central vascular ligation (CVL)" together with "CME" was used in the original article reported by Hohenberger et al. (2009). On the other hand, Japanese D3 dissection is defined as the removal of all regional lymph node stations depending on the location of the tumor, which includes the removal of main lymph nodes around the origin of feeding arteries.

Next, the length of the resected bowel is different. In Japan, as previously mentioned, the length of the bowel to be resected basically follows the "10 cm rule"; however, it varies depending on the location of feeding artery (Fig. 7.3). On the other hand, the CME procedure removes a broader area because it includes the removal of the nearby vascular arcade beyond the "10 cm" margin. West et al. compared the resected colon cancer specimen treated with CME in German centers and Japanese D3 lymphadenectomy in Japanese centers. Japanese D3 specimens were significantly shorter, contained a smaller area of mesentery, and a smaller number of harvested lymph nodes ($P<0.001$, $P<0.001$, and $P<0.001$, respectively) compared to CME specimens. The distance from the high vascular tie to the bowel wall was similar between the two (West et al., 2012). Again, under the concept of Japanese D3 lymphadenectomy, only the primary feeding arteries entering the regional pericolic area are concerned.

A third difference is the extent of lymphadenectomy in and beyond the main lymph node area. For example, the extramesenteric lymph nodes, such as gastroepiploic arcade and the nodes along the head of the pancreas are resected in the CME procedure, when tumor is located at hepatic flexure (Hohenberger et al., 2009). In Japan, resection of the gastroepiploic and infrapyloric lymph nodes was performed in daily practice at specialized centers, taking into account the possibility of cancer spread to the extramesocolic lymphatic network around the gastrocolic trunk. However, resection of these extramesenteric lymph nodes is not performed recently because analysis of a Japanese nationwide database has shown that metastases to these extramesenteric nodes are rare (Hashiguchi et al., 2020). When the tumor is not invading the pancreas or duodenum, a Kocher maneuver of mobilizing the duodenum and the pancreatic head to expose the origin of the superior mesenteric artery is usually not performed in Japan.

7.2 Tumor node metastasis versus Japanese lymph node classification

The Union for International Cancer Control (UICC) TNM classification and TNM in the JCCRC (Hashiguchi et al., 2020) are almost the same, except for two points. The first point is the subclassification of T1a and T1b. In the Japanese classification, T1 is subclassified into T1a and T1b, depending on whether the tumor invasion is 1000 μm or more or not. The reason is that one of the indications of curative endoscopic resection is tumor depth less than 1000 μm.

The biggest difference is "N3." The definition of N3 is "Metastasis in the main lymph node (s)". In low rectal cancer, it is defined as "metastasis in the main and/or lateral lymph node (s)." A retrospective analysis of 860 Stage III colon cancer patients revealed the prognostic significance of the number of invaded lymph nodes, but not their anatomical location. However, some groups have reported that colorectal cancer patients with invaded "main" lymph nodes have worse survival. The percentage of metastatic main lymph nodes is reported to be 1%−8% (Huh et al., 2012; Kanemitsu et al., 2006; Kanemitsu et al., 2013; Kataoka et al., 2019). Retrospective analyses of a Japanese multi-institutional database also reported the prognostic impact of main lymph node metastases, which may be different according to the location of tumor. Right-sided tumors more frequently invaded main lymph nodes than left-sided lesions (8.5% vs 3.7%; $P < .001$) (Kataoka et al., 2020). Colon cancer patients with main lymph node metastases showed worse survival, but the hazard ratio (HR) was much higher for left-sided compared to right-sided colon cancer (HR 2.89 vs 1.67).

Lateral lymph node metastasis is also classified as N3, not as M1. A Japanese nationwide multiinstitutional study showed that 5-year overall survival of patients with lower rectal cancer and positive lateral lymph node metastases was similar to those with stage N2b disease (29% vs 32%, $P = .33$), and better than patients with distant metastases (29% vs 24%, $P = .024$) (Akiyoshi et al., 2012). Since lateral lymph node metastasis is considered as regional disease which can be managed with surgery, the presence of metastasis in lateral lymph nodes is categorized with "N3."

Metastasis to main lymph nodes does not lead to the metastases to intermediate and/or pericolic lymph nodes. Biologically, two hypotheses have been proposed regarding lymph node spreading pattern: the Halstedian model and Fisher model. The Halstedian model assumes that lymphatic spread follows a well-defined temporal and anatomical path, from the primary tumor to pericolic nodes, next to intermediate nodes, subsequently to main nodes, and eventually to other organs (Halsted, 1894). This model presupposes that (1) progression along the metastatic pathway can only arise when the previous lymphatic nodal barrier is breached and (2) efforts to remove all cancer invaded nodes may render the patient disease-free. Head and neck cancers are likely to follow this model. The CME and Japanese D3 lymphadenectomy come from this concept. On the other hand, the Fisher model assumes that lymphatic as well as hematogenous metastasis occurs early and in parallel (Fisher, 2008). When the cancer follows this model, extended lymphadenectomy to remove all invaded lymph nodes are very unlikely to affect survival. Breast cancer is likely to follow this model. Several genomic analyses using phylogenetic trees suggested that lymphatic spread and metastasis to distant organs in colon cancer occur in parallel and simultaneously from the onset of primary tumor growths (Naxerova et al., 2017; Ulintz et al., 2018; Wei et al., 2017). However, true behavior of lymph node spreading pattern in colon cancer still remains unknown. Currently, according to the table of TNM in the JCCRC, N3 is categorized in the same box with N2b.

According to the largest analysis using a Japanese multiinstitutional database to evaluate the lymph node spreading pattern, a "skipped" lymph node spreading pattern (when one or two nodal stations (pericolic and/or intermediate) was negative and the more centrally located nodal stations (intermediate or main) are positive) was present in right-sided and left-sided colon cancer were 13.7% and 9.0%, respectively (Kataoka et al., 2020). Specifically, skipped lymph node spreading pattern with the main lymph node positive was found in right-sided and left-sided colon cancer are 4.4% and 1.6%, respectively (Kataoka et al., 2020). So far, the prognostic significance of this skipped lymph node spreading pattern remains unknown. In the series mentioned before, the skipped pattern was associated with a better disease-free survival for left-sided colon cancer, but not for right-sided colon cancer. For further assessment regarding this clinical question, the way of examination for lymph nodes needs to be reconsidered. When lymph nodes are evaluated pathologically, they are cut into half and only one slice is evaluated. The possibility of micrometastases is not taken into account. A Japanese group reported that 17.6% of pathological stage II colon cancer had micrometastases, using one-step nucleic acid amplification (OSNA) of cytokeratin 19 (CK19) mRNA, which is equivalent to a 2-mm-interval histopathologic examination of lymph nodes of colorectal cancer (Yamamoto et al., 2016). We are currently trying to reevaluate the lymph node spreading pattern, taking into account the micrometastases using OSNA (supported by KAKENHI 21K15494).

7.3 Conclusion

The overview of the UICC TNM staging system and JCCRC system was presented in this chapter. Further discussion and evidence are required to deal with "N3," in order to incorporate JCCRC system into UICC TNM staging system. Further insight into lymph node spreading pattern may help to clarify the prognostic role of not only central lymph node metastases but also lateral lymph node metastases.

References

Akiyoshi, T., Watanabe, T., Miyata, S., et al. (2012). Results of a Japanese nationwide multi-institutional study on lateral pelvic lymph node metastasis in low rectal cancer: Is it regional or distant disease? *Annals of Surgery, 255*, 1129–1134.

Brierley, J. D., Gospodarowicz, M. K., & Wittekind, C. (2017). *TNM classification of malignant tumours*. John Wiley & Sons.

Fisher, B. (2008). Biological research in the evolution of cancer surgery: a personal perspective. *Cancer Research, 68*, 10007–10020.

Halsted, W. S. I. (1894). The Results of Operations for the Cure of Cancer of the Breast Performed at the Johns Hopkins Hospital from June, 1889, to January, 1894. *Annals of Surgery, 20*, 497–555.

Hashiguchi, Y., Muro, K., Saito, Y., et al. (2020). Japanese Society for Cancer of the Colon and Rectum (JSCCR) guidelines 2019 for the treatment of colorectal cancer. *International Journal of Clinical Oncology / Japan Society of Clinical Oncology, 25*, 1–42.

Heald, R. J., Husband, E. M., & Ryall, R. D. (1982). The mesorectum in rectal cancer surgery—The clue to pelvic recurrence? *The British Journal of Surgery, 69*, 613–616.

Heald, R. J., & Ryall, R. D. (1986). Recurrence and survival after total mesorectal excision for rectal cancer. *Lancet, 1*, 1479–1482.

Hida, J., Yasutomi, M., Maruyama, T., et al. (1997). The extent of lymph node dissection for colon carcinoma: the potential impact on laparoscopic surgery. *Cancer, 80*, 188–192.

Hohenberger, W., Weber, K., Matzel, K., et al. (2009). Standardized surgery for colonic cancer: Complete mesocolic excision and central ligation—technical notes and outcome. *Colorectal Disease: The Official Journal of the Association of Coloproctology of Great Britain and Ireland, 11*, 354–364, discussion 364-355.

Huh, J. W., Kim, Y. J., & Kim, H. R. (2012). Distribution of lymph node metastases is an independent predictor of survival for sigmoid colon and rectal cancer. *Annals of Surgery, 255*, 70–78.

Kanemitsu, Y., Hirai, T., Komori, K., & Kato, T. (2006). Survival benefit of high ligation of the inferior mesenteric artery in sigmoid colon or rectal cancer surgery. *The British Journal of Surgery, 93*, 609–615.

Kanemitsu, Y., Komori, K., Kimura, K., & Kato, T. (2013). D3 lymph node dissection in right hemicolectomy with a no-touch isolation technique in patients with colon cancer. *Diseases of the Colon and Rectum, 56*, 815–824.

Kataoka, K., Beppu, N., Shiozawa, M., et al. (2020). Colorectal cancer treated by resection and extended lymphadenectomy: patterns of spread in left- and right-sided tumours. *The British Journal of Surgery, 107*, 1070–1078.

Kataoka, K., Ysebaert, H., Shiozawa, M., et al. (2019). Prognostic significance of number vs location of positive mesenteric nodes in stage iii colon cancer. *European Journal of Surgical Oncology: the Journal of the European Society of Surgical Oncology and the British Association of Surgical Oncology, 45*, 1862–1869.

Kobayashi, H., Mochizuki, H., Kato, T., et al. (2009). Outcomes of surgery alone for lower rectal cancer with and without pelvic sidewall dissection. *Diseases of the Colon and Rectum, 52*, 567–576.

Naxerova, K., Reiter, J. G., Brachtel, E., et al. (2017). Origins of lymphatic and distant metastases in human colorectal cancer. *Science (New York, N.Y.), 357*, 55–60.

Paquette, I. M., Madoff, R. D., Sigurdson, E. R., & Chang, G. J. (2018). Impact of proximal vascular ligation on survival of patients with colon cancer. *Annals of Surgical Oncology, 25*, 38–45.

Shiozawa, M., Ueno, H., Shiomi, A., et al. (2021). Study protocol for an International Prospective Observational Cohort Study for Optimal Bowel Resection Extent and Central Radicality for Colon Cancer (T-REX study). *Japanese Journal of Clinical Oncology, 51*, 145–155.

Sugihara, K., Kobayashi, H., Kato, T., et al. (2006). Indication and benefit of pelvic sidewall dissection for rectal cancer. *Diseases of the Colon and Rectum, 49*, 1663–1672.

Ulintz, P. J., Greenson, J. K., Wu, R., et al. (2018). Lymph node metastases in colon cancer are polyclonal. *Clinical Cancer Research: An Official Journal of the American Association for Cancer Research, 24*, 2214–2224.

Wei, Q., Ye, Z., Zhong, X., et al. (2017). Multiregion whole-exome sequencing of matched primary and metastatic tumors revealed genomic heterogeneity and suggested polyclonal seeding in colorectal cancer metastasis. *Annals of Oncology: Official Journal of the European Society for Medical Oncology / ESMO, 28*, 2135–2141.

West, N. P., Kobayashi, H., Takahashi, K., et al. (2012). Understanding optimal colonic cancer surgery: Comparison of Japanese D3 resection and European complete mesocolic excision with central vascular ligation. *Journal of Clinical Oncology: Official Journal of the American Society of Clinical Oncology, 30*, 1763–1769.

Yamamoto, H., Tomita, N., Inomata, M., et al. (2016). OSNA-assisted molecular staging in colorectal cancer: A prospective multicenter trial in Japan. *Annals of Surgical Oncology, 23*, 391–396.

CHAPTER 8

Detection and significance of micrometastases and isolated tumor cells in lymph nodes of colorectal cancer resections

Anne Hoorens
Department of Pathology, Ghent University Hospital, Ghent University, Ghent, Belgium

8.1 Definition of micrometastases and isolated tumor cells

8.1.1 AJCC/UICC definition

Initially, the terms micrometastases and occult tumor cells were generally used for tiny metastases that are not detectable by routine histological examination, but where ancillary techniques such as immunohistochemistry (IHC) or reverse transcription-polymerase chain reaction (RT-PCR) are required for detection. When this concept was subsequently introduced in the TNM classification first in 2002, however, a definition for lymph-node micrometastases was used based on the size of the metastatic focus, and lymph-node micrometastases were in addition further divided into several categories according to both the size and the method of detection. As such, lymph-node micrometastases are actually defined, in the eighth edition of the TNM classification ("AJCC Cancer Staging Manual" published in 2017 and initiated in 2018, "UICC TNM Classification of Malignant Tumours" published and initiated in 2017) as a cluster of tumor cells measuring greater than 0.2 mm, but not more than 2.0 mm (>0.2 mm and ≤2.0 mm), while isolated tumor cells (ITCs) are single tumor cells or small clusters of tumor cells measuring not more than 0.2 mm in greatest diameter (≤0.2 mm) (Amin et al., 2017; Brierley et al., 2017). Definitions of ITCs may, however, vary by tumor site. In breast cancer an additional criterion has been proposed to include a cluster of fewer than 200 cells in a single histological section of a lymph node (Amin et al., 2017; Brierley et al., 2017). A three-dimensional 0.2 mm cluster contains approximately 1000 tumor cells. Thus, if more than 200 individual tumor cells are identified as single dispersed tumor cells or as a nearly confluent elliptical or spherical focus in a single histological section of a lymph node, there is a high probability that more than 1000 cells are present in the lymph node. In these situations the node may be classified as containing micrometastases (Amin et al., 2017). Others have proposed for other tumor sites, including colon cancer, that a cluster should have only 20 cells or fewer to be considered as ITC (Amin et al., 2017; Brierley et al., 2017). Moreover, uncertainty exists regarding the addition of abnormalities in the case of micrometastases and ITCs (either a single focus in a single node, multiple foci within a single or multiple nodes). In this regard the new AJCC manual incorporates supplemental figures in the chapter on breast cancer that reinforces the concept that ITC clusters and micrometastases are more likely to be present as multiple tumor deposits, either in proximity to one another or dispersed in different locations within the lymph node, rather than as single tumor deposits. The size of a tumor deposit in a lymph node is determined by measuring the largest dimension of any group of cells that are touching one another (confluent or contiguous tumor cells), regardless of whether the deposit is confined to the lymph node or extends outside the node (extra-nodal extension) (Amin et al., 2017). When a tumor deposit has induced a fibrous (desmoplastic) stromal reaction, the combined contiguous dimension of tumor cells and fibrosis determines the size of the metastasis, except following neoadjuvant therapy (Amin et al., 2017). It is emphasized that when multiple tumor deposits are present within a lymph node, whether ITCs, micrometastases, or macro-metastases, only the size of the largest contiguous tumor deposit in this lymph node is used to classify that node, and the collective summary of all lymph nodes is used to determine the final pN category; neither the sum of all individual tumor deposit sizes nor the area in which the tumor deposits are distributed is used for pN (Amin et al., 2017). In case of applying the criterion of tumor cell clusters for ITCs (e.g., a cluster of fewer than 200 cells in a single histological section of a lymph node in breast cancer), tumor cells in different levels of the tissue block or in different cross- or longitudinal sections of a lymph node should not be added together (Amin et al., 2017).

8.1.2 Methods of detection

ITCs or lymph-node micrometastases may be detected on routine hematoxylin and eosin (H&E) staining, or by specialized histopathological techniques such as IHC for cytokeratins in case of carcinomas. Cytokeratin is a component of the cytoskeleton of epithelial cells that is not present in normal lymph nodes. Therefore IHC using an antibody that recognizes cytokeratin enables the detection of minute deposits of carcinoma cells in lymph nodes. Monoclonal anticytokeratin antibodies are most frequently used and are recommended, because of the higher risk of false-positive findings when using polyclonal antibodies. The use of highly colon-specific cytokeratin 20 (CK20) antibodies might result in an underestimation of the presence of occult tumor cells due to dedifferentiation. Therefore the use of generic monoclonal cytokeratin antibodies might be more suitable (e.g., CAM5.2) (Sloothaak et al., 2014) (Fig. 8.1). Weak staining for cytokeratin can occur in dendritic reticulum cells and should be distinguished from positive tumor cells (Franke & Moll, 1987). Nonmorphologic techniques, such as flow cytometry and molecular analysis, however, can sometimes identify minimal deposits of cancer cells in lymph nodes not detectable by microscopy, which may also be considered as

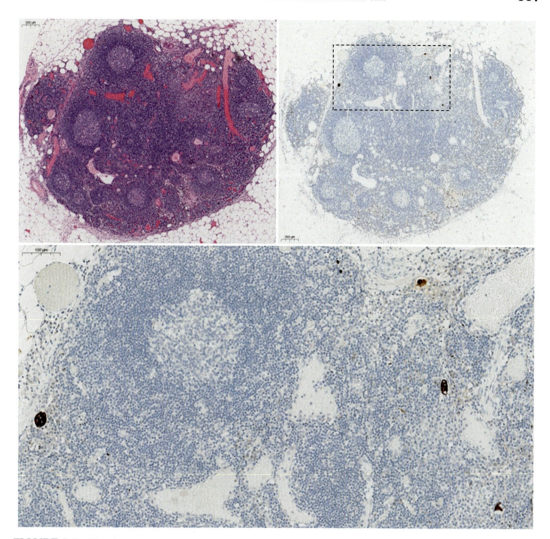

FIGURE 8.1 This figure is a microscopic image of an H&E staining (upper left image) and an immunohistochemical staining for CAM5.2 (generic monoclonal cytokeratin antibody - upper right) on a tissue section of a lymph node from a colectomy resection specimen. We recognize small clusters of CAM5.2 positive cells (brown labeled). Lower image: higher magnification of the rectangle in the upper right image. The scale bar on the figure shows that these clusters of epithelial cells are much smaller than 200 μm, consistent with the AJCC/UICC definition of isolated tumour cells. These brown labeled epithelial cells are located in or near the marginal sinus, but also deeper in the lymph node. Counterstaining is with haematoxylin with blue staining of all nuclei.

ITCs, while according to the definition of the TNM classification, micrometastases can only be detected by histologic examination (Amin et al., 2017; Brierley et al., 2017). The method used to detect ITCs should always be recorded as IHC has a lower false-positive rate than nonmorphological methods.

8.1.3 Biological significance

The formation of metastases is complex and it is estimated that only a very small percentage of circulating tumor cells (<0.05%) survive to initiate a metastatic focus (Abati & Liotta, 1996). Micrometastases can only develop if there has been arrest, implantation, and proliferation of ITCs in the organ involved (Hermanek et al., 1999). For this reason, AJCC and UICC guidelines suggest since the introduction of micrometastases in the cancer staging system in 2002, to distinguish micrometastases from ITCs. According to their definition, ITCs are usually found in the subcapsular nodal sinuses but may be seen within the nodal parenchyma (Amin et al., 2017). They usually do not show evidence of metastatic activity (e.g., proliferation or stromal reaction) or penetration of vascular structures or lymphatic sinus walls (Amin et al., 2017; Brierley et al., 2017). Because ITCs may represent in-transit tumor cells that are not proliferating within the node, lymph nodes with only ITCs are categorized as pN0, with exception of melanoma and Merkel cell carcinoma where lymph nodes with ITCs are considered pN1 or higher, depending on the number of involved nodes (Amin et al., 2017; Brierley et al., 2017). Not everybody, however, agrees on the actual definition of ITCs. Some authors believe that the formation of tumor cell clusters, even if only up to 20 cells, indicates that tumor cells have entered a lymph node and replicated and that they are not merely dormant cells (Scheunemann et al., 1999). Cases in which micrometastases are detected, but no metastases larger than 2 mm (no macro-metastases), in contrast, are classified as pN1 or higher, depending on the number of nodes involved by micrometastases. For certain tumor sites (e.g., breast cancer), cases in which at least one micrometastasis is detected, but no macro-metastases, are classified as pN1 denoted by the addition of the symbol "mi" in parenthesis after pN1 (pN1(mi)), regardless of the number of nodes involved (Amin et al., 2017; Brierley et al., 2017) (Table 8.1). In this case the number of nodes involved by micrometastases should be noted (Amin et al., 2017).

TABLE 8.1 Recommendations for reporting ITCs and micrometastases in lymph nodes according to TNM classification

pN0	No regional lymph-node metastasis histologically, no examination for ITCs (isolated tumor cells)
pN0(i−)	No regional lymph-node metastasis histologically, negative morphological findings for ITCs
pN0(i+)	No regional lymph-node metastasis histologically, positive morphological findings for ITCs
pN0(mol−)	No regional lymph-node metastasis histologically, negative nonmorphological findings for ITCs
pN0(mol+)	No regional lymph-node metastasis histologically, positive nonmorphological findings for ITCs
pN0(i−)(sn)	No sentinel lymph-node metastasis histologically, negative morphological findings for ITCs
pN0(i+)(sn)	No sentinel lymph-node metastasis histologically, positive morphological findings for ITCs
pN0(mol−)(sn)	No sentinel lymph-node metastasis histologically, negative nonmorphological findings for ITCs
pN0(mol+)(sn)	No sentinel lymph-node metastasis histologically, positive nonmorphological findings for ITCs
pN1(mi)[a]	Only micrometastasis in regional lymph nodes (can only be detected by histological examination)

[a]*(mi) is only applicable for certain organs*

Adapted from Amin, M. B., Edge, S. B., Greene, F. L., Byrd, D. R., Brookland, R. K., Washington, M. K., Gershenwald, J. E., Compton, C. C., Hess, K. R., Sullivan, D. C., Jessup, J. M., Brierley, J. D., Gaspar, L. E., Schilsky, R. L., Balch, C. M., Winchester, D. P., Asare, E.A., Madera, M., Gress, D.M., & Meyer, L.R. (2017). American Joint Committee on Cancer (AJCC) Cancer Staging Manual (8th ed.); Brierley, J. D., Gospodarowicz, M. K., & Wittekind, C. (2017). UICC (Union for International Cancer Control) TNM Classification of Malignant Tumours (8th ed.).

8.1.4 Reporting

As already emphasized, both ITCs and lymph-node micrometastases may be detected on H&E or IHC, and nonmorphologic techniques (e.g., flow cytometry, RT-PCR to detect tumor-specific RNA) may in addition identify minimal deposits of cancer cells in lymph nodes not visible in microscopy. Cases with tumor deposits only detected by nonmorphological techniques are classified as clinically node negative and are identified with the symbol "mol," for molecular, in parenthesis after pN0, pN0(mol+) versus pN0(mol−). If ITCs are detected on H&E or IHC, they are reported with the symbol "i," pN0(i+) versus pN0(i−) (Table 8.1) (Amin et al., 2017; Brierley et al., 2017). The number of ITC-involved lymph nodes should be noted (Amin et al., 2017; Brierley et al., 2017). Actually, however, additional use of special/ancillary techniques is not recommended for routine examination of lymph nodes. Sentinel lymph nodes (SLNs) examined for ITCs in the absence of complete dissection of the nodal basin similarly can be classified as pN0(i−)(sn), pN0(i+)(sn), pN0(mol−)(sn), or pN0(mol+)(sn) (Amin et al., 2017; Brierley et al., 2017) (Table 8.1).

8.1.5 Summary

The tabulation of positive lymph nodes recognizes only the largest tumor deposit in each lymph node, and the sum of the number of lymph nodes with macro-metastases and the number of lymph nodes with micrometastases is equivalent to the total number of lymph nodes containing metastatic tumor deposits. Lymph nodes with ITCs only are tabulated in the report but do not contribute to overall pN classification (Amin et al., 2017). Cases with micrometastases, but without macro-metastases, may be indicated as pN1(mi) for certain tumor sites, stating the number of involved lymph nodes. Cases with ITCs as the only form of nodal involvement are classified as pN0 (except for melanoma and Merkel cell carcinoma), again with mention of the number of nodes involved by ITCs. This is consistent with the general rule of the TNM classification that if there is doubt concerning the correct T, N, or M category to which a particular case should be assigned, then the lower (i.e., less advanced) category should be chosen. Cases in which ITCs represent the only form of nodal involvement should be analyzed separately when, for example, investigating tumor relapse, progression-free survival, and/or overall survival.

8.2 Micrometastases and isolated tumor cells in colorectal cancer

8.2.1 Implications in colorectal cancer

The concept and the implications of ITCs and micrometastases versus standard regional lymph-node metastases (macro-metastases) are best studied in breast cancer, where it was introduced through the sixth edition of the "TNM Classification of Malignant Tumours" of the International Union Against Cancer (UICC) and the sixth edition of the "Cancer Staging Manual" of the American Joint Committee on Cancer (AJCC), 2002. In the seventh and eighth edition of the TNM classification, it became part of the introduction session with general definitions used throughout and as such may in principle hold for metastases of all tumors, with the exception of melanoma and Merkel cell carcinoma, as already discussed. This concept regarding the staging of ITCs and micrometastases is actually, however, meant as a guideline for uniformity and consistency in practice in reporting information for future evaluation of their prognostic value and possible improvement of the TNM system. It continues to evolve, and further study is warranted to determine the significance of micrometastases and ITCs in the different tumor types. As for analysis, it is necessary to register not only positive but also negative findings (Table 8.1) (Hermanek et al., 1999). When introduced in the staging system in 2002, there was debate regarding whether the identification of micrometastases and ITCs was of clinical significance and whether it should be incorporated into staging systems. Some pointed to a large body of evidence showing that tumor cells can be found in a variety of sites with no clinical impact on patient's prognosis, whereas others found that the finding of viable tumor cells should be included in all staging systems (Hermanek et al., 1999; Scheunemann et al., 1999). Moreover, it should be clearly noted that there is not yet hard evidence to justify general upstaging of micrometastases while "not upstaging" of ITCs, as the distinction between the two categories was made arbitrarily at 0.2 mm and not on the basis of measured survival differences. Also, the statement that ITCs do not typically show evidence of metastatic activity, for example, proliferation or stromal reaction, is not evidence based. Because of these ambiguities, pathological and oncological practices differ significantly, even between countries in Western Europe. Germany and Switzerland, for example, adopted upon its introduction the TNM 7 classification for colon cancer, classifying ITCs as pN0 and offering adjuvant chemotherapy to patients with micrometastases staged as pN1. Conversely, in the United Kingdom, colon cancer during the same time period was staged according to the TNM 5 classification. Any metastatic deposit in a lymph node is staged as pN1, suggesting adjuvant chemotherapy for patients with ITCs as well as for patients with micrometastases. In the Netherlands, colon cancer at that moment was also staged according to the TNM 5 classification, but ITCs and micrometastases were staged as pN0, not suggesting adjuvant chemotherapy for these patients (Weixler et al., 2017).

In many organs, including the gastrointestinal tract, and more in particular in colorectal carcinoma, the prognostic significance of ITCs and micrometastases in regional lymph nodes, and evidence for down-staging ITC's, actually remains controversial. Hence, pathologists in daily practice for the moment frequently consider both micrometastases and ITCs in regional lymph nodes in colorectal cancer still as standard positive lymph nodes (Jin & Frankel, 2018). Search for associations between micrometastases and ITCs and clinical outcomes is important and several studies have been performed. As shown below, there is considerable evidence that colon cancer patients with micrometastases will indeed have a prognosis similar to patients with macro-metastases. However, for cases in which ITCs represent the only form of nodal involvement it is not yet clear if this means that there is local disease and that they are not a marker of systemic disease, but cured by comprehensive lymph-node dissection or that they indicate that the disease is beyond cure with local treatment and that adjuvant treatment is indicated.

8.2.2 Occult disease in lymph nodes in colorectal cancer

The TNM staging system discriminates nodal negative (stage I/II) from nodal positive (stage III) colon cancer. Adjuvant chemotherapy is usually reserved to stage III disease as it has been shown to improve survival of patients with locoregional and metastatic colon cancer, while colorectal cancer patients with node-negative disease are usually cured by surgery alone. Unfortunately, about 20% of patients of the prognostic favorable stage I/II colorectal cancer patients, with histologically negative lymph nodes on conventional histopathologic analysis, will develop disease recurrence within 5 years after diagnosis, despite node-negative status, and up to 20%–30% of patients with stage I/II colorectal cancer will eventually die from local tumor relapse or overwhelming metastatic disease (Nissan et al., 2012). Recurrence may theoretically be attributed to incomplete resection of tumor-bearing regional lymph nodes, aggressive tumor biology with occult residual disease spreading to extra-nodal sites, and/or occult nodal metastatic disease, including ITCs or micrometastases in regional lymph nodes, that has not been detected by standard histopathologic analysis and that might play a role in systemic tumor spread. In fact, such patients might not have been lymph-node negative but under-staged instead. Studies have found only small improvements in survival (estimated to be between 2% and 4%) with the addition of adjuvant chemotherapy in patients with high-risk stage II disease with one or several of the following features: pT4 tumor, poor differentiation of the primary tumor, fewer than 12 lymph nodes analyzed, presence of lymphovascular invasion, localized perforation, and/or bowel obstruction (Benson et al., 2004; Weitz et al., 2005). This has to be weighed with the possible side effects of the chemotherapy treatment. While uncertainty persists regarding the explanation of the high disease recurrence rate and poor survival in node-negative colon cancer, there is increasing evidence that this can be at least partially explained by occult disease in resected lymph nodes detected by using IHC and molecular detection techniques such as RT-PCR, that are missed during standard histopathological analysis (Iddings et al., 2006; Nicastri et al., 2007; Sirop et al., 2011; Mescoli et al., 2012; Nissan et al., 2012).

Rahbari et al. (2012) showed in a *meta*-analysis of 39 studies that detection of occult tumor cells in lymph nodes in patients found to be node negative by standard histopathology with stage I/II colorectal cancer is associated with disease recurrence and poorer survival. The authors included, however, a high percentage of studies that detected occult tumor cells by molecular techniques and only a minority of the included studies distinguished micrometastases from ITCs and a subgroup analysis to validate the prognostic features of micrometastases and ITCs separately was not possible (Sloothaak et al., 2014). Moreover, concerning the biological significance and clinical meaning of micrometastases and ITCs, the literature is unclear and contradictory, suffering from differences in study designs and methods of detection, such as the sample size, tumor stage of patients, number of retrieved lymph nodes, use of IHC or molecular techniques, types of antibodies used, and, particularly, the definition of micrometastasis and ITCs. Some authors have defined ITCs as individual tumor cells in the subcapsular nodal sinuses, while others considered clumps of up to 20 tumor cells as ITCs. On the other hand, clusters of up to 10–20 tumor cells have been defined in some studies as micrometastases.

A more recent systematic review and *meta*-analysis of five studies evaluated the prognostic value in colorectal cancer of micrometastases and ITCs separately, using the TNM definition (Sloothaak et al., 2014). Sloothaak et al. concluded that in patients with stage I/II colorectal cancer, disease recurrence was significantly increased in the presence of micrometastases in comparison with absent occult tumor cells. This was even more pronounced in patients with colon cancer compared to rectal cancer. In contrast, detection of ITCs as the only form of nodal involvement, using the TNM definition, was not associated with an increased risk of cancer recurrence in patients with stage I/II colorectal cancer and pN0 was considered justified.

Only the presence of micrometastases in colorectal cancer, not ITCs, thus appears to be a significant poor prognostic factor. Although micrometastases may be designated as pN1 (mi) per UICC and AJCC eighth edition, based on the *meta*-analysis of Sloothaak et al. that demonstrated that they are associated with poor prognosis, it is recommended in the context of colorectal cancer to consider them as positive and denote as pN1, to avoid confusion and prevent patients from being undertreated (Loughrey et al., 2020; Sloothaak et al., 2014; Weiser, 2018). Regional lymph nodes with micrometastases in colorectal cancer are best considered as standard positive lymph nodes with the corresponding number, as pathologists have considered these to be positive lymph nodes in the past (Jin & Frankel, 2018; Loughrey et al., 2020). Accordingly, cases in which ITCs represent the only form of nodal involvement should be classified as pN0, and their presence does not elevate disease to stage III as they remain of controversial prognostic value (Loughrey et al., 2020; Weiser, 2018). It is advised to give a comment in the report on the presence of ITCs and optional designation as pN0(i+) or pN0(mol+) (Loughrey et al., 2020).

In a recent observational and prospective study of 105 patients where about one-quarter of patients were upstaged from pN0, 18 patients with ITCs and 1 patient with micrometastases, upstaged patients did not have a poorer prognosis than patients without node involvement. This suggests that detection of ITCs indeed is not a factor in the poor prognosis of some of these patients (Estrada et al., 2017). Mescoli et al., however, found different results and showed in a small retrospective study of pN0 colorectal carcinomas, using IHC, that cancer relapse was significantly associated with ITCs in regional lymph nodes (Mescoli et al., 2012). In addition, in a multicenter prospective trial, nearly 200 patients with Stage I or II disease had negative nodes based on standard H&E staining but were found to have tumor cells ≤0.2 mm in diameter that stained positively with a pan-cytokeratin antibody (pN0(i+)) (Nissan et al., 2012). These patients had a 10% decrease in overall survival compared to patients with pN0 disease (pN0(i−)). This decrease in survival occurred in patients with T3–T4 primary tumors, but not in those with T1 or T2 primary tumors (Nissan et al., 2012). Weixler et al. detected ITCs (pN0(i+)) in locoregional lymph nodes in 31% of patients in a study comprising 74 stage I/II colon cancer patients. The presence of ITCs is associated significantly with worse disease-free and overall survival (Weixler et al., 2016). These data, however, were considered insufficient to be implemented in diagnostic practice and further studies are awaited to clarify the real biological significance of ITCs as the benefit of adjuvant systematic therapy in these patients has yet to be convincingly demonstrated. Besides, a prospective randomized trial demonstrated that patients with optimally staged node-negative colon cancer, ≥12 lymph nodes and pN0(i−), are unlikely to benefit from adjuvant chemotherapy, as 97% remained disease free after primary tumor resection (Protic et al., 2015). Moreover, these low-risk patients thus would not benefit but might suffer injury from excessive adjuvant chemotherapy treatment (Zhao et al., 2016).

8.2.3 Recommendations for standardized histopathological analysis of lymph nodes

In relation to micrometastasis and ITCs additional major problems remain. Data from multicenter studies are mostly based on "routine examination of lymph nodes." The worldwide standard of routine histopathological assessment of colon cancer lymph nodes most often still represents single-level sectioning with microscopic examination of 1 or 2 H&E stained sections of one single cross-section of each discovered lymph node, resulting in a substantial risk of missing metastases in other parts of the lymph nodes that are not processed. To increase detection of lymph-node metastases recent College of American Pathologists guidelines state that grossly positive lymph nodes may be partially submitted for microscopic confirmation of metastases but recommend that all grossly negative or equivocal lymph nodes should be submitted entirely (Burgart et al., 2021). However, this method still limits analysis to <1% of the lymph node and suffers from the inability to detect small tumor cell aggregates (Bilchik et al., 2007). It implies a relevant risk of sampling bias and under-staging, especially concerning detection of micrometastases. A single cell in one section may be a >0.2 mm deposit at a deeper plain of sectioning. Moreover, detection of ITCs generally requires either step sectioning combined with IHC staining or with molecular methods, hence ITCs are often missed in standard analysis (Nissan et al., 2012). Identification of micrometastases and ITCs is particularly critical for patients with stage I/II colorectal cancer, to identify patients that might not be cured by surgery alone. To improve the accuracy of nodal staging in these patients, it would be necessary to perform a much more detailed analysis of all lymph nodes recovered from the specimen. The use of ultra-staging techniques such as serial sectioning, combined with IHC or RT-PCR, on all lymph nodes harvested in a lymphadenectomy specimen is however not feasible in a routine laboratory, as it would be impractical, time consuming, labor intensive, and expensive. Applying this to a selected sample of high-risk nodes could, however, improve staging accuracy in a cost-efficient fashion. The SLN concept can offer a solution as it aims to enable the pathologist to analyze more meticulously one or a few lymph nodes harboring the highest risk of metastatic disease.

8.3 Micrometastases and isolated tumor cells in sentinel lymph-node biopsy in colorectal cancer

8.3.1 Sentinel lymph-node mapping with ultra-staging

The concept of SLN biopsy is centered on the premise that lymphatic drainage from a primary tumor occurs in an orderly and progressive fashion, with the SLN representing the first draining regional lymph node that receives direct afferent lymphatic drainage from the primary tumor site, and the most likely site of metastatic disease when regional lymph nodes become involved. More than one SLN may be present. The status of the SLN(s) accurately predicts the status of the remaining regional lymph nodes. If it contains metastatic tumor this indicates that other lymph nodes may contain tumor. If not, other lymph nodes are unlikely to contain tumor. SLNs are identified by lymphatic mapping, as evidenced by nodes that concentrate a vital dye or radiolabelled colloidal material injected near the area of the primary tumor or in the involved organ. When the lymph nodes with the highest risk of metastasis are identified, the possibility arises to investigate them with ancillary techniques. IHC is preferred, as was shown to have a lower false-positive rate than nonmorphological methods (Hermanek et al., 1999). Nonvital cells, shed debris or false positively stained cells (in up to 25% of patients) can be falsely identified as clinically relevant occult tumor cells with nonmorphological methods and discrimination between micrometastases and ITCs cannot be made. IHC allows histological confirmation as well as subclassification of small lesions. This is not possible if RT-PCR is performed. Development of SLN mapping with ultra-staging of the SLN(s), moreover, has brought to the forefront the investigation of the influence of nodal ITCs and micrometastases. However, also in the context of SLN procedure with ultra-staging, major problems remain. As already emphasized, there is no guidance on how these lymph nodes should be sectioned and examined, which makes interlaboratory comparisons difficult. Standardization of histopathological analysis of lymph nodes is important in future clinical trials to reveal the real importance of this 0.2 mm cut-off value. Systematic techniques in SLNs that allow detection of metastases of different sizes require a disciplined approach in the pathology laboratory and power of detection should be balanced against cost. The most cost-effective approach is stepwise. SLNs with grossly visible metastases do not require special treatment. If grossly negative, the first step would be one or two microsections from the face of the block. If negative for metastases, a systematic serial sectioning approach will follow. Complete nodes too small to bisect (<3 mm) may be processed together in one cassette and should be examined completely in H&E stained sections at histological levels at maximum 200 μm apart, which in theory will give 100% accuracy of identifying micrometastases >0.2 mm in diameter. Nodes larger than 3 mm may be bisected through the plane of the hilum and both halves embedded or, better, may be sliced in a plane perpendicular to their longitudinal axis at 2 mm intervals with all slices processed (Meyer, 1998). All parts of any node cut through during dissection in the laboratory should be processed together in a cassette separate from other nodes or labeled with different colors to identify the different nodes within one cassette. Consecutive sections with H&E staining as well as IHC for cytokeratin should be performed.

8.3.2 Sentinel lymph-node biopsy in colorectal cancer

Lymphatic mapping with focused analysis of the SLN(s) was first developed in melanoma and subsequently used in breast cancer, where it can avoid unnecessary complete lymph-node dissection with its associated complications of lymphedema and wound problems. As such, sentinel node biopsy is today a standard procedure in many tumors (melanoma, breast carcinoma, gynecological and urological malignancies, and thyroid gland carcinoma) to detect patients that may benefit from regional lymphadenectomy. The possibility for other nodes to be positive when the sentinel node is negative is generally less than 1% (Gipponi, 2005). However, intraoperative lymphatic mapping in gastrointestinal malignancies is more complex, and aberrant lymphatic drainage, defined as lymphatic drainage identified outside the usual resection margin with location of the SLN more than 5 cm from the macroscopic border of the tumor, is much more frequent in colon cancer compared to other organs (Dragan et al., 2009). There is also a higher frequency of skip metastases (metastatic non-SLNs without SLN involvement) compared to other organs (Tsioulias et al., 2000). All patients with colon cancer, however, undergo en bloc lymphadenectomy with segmental resection and the aim is not to change the extent of lymph-node resection, as lymphadenectomy attending colon resection confers no significant morbidity, but it allows to identify SLN(s) for more rigorous histopathologic analysis. Although the potential of lymphatic mapping to enhance the nodal count as well as to increase the adequacy of resection of the colon site has been investigated, the SLN procedure in colorectal cancer in most studies had no purpose of altering the surgical management, but it made possible a more focused and cost-effective pathologic evaluation of nodal disease (Saha et al., 2018). In this context, lymphatic mapping can be done in vivo as well as ex vivo (Wood et al., 2001).

In the experimental setting the use of the SLN procedure in colorectal cancer has been studied by several groups and was found to have good identification and accuracy rates, which further improve with increasing experience (Qiao, 2020). It was shown that one or more SLN can be detected in 90%–100% of patients (Qiao, 2020; van der Pas et al., 2011; van der Zaag et al., 2012; Weitz et al., 2005). Rectal cancer appeared to have a significantly lower detection rate compared to colon cancer (Qiao, 2020). The fat and bulky mesorectum might be one of the reasons hampering SLN detection (Sommariva et al., 2010). In addition, radiotherapy, an essential treatment for advanced rectal cancer, can alter lymphatic flow and induce fibrosis, resulting in a significant decrease of lymph nodes detected within the surgical resection specimen (Sermier et al., 2006). Investigations with focused analysis of lymph nodes, or ultra-staging, by step sectioning of the SLN(s) combined with cytokeratin IHC or RT-PCR to increase tumor cell identification, have shown promising results in upstaging colon cancer and predicting recurrence. Three systematic reviews and *meta*-analyses including, respectively, 52, 57, and 68 studies, investigating the diagnostic performance of SLN mapping in colorectal cancer reported a sensitivity ranging from 70% to 90% (Qiao, 2020; van der Pas et al., 2011; van der Zaag et al., 2012). In these studies the sensitivity of the SLN procedure is calculated as TP/(TP + FN) (TP = true positives, that is, all positive SLNs with or without ultra-staging techniques, FN = false negatives, that is, any tumor-negative SLN in the presence of a tumor positive non-SLN). Most importantly, a mean upstaging in node-negative patients varying between 15% and 22% is reported, with a true upstaging, defined as micrometastases (excluding SLNs with only ITCs) between 8% and 14%, compared to conventional histopathological analysis. Upstaged patients were associated with worse overall survival compared with node-negative patients (Qiao, 2020; van der Pas et al., 2011; van der Zaag et al., 2012). In addition, upstaged patients had a lower 5-year disease-free survival rate than node-negative patients (Qiao, 2020). A *meta*-analysis of 11 studies using a newer tracer (indocyanine green) reported a pooled sensitivity and specificity for detection of metastatic lymph nodes of 64% and 65% respectively (Villegas-Tovar et al., 2020). The technique used for SLN assessment does not appear to have a major influence on the sensitivity, as originally thought (Qiao, 2020; Villegas-Tovar et al., 2020). Sensitivity appeared similar for colon carcinoma and rectal carcinoma (Qiao, 2020).

The use of the SLN procedure, in addition, made it possible in study context to select patients, normally considered node-negative, who might potentially benefit from adjuvant chemotherapy. A multicenter randomized controlled trial using ex vivo lymphatic mapping and ultra-staging with step sectioning of the SLN(s) followed by cytokeratin IHC in nonmetastatic colon cancer demonstrated a significantly increased nodal positivity compared with standard histopathology alone (57.3 vs 38.7% respectively) (Stojadinovic et al., 2007). Moreover, 11% of the node-negative patients randomized to nodal mapping with ultra-staging were found to have nodal micrometastases and otherwise would not have received adjuvant systemic chemotherapy (Stojadinovic et al., 2007). Patients had improved disease-free survival in early colorectal cancer (Nissan et al., 2012). In a prospective study, Saha et al. showed that SLN mapping is highly successful, easily reproducible, and finds micrometastases in over 15% of patients, which could have been missed by conventional histopathological examination. These patients when treated with adjuvant chemotherapy have similar survival as those with node-negative disease. Similarly, patients without any nodal metastases after SLN mapping and ultra-staging may be considered as true node-negative disease and may avoid further adjuvant chemotherapy (Saha et al., 2018).

After more than one decade of publications referring to colorectal cancer, there still is, however, no consensus on the usefulness of SLN mapping in daily clinical practice. Results from studies investigating SLN procedure are difficult to compare because of a tremendous clinical diversity seen across studies regarding patient selection, SLN mapping techniques, pathological analysis, and surgeon experience (Qiao, 2020). Consensus and standardization of the SLN procedure are required as the accuracy of detecting micrometastases and ITCs in SLN(s) depends on both surgical and pathological techniques. If the SLN patients are found in the high-risk group of stage I/II colorectal cancer patients, then it will not add much. Especially important is the low-risk group. Further studies must clarify the clinical impact of these findings in this group of low-risk colon cancer patients not only in terms of prognosis but also investigate the indication of adjuvant therapy. It will probably be very difficult in this small subset of patients to find out if there is survival benefit with adjuvant therapy (Thomas et al., 2006).

8.4 Micrometastases and isolated tumor cells after neoadjuvant therapy

Neoadjuvant therapy is standard of care in locally advanced rectal cancer and is also administered in some instances in colon cancer, for example, in the case of liver first resection in colon cancer with limited/isolated liver metastases. After neoadjuvant treatment, finding viable tumor cells is essential to classify a lymph node as tumor positive. Acellular mucin pools found in lymph nodes from colorectal cancer resection specimens after neoadjuvant therapy are considered negative. Evaluation of therapy response should not be limited to the primary tumor but should also comment on the presence or absence of signs of regression within nodal tissue, to allow correlation with initial staging MRI (Loughrey et al., 2020). The significance of micrometastases and ITCs in the context of neoadjuvant therapy in colorectal cancer as compared to macro-metastases, however, has not yet been investigated. It is advocated that colon and rectal cancer should be considered as different diseases and analyzed separately in the study context. A recent retrospective study showed a relatively high incidence of ITCs in lateral lymph nodes in rectal cancer patients by IHC for cytokeratin 20, with an important impact on overall recurrence. Caution is, however, warranted and further larger prospective studies are needed to confirm this finding (Yang et al., 2019).

8.5 Differential diagnosis of micrometastases and isolated tumor cells

Finally, it needs to be emphasized that ITCs and micrometastases in lymph nodes in colorectal cancer patients must be distinguished from benign mesothelial inclusions and Müllerian remnants, which also stain with generic monoclonal cytokeratin antibodies. Müllerian glandular inclusions are predominantly observed in pelvic and para-aortic lymph nodes, but mesothelial inclusions tend to occur in the sinus of mediastinal as well as abdominal lymph nodes. Alertness and use of IHC, including PAX-8 and WT-1, in case of doubt, can prevent incorrect diagnosis of adenocarcinoma cells involving a lymph node (Gallan & Antic, 2016; Reich et al., 2000).

8.6 Conclusion

There is considerable evidence that colon cancer patients with micrometastases will have a prognosis similar to patients with macro-metastases. Nodes with micrometastases thus should be considered as involved by cancer in colon carcinoma. Reporting the number of lymph nodes involved by micrometastases is appropriate, particularly in the study context. The prognostic value of ITCs, however, is less clear and remains debated. Some recent studies suggest that the presence of ITCs associates significantly with worse disease-free and overall survival. These data are however up to now considered insufficient to be implemented in diagnostic practice and further studies are awaited to clarify the real biological significance of ITCs and proof of benefit of adjuvant systematic therapy in these patients. To enable comparisons of treatment results, a finding of ITCs should actually not be considered in the TNM. For future evaluation of the prognostic significance of ITCs by both retrospective and prospective studies and for possible improvement of the TNM system, they should however be documented withstanding that they are classified according to uniform criteria as proposed by the TNM. The use of an inconsistent nomenclature in many studies renders a comparison among published studies difficult. For analysis, it is necessary to register not only positive but also negative findings (pN0(i+), pN0(i−), pN0(mol+), and pN0(mol−)). The same holds for ITCs in the general circulation (bone marrow, other distant sites, and blood). It is advised in cases in which ITCs represent the only form of nodal involvement, that the number of lymph nodes involved by ITCs and the method of detection is clearly stated in a comment section or elsewhere in the report, as this may be useful in future retrospective studies. Routine assessment of regional lymph nodes, however, is limited to conventional pathologic techniques (gross assessment and histologic examination), and data are currently insufficient to recommend special measures to detect ITCs. Thus neither multiple levels of paraffin blocks nor the use of special/ancillary techniques such as IHC are actually recommended for routine examination of regional lymph nodes.

The TNM staging system discriminates nodal negative (stage I/II) from nodal positive (stage III) colon cancer. Adjuvant chemotherapy is usually reserved to stage III disease as it has been shown to improve survival of patients with locoregional and metastatic colon cancer, while colorectal cancer patients with node-negative disease are usually cured by surgery alone. However, there remains a subgroup of patients undergoing curative resections with apparently no evidence of nodal disease, who unfortunately succumb to relapse. Such patients might not have been lymph-node negative but under-staged. If these patients could be identified earlier with a more sensitive nodal examination, they could benefit from adjuvant chemotherapy. There is growing evidence that SLN mapping with ultra-staging may lead to upstaging of a considerable number of patients and that these patients show a significantly lower disease-free survival and overall survival compared to true node-negative patients. Further studies must clarify the clinical impact of these findings in this group of low-risk colon cancer patients and investigate if there is survival benefit with adjuvant therapy. Hence, SNL mapping with ultra-staging is actually only considered justified in the study context.

References

Abati, A., & Liotta, L. A. (1996). Looking forward in diagnostic pathology: The molecular superhighway. *Cancer*, 78(1), 1–3. Available from https://doi.org/10.1002/(SICI)1097-0142(19960701)78:1 < 1::AID-CNCR1 > 3.0.CO;2-S, 8646703.

Amin, M. B., Edge, S. B., Greene, F. L., Byrd, D. R., Brookland, R. K., Washington, M. K., Gershenwald, J. E., Compton, C. C., Hess, K. R., Sullivan, D. C., Jessup, J. M., Brierley, J. D., Gaspar, L. E., Schilsky, R. L., Balch, C. M., Winchester, D. P., Asare, E. A., Madera, M., Gress, D. M., & Meyer, L. R. (2017). *American Joint Committee on Cancer (AJCC) Cancer Staging Manual*, (8th edn.). New York: Springer.

Benson, A. B., Schrag, D., Somerfield, M. R., Cohen, A. M., Figueredo, A. T., Flynn, P. J., Krzyzanowska, M. K., Maroun, J., McAllister, P., Van Cutsem, E., Brouwers, M., Charette, M., & Haller, D. G. (2004). American Society of Clinical Oncology recommendations on adjuvant chemotherapy for stage II colon cancer. *Journal of Clinical Oncology*, 22(16), 3408–3419. Available from https://doi.org/10.1200/JCO.2004.05.063.

Bilchik, A. J., Hoon, D. S. B., Saha, S., Turner, R. R., Wiese, D., DiNome, M., Koyanagi, K., McCarter, M., Shen, P., Iddings, D., Chen, S. L., Gonzalez, M., Elashoff, D., & Morton, D. L. (2007). Prognostic impact of micrometastases in colon cancer: Interim results of a prospective multicenter trial. *Annals of Surgery*, 246(4), 568–575. Available from https://doi.org/10.1097/SLA.0b013e318155a9c7, 17893493.

Brierley, J. D., Gospodarowicz, M. K., & Wittekind, C. (2017). *UICC (Union for International Cancer Control) TNM Classification of Malignant Tumours* (8th edn.). Oxford: Wiley-Blackwell.

Burgart, L.J, Chopp, W.V., & Jain, D. (2021). *Protocol for the examination of resection specimens from patients with primary carcinoma of the colon and rectum. Version: 4.2.0.1. Protocol posting date:* November 2021. College of American Pathologists.

Dragan, R., Nebojsa, M., Dejan, S., Ivan, P., Dragos, S., Damir, J., Predrag, S., & Vladan, Z. (2009). Clinical application of sentinel lymph node biopsy for staging, treatment and prognosis of colon and gastric cancer. *Hepato-Gastroenterology*, 56(96), 1606–1611, 20214202.

Estrada, O., Pulido, L., Admella, C., Hidalgo, L. A., Clavé, P., & Suñol, X. (2017). Sentinel lymph node biopsy as a prognostic factor in non-metastatic colon cancer: A prospective study. *Clinical and Translational Oncology*, 19(4), 432–439. Available from https://doi.org/10.1007/s12094-016-1543-8, 27541595.

Franke, W. W., & Moll, R. (1987). Cytoskeletal components of lymphoid organs. *Differentiation*, 36(2), 145–163. Available from https://doi.org/10.1111/j.1432-0436.1987.tb00189.x, 2452110.

Gallan, A. J., & Antic, T. (2016). Benign müllerian glandular inclusions in men undergoing pelvic lymph node dissection. *Human Pathology*, 57, 136–139. Available from https://doi.org/10.1016/j.humpath.2016.07.003, 27438608.

Gipponi, M. (2005). Applicazioni cliniche del linfonodo sentinella nella stadiazione e nel trattamento dei tumori solidi. *Minerva Chirurgica*, 60(4), 217–233, 16166921.

Hermanek, P., Hutter, R. V. P., Sobin, L. H., & Wittekind, C. (1999). Classification of isolated tumor cells and micrometastasis. *Cancer*, 2668–2673, 10594862.

Iddings, D., Ahmad, A., Elashoff, D., & Bilchik, A. (2006). The prognostic effect of micrometastases in previously staged lymph node negative (N0) colorectal carcinoma: A *meta*-analysis. *Annals of Surgical Oncology*, 13(11), 1386–1392. Available from https://doi.org/10.1245/s10434-006-9120-y, 17009147.

Jin, M., & Frankel, W. L. (2018). Lymph node metastasis in colorectal cancer. *Surgical Oncology Clinics of North America*, 27(2), 401–412. Available from https://doi.org/10.1016/j.soc.2017.11.011, 17009147.

Loughrey, M. B., Arends, M., Brown, I., Burgart, L. J., Cunningham, C., Flejou, J. F., Kakar, S., Kirsch, R., Kojima, M., Lugli, A., Rosty, C., Sheahan, K., West, N. P., Wilson, R., & Nagtegaal, I. D. (2020). *Colorectal cancer histopathology reporting guide*. Sydney, Australia: International Collaboration on Cancer Reporting.

Mescoli, C., Albertoni, L., Pucciarelli, S., Giacomelli, L., Russo, V. M., Fassan, M., Nitti, D., & Rugge, M. (2012). Isolated tumor cells in regional lymph nodes as relapse predictors in stage I and II colorectal cancer. *Journal of Clinical Oncology*, 30(9), 965–971. Available from https://doi.org/10.1200/JCO.2011.35.9539, 22355061.

Meyer, J. S. (1998). Sentinel lymph node biopsy: Strategies for pathologic examination of the specimen. *Journal of Surgical Oncology*, 69(4), 212–218. Available from https://doi.org/10.1002/(sici)1096-9098(199812)69:4 < 212::aid-jso4 > 3.0.co;2-v, 9881937.

Nicastri, D. G., Doucette, J. T., Godfrey, T. E., & Hughes, S. J. (2007). Is occult lymph node disease in colorectal cancer patients clinically significant? A review of the relevant literature. *Journal of Molecular Diagnostics*, 9(5), 563–571. Available from https://doi.org/10.2353/jmoldx.2007.070032.

Nissan, A., Protic, M., Bilchik, A. J., Howard, R. S., Peoples, G. E., & Stojadinovic, A. (2012). United States Military Cancer Institute Clinical Trials Group (USMCI GI-01) randomized controlled trial comparing targeted nodal assessment and ultrastaging with standard pathological evaluation for colon cancer. *Annals of Surgery*, 256(3), 412–427. Available from https://doi.org/10.1097/SLA.0b013e31826571c8.

Protic, M., Stojadinovic, A., Nissan, A., Wainberg, Z., Steele, S. R., Chen, D. C., Avital, I., & Bilchik, A. J. (2015). Prognostic effect of ultra-staging node-negative colon cancer without adjuvant chemotherapy: A Prospective National Cancer Institute-Sponsored Clinical Trial. *Journal of the American College of Surgeons*, 221(3), 643–651. Available from https://doi.org/10.1016/j.jamcollsurg.2015.05.007.

Qiao, L. (2020). Sentinel lymph node mapping for metastasis detection in colorectal cancer: a systematic review and meta-analysis. *Revista española de enfermedades digestivas: organo oficial de la Sociedad Española de Patología Digestiva*, 112(9), 722–730. Available from https://doi.org/10.17235/reed.2020.6767/2019.

Rahbari, N. N., Bork, U., Motschall, E., Thorlund, K., Büchler, M. W., Koch, M., & Weitz, J. (2012). Molecular detection of tumor cells in regional lymph nodes is associated with disease recurrence and poor survival in node-negative colorectal cancer: A systematic review and *meta*-analysis. *Journal of Clinical Oncology*, 30(1), 60–70. Available from https://doi.org/10.1200/JCO.2011.36.9504.

Reich, O., Tamussino, K., Haas, J., & Winter, R. (2000). Benign mullerian inclusions in pelvic and paraaortic lymph nodes. *Gynecologic Oncology*, 78(2), 242–244. Available from https://doi.org/10.1006/gyno.2000.5867.

Saha, S., Elgamal, M., Cherry, M., Buttar, R., Pentapati, S., Mukkamala, S., Devisetty, K., Kaushal, S., Alnounou, M., Singh, T., Grewal, S., Eilender, D., Arora, M., & Wiese, D. (2018). Challenging the conventional treatment of colon cancer by sentinel lymph node mapping and its role of detecting micrometastases for adjuvant chemotherapy. *Clinical and Experimental Metastasis*, 35(5–6), 463–469. Available from https://doi.org/10.1007/s10585-018-9927-5.

Scheunemann, P., Izbicki, J. R., & Pantel, K. (1999). Tumorigenic potential of apparently tumor-free lymph nodes. *New England Journal of Medicine*, 340(21), 1687. Available from https://doi.org/10.1056/NEJM199905273402116.

Sermier, A., Gervaz, P., Egger, J. F., Dao, M., Allal, A. S., Bonet, M., & Morel, P. (2006). Lymph node retrieval in abdominoperineal surgical specimen is radiation time-dependent. *World Journal of Surgical Oncology*, 4. Available from https://doi.org/10.1186/1477-7819-4-29.

Sirop, S., Kanaan, M., Korant, A., Wiese, D., Eilender, D., Nagpal, S., Arora, M., Singh, T., & Saha, S. (2011). Detection and prognostic impact of micrometastasis in colorectal cancer. *Journal of Surgical Oncology*, 103(6), 534–537. Available from https://doi.org/10.1002/jso.21793.

Sloothaak, D. A. M., Sahami, S., Van Der Zaag-Loonen, H. J., Van Der Zaag, E. S., Tanis, P. J., Bemelman, W. A., & Buskens, C. J. (2014). The prognostic value of micrometastases and isolated tumour cells in histologically negative lymph nodes of patients with colorectal cancer: A systematic review and *meta*-analysis. *European Journal of Surgical Oncology*, 40(3), 263–269. Available from https://doi.org/10.1016/j.ejso.2013.12.002.

Sommariva, A., Donisi, P. M., Gnocato, B., Vianello, R., Stracca Pansa, V., & Zaninotto, G. (2010). Factors affecting false-negative rates on ex vivo sentinel lymph node mapping in colorectal cancer. *European Journal of Surgical Oncology*, 36(2), 130–134. Available from https://doi.org/10.1016/j.ejso.2009.06.007.

Stojadinovic, A., Nissan, A., Protic, M., Adair, C. F., Prus, D., Usaj, S., Howard, R. S., Radovanovic, D., Breberina, M., Shriver, C. D., Grinbaum, R., Nelson, J. M., Brown, T. A., Freund, H. R., Potter, J. F., Peretz, T., & Peoples, G. E. (2007). Prospective randomized study comparing sentinel lymph node evaluation with standard pathologic evaluation for the staging of colon carcinoma: Results from the United States Military Cancer Institute Clinical Trials Group study GI-01. *Annals of Surgery*, 245(6), 846–857. Available from https://doi.org/10.1097/01.sla.0000256390.13550.26.

Thomas, K. A., Lechner, J., Shen, P., Waters, G. S., Geisinger, K. R., & Levine, E. A. (2006). Use of sentinel node mapping for cancer of the colon: "To map or not to map". *The American Surgeon*, 72(7), 606–612, 16875082.

Tsioulias, G. J., Wood, T. F., Morton, D. L., & Bilchik, A. J. (2000). Lymphatic mapping and focused analysis of sentinel lymph nodes upstage gastrointestinal neoplasms. *Archives of Surgery*, 135(8), 926–932. Available from https://doi.org/10.1001/archsurg.135.8.926.

van der Pas, M. H. G. M., Meijer, S., Hoekstra, O. S., Riphagen, I. I., de Vet, H. C. W., Knol, D. L., van Grieken, N. C. T., & Meijerink, W. J. H. J. (2011). Sentinel-lymph-node procedure in colon and rectal cancer: A systematic review and *meta*-analysis. *The Lancet Oncology*, 12(6), 540–550. Available from https://doi.org/10.1016/S1470-2045(11)70075-4.

van der Zaag, E. S., Bouma, W. H., Tanis, P. J., Ubbink, D. T., Bemelman, W. A., & Buskens, C. J. (2012). Systematic review of sentinel lymph node mapping procedure in colorectal cancer. *Annals of Surgical Oncology*, 19(11), 3449–3459. Available from https://doi.org/10.1245/s10434-012-2417-0.

Villegas-Tovar, E., Jimenez-Lillo, J., Jimenez-Valerio, V., Diaz-Giron-Gidi, A., Faes-Petersen, R., Otero-Piñeiro, A., De Lacy, F. B., Martinez-Portilla, R. J., & Lacy, A. M. (2020). Performance of Indocyanine green for sentinel lymph node mapping and lymph node metastasis in colorectal cancer: A diagnostic test accuracy *meta*-analysis. *Surgical Endoscopy*, 34(3), 1035–1047. Available from https://doi.org/10.1007/s00464-019-07274-z.

Weiser, M. R. (2018). AJCC 8th edition: Colorectal cancer. *Annals of Surgical Oncology*, 25(6), 1454–1455. Available from https://doi.org/10.1245/s10434-018-6462-1.

Weitz, J., Koch, M., Debus, J., Höhler, T., Galle, P. R., & Büchler, M. W. (2005). Colorectal cancer. *Lancet*, 365(9454), 153–165. Available from https://doi.org/10.1016/S0140-6736(05)17706-X.

Weixler, B., Viehl, C. T., Warschkow, R., Guller, U., Ramser, M., Sauter, G., & Zuber, M. (2017). Comparative analysis of tumor cell dissemination to the sentinel lymph nodes and to the bone marrow in patients with nonmetastasized colon cancer a prospective multicenter study. *JAMA Surgery*, 152(10), 912–920. Available from https://doi.org/10.1001/jamasurg.2017.1514.

Weixler, B., Warschkow, R., Güller, U., Zettl, A., von Holzen, U., Schmied, B. M., & Zuber, M. (2016). Isolated tumor cells in stage I & II colon cancer patients are associated with significantly worse disease-free and overall survival. *BMC Cancer*, 16(1). Available from https://doi.org/10.1186/s12885-016-2130-7.

Wood, T. F., Saha, S., Morton, D. L., Tsioulias, G. J., Rangel, D., Hutchinson, W., Foshag, L. J., & Bilchik, A. J. (2001). Validation of lymphatic mapping in colorectal cancer: In vivo, ex vivo, and laparoscopic techniques. *Annals of Surgical Oncology*, 8(2), 150–157. Available from https://doi.org/10.1245/aso.2001.8.2.150.

Yang, X., Hu, T., Gu, C., Yang, S., Jiang, D., Deng, X., Wang, Z., & Zhou, Z. (2019). The prognostic significance of isolated tumor cells detected within lateral lymph nodes in rectal cancer patients after laparoscopic lateral lymph node dissection. *Journal of Laparoendoscopic and Advanced Surgical Techniques*, 29(11), 1462–1468. Available from https://doi.org/10.1089/lap.2019.0489.

Zhao, W., Chen, B., Guo, X., Wang, R., Chang, Z., Dong, Y., Song, K., Wang, W., Qi, L., Gu, Y., Wang, C., Yang, D., & Guo, Z. (2016). A rank-based transcriptional signature for predicting relapse risk of stage II colorectal cancer identified with proper data sources. *Oncotarget*, 7(14), 19060–19071. Available from https://doi.org/10.18632/oncotarget.7956.

9

Anatomical and temporal patterns of lymph node metastasis in colorectal cancer

Mathieu J.R. Struys and Wim P. Ceelen

Department of GI Surgery, Ghent University Hospital and Department of Human Structure and Repair, Ghent University, Ghent, Belgium

9.1 Introduction

Colon cancer (CC) causes over 200,000 deaths per year in Europe, and over 800,000 worldwide (Ferlay et al., 2018; Sharma, 2020). Surgery is the mainstay of treatment of nonmetastatic CC. Since the 19th century, little has changed in the general surgical approach: a segment of the colon is removed with the adjacent mesentery. However, debate persists on the ideal extent of surgery: how much of the colon and mesentery should be removed? What is the ideal extent of lymphadenectomy?

This debate is closely linked with the uncertainty that surrounds the mechanisms of lymphatic spread in solid cancer. The Halstedian view assumes that lymphatic spread follows a well-defined temporal and anatomical path, from the primary tumor to nearby lymph nodes (LNs), henceforth to intermediate nodes, subsequently to central nodes, and eventually to distant organs such as the liver (Klein, 2009). This model presupposes that (1) progression along the metastatic pathway can only arise when the previous lymphatic nodal barrier is breached; (2) distant metastasis occurs late in the natural history of the disease; and (3) efforts to remove all cancer-invaded nodes may render the patient disease-free. This model led to mutilating, locally aggressive treatment of breast cancer aiming to eradicate tumor at the primary site, as well as in all draining regional LNs, and soon became a paradigm for cancer surgery in general.

In breast cancer, however, the Halsted model failed to explain the frequent observation of late distant metastasis arising in patients in whom the primary tumor and local LNs were successfully treated. This observation underlies the theory originally proposed by Geoffrey Keynes and developed by Bernard Fisher in cancer of the breast, which assumes that lymphatic as well as hematogenous metastasis occur early, and at random (Hellman, 2005). If the Fisher model holds, lymphadenectomy is only a staging procedure, and efforts to maximize LN retrieval are unlikely to affect the natural history of the disease (Fisher, 2008). As a consequence, breast cancer surgery has evolved to ever less invasive and extensive undertakings. In reality, and as proposed by Hellman (1994), the Halsted and Fisher models likely represent two extremes of a continuous spectrum of disease behavior, ranging from those whose cancer will remain localized to those who have disseminated disease before clinical detection.

In cancer of the colon the temporal and anatomical patterns of lymphatic spread are incompletely characterized. Also, whether cancer-invaded LNs govern metastatic behavior or merely indicate an unfavorable disease biology is an unsettled question. As a consequence, the ideal extent of lymphadenectomy remains a matter of debate between those favoring standard techniques and proponents of extensive removal of LNs during surgical treatment [complete mesocolic excision (CME) and D3 lymphadenectomy]. In this chapter, we address recent insights in the pathogenesis of CC that may contribute to define the ideal extent of surgery.

9.2 Mechanisms of lymphatic spread in colon cancer

The mechanisms that drive lymphatic metastasis in CC are only partially known. Cancer-associated lymphangiogenesis is driven by a range of soluble factors including vascular endothelial growth factors (VEGF-C/D and VEGFR-2/3), neuroplilin-2, fibroblast growth factor, hepatocyte growth factor, and noncoding RNAs (Md Yusof et al., 2020). While the elevated solid and fluid pressures inside the cancer stroma lead to collapse of intratumoral lymphatic endothelial vessels, they proliferate at the margin of the tumor, and a high density of lymphatic microvessels correlates with disease-free survival and local recurrence (LR; Huang & Chen, 2017).

In preclinical models, it was observed that the primary tumor secretes soluble factors that "prime" the nodal microenvironment as a premetastatic niche before disseminated cancer cells arrive (Jones et al., 2018; Liu & Cao, 2016). Premetastatic niche formation is characterized by diverse mechanisms including local tissue inflammation, immune suppression, stromal cell activation, and ECM remodeling (Lafitte et al., 2019). In a recent mouse colorectal cancer model, tumor cell-derived exosomes carrying IRF-2 induced the release of VEGF-C by macrophages, resulting in lymphatic network remodeling in LNs (Sun et al., 2019). At present, the existence of a similar mechanism of premetastatic niche formation in human colorectal cancer pathogenesis is unknown.

9.3 Temporal patterns of metastasis in colon cancer

The dynamic evolution of metastatic spread is of major importance for the decision of which therapeutic strategy in CC to apply, and at which timepoint. Specifically, locoregional therapy will be less effective in patients with early systemic spread. In rectal cancer, this may explain the observation that despite increasingly effective locoregional control using radiotherapy, little progress was made in distant metastasis-free survival, and this has motivated recent strategies of upfront or concurrent systemic treatment in these patients (Gilshtein et al., 2020). In CC, recent data suggest that at least in a proportion of patients, distant metastatic spread occurs early and is independent from LN metastasis.

9.3.1 Evidence from circulating and tissue biomarkers

Circulating tumor cells (CTCs) in the peripheral blood of colorectal cancer patients have been found in every stage of the disease, independently of methods and marker(s) used (Torino et al., 2013). Bork et al. (2015) found a detection rate of CTC of 4.9%, 10.5%, and 8.3% in patients with stages I, II, and III disease, respectively. In recent metaanalyses the presence of CTCs in nonmetastatic CC was found to predict worse survival in node negative as well as in stage III disease (Lu et al., 2017; Tan & Wu, 2018). Reinert et al. (2019) examined the presence and prognostic significance of circulating tumor DNA (ctDNA) in stages I—III CRC. They found ctDNA in 40%, 92%, and 90% of patients with stages I, II, and III disease, respectively, and ctDNA-positive patients were more than 40 times more likely to experience disease recurrence compared to ctDNA-negative patients (HR 43.5, 95% CI 9.8—193.5, $P < .001$) Similarly, the presence of micrometastases in colorectal LNs was found to be associated with worse outcome in several metaanalyses (Eiholm & Ovesen, 2010; Rahbari et al., 2012; Sloothaak et al., 2014).

9.3.2 Evidence from genomic and phylogenetic studies

Important insights into the dynamics of cancer metastasis can be generated from genomic and phylogenetic studies. Cancer is generally regarded as a disease resulting from an accumulation of genomic aberrations, driven by Darwinian evolutionary principles of diversification and selection for mutations that promote tumor cell proliferation and survival. Modern genomic analyses allow to reconstruct and build evolutionary trees to understand tumor progression. In colorectal cancer, two opposite theoretical scenarios exist (Fig. 9.1). Traditionally, it was assumed that as a cancer mass grows, an individual cell may acquire the mutations that provide a selective advantage and at the same time allow it to metastasize to nearby LNs. This subclone subsequently traverses and colonizes anatomically more distant LNs until a distant metastasis is established (Fig. 9.1A). However, recent data based on clinical samples from CC patients suggest that different metastatic subclones exist from an early stage of the primary tumor, and these subclones give rise to regional and systemic polyclonal metastases (Fig. 9.1B).

In a landmark study, Naxerova et al. (2017) studied the evolutionary origin of lymphatic and distant human colorectal metastases using phylogenetic trees based on the detection of somatic variants in hypermutable DNA regions. They found that in 65% of cases, lymphatic and distant metastases arose from independent subclones in the primary tumor, whereas in 35% of cases they shared a common subclonal origin, suggesting that in the majority of CC patients metastatic progression is probably not sequential in nature.

Ulintz et al. (2018) performed a detailed genetic analysis of primary tumors and paired LN metastases from CC and found that the primary tumor regions and matched LN

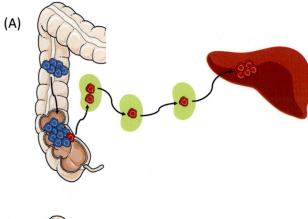
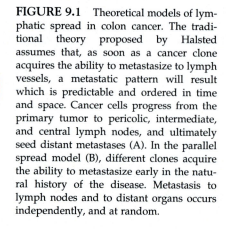
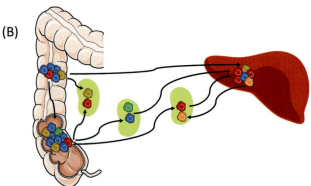

FIGURE 9.1 Theoretical models of lymphatic spread in colon cancer. The traditional theory proposed by Halsted assumes that, as soon as a cancer clone acquires the ability to metastasize to lymph vessels, a metastatic pattern will result which is predictable and ordered in time and space. Cancer cells progress from the primary tumor to pericolic, intermediate, and central lymph nodes, and ultimately seed distant metastases (A). In the parallel spread model (B), different clones acquire the ability to metastasize early in the natural history of the disease. Metastasis to lymph nodes and to distant organs occurs independently, and at random.

metastases were each polyclonal, while the clonal populations differed from one LN to another, suggesting multiple "waves" of seeding from the primary tumor to the nodal basin rather than a stepwise progression. Dang et al. (2020) confirmed the potential of polyclonal spread and clonal heterogeneity in CRC metastases. In addition, they identified metastasis-seeding clones that were not identified in any primary region in four cases, suggesting the existence of a metastasis-seeding-metastasis model. Similarly, Wei et al. showed that metastatic tumors inherited multiple genetically distinct subclones from the primary tumor, again supporting a polyclonal seeding mechanism. Interestingly, analysis of a patient with samples of primary, metastatic, and LN tumors supported a model of synchronous parallel dissemination from the primary to metastatic sites, which was not mediated through LN (Wei et al., 2017).

Zhao et al. (2016) sequenced normal, primary, and matched metastatic tumor tissues from 13 types of cancer (including one patient with CRC) and found an overwhelming nonlinear pattern of tumor progression, with early origins of metastatic lineages. Strikingly, they found that in nearly 90% of patients, genetic divergence of the first metastatic lineage had already occurred by the time of initial cancer diagnosis, while some lineages diverged years earlier. Alves et al. used a Bayesian phylogenetics approach to reconstruct evolutionary pathways based on whole-exome sequencing of a single primary CRC and multiple metastatic sites (Alves et al., 2019; Bao et al., 2018). They estimated that the age of the primary tumor was approximately 6.5–7 years before clinical diagnosis, and metastatic spread occurred early (estimated at 4.2 years before diagnosis) and in multiple directions radiating outward from the primary tumor. Biogeographic reconstruction showed direct metastatic seeding to the liver, without apparent early involvement of the lymphatic system. Interestingly, the model suggested apparent *retrograde* seeding from liver metastases to mesocolic LNs.

Reiter et al. (2020) investigated patterns of inter and intrametastatic heterogeneity of LN and distant organ metastases from CRC. Interestingly, they found that LN and distant metastases displayed considerably different levels of genetic diversity: LN metastases were polyphyletic and polyclonal, and develop through a wider evolutionary bottleneck than distant metastases, while distant metastases were less polyclonal than and typically form monophyletic groups, suggesting that LN and distant metastases develop through fundamentally different evolutionary mechanisms of note, the chronological time of metastatic dissemination cannot readily be inferred from phylogenetic studies since the divergence time on the phylogenetic tree is not equivalent to the time of metastatic dissemination (Hu & Curtis, 2020). To address this shortcoming, Hu et al. (2019) developed a spatial computational model of tumor growth and metastasis combined with exome sequencing on paired primary CRC and liver metastases. They found that only limited driver gene heterogeneity existed between primary tumors and paired metastases, suggesting that few additional genomic drivers were required for liver metastasis to occur. Strikingly, computational models from that study estimated that in 83% (19/23) of primary/metastasis pairs, metastatic dissemination occurred early, when the primary CRC was below the limit of clinical detection and typically consisted of $<10^6$ cells. In a follow-up study by the same group, genomic data from colorectal, breast, and lung cancers were analyzed, and it was estimated that metastatic dissemination in CC occurs 4 years prior to diagnosis of the primary tumor.

9.3.3 Evidence from growth rate of primary and metastatic colorectal tumors

When the appearance of systemic metastases would be the final step in a metastatic cascade encompassing sequential nodal stations, it would be expected that distant metastases occur late relative to the detection of the primary tumor. In fact, the available data suggest that the estimated growth rates of primary CC and distant metastases are comparable. Using serial imaging, Finlay et al. (1988) calculated that the age of colorectal liver metastases at the time of surgery is approximately 3.5 years, suggesting that they originate almost in parallel with the primary tumor. Similarly, mathematical models suggest that distant metastases from CRC are seeded before surgery of the primary, depending on its timing. With a primary tumor of 1.6 cm, the probability of subclinical synchronous metastases was estimated at 1%, whereas with a diameter of 12.4 cm, this probability was 99% (Avanzini & Antal, 2019). Taken together, the results from the above (phylo)genetic studies suggest that metastatic spread occurs early in the natural history of CC, and that LN distant metastases develop through independent, evolutionary different mechanisms.

9.3.4 Evidence from autopsy findings in metastatic colorectal cancer

Knijn and coworkers reported a highly interesting study that investigated the metastatic pattern in an autopsy series of 1393 colorectal cancer patients. They found that the incidence of liver and lung metastasis was comparable between node-negative and node-positive patients, suggesting that dissemination to distant organs occurred independently of lymphatic spread (Akgun et al., 2018; Knijn et al., 2016).

9.3.5 Evidence from clinical studies

Important insights can be obtained from clinical studies that have examined the survival benefit of extensive or additional removal of LNs. Assuming a linear, stepwise model of metastatic spread, extensive removal of LNs may remove all disseminated cancer cells, before they breach the final lymphatic barrier and reach the systemic circulation. However, multiple clinical studies argue against this assumption.

9.3.5.1 Segmental resection versus hemicolectomy

In a French multicenter prospective trial, Rouffet et al. (1994) randomly allocated 260 CC patients to either a left segmental colectomy or a left hemicolectomy. Only the length of tumor-free margins of colon removed was significantly greater after left hemicolectomy. Survival in both groups was, however, similar. La Torre et al. (2014) compared segmental with standard colectomy in patients with high-risk T1 CC in a matched case-control study. They found a higher specimen length and lymph node count (LNC) in the standard colectomy group. However, no differences were found in the proportion of node-positive patients, disease-free survival, or OS. Degiuli et al. (2020) compared segmental resection with extended right or left hemicolectomy in 1304 patients with cancer of the splenic flexure. In that retrospective study, no differences were found in either overall survival or progression-free survival. Two recent metaanalyses concluded that segmental resection yields comparable outcomes compared to more extensive resection in surgery for splenic flexure carcinoma of the colon (Hajibandeh et al., 2020; Wang et al., 2020). These data suggest that, in the absence of obstructive disease or the presence of hereditary nonpolyposis CC (Lynch syndrome) or cancer originating in patients with long-standing inflammatory large bowel disease, segmental resection does not compromise long-term oncological results compared to more extensive resection.

9.3.5.2 Extra-mesenteric lymph node dissection

In a retrospective single-center study, Tagliacozzo and coauthors compared extended mesenteric excision (up to the origin of the mesenteric trunk combined with retropancreatic lymphadenectomy) with standard right hemicolectomy (Tagliacozzo & Tocchi, 1997). Although radical resection resulted in a significantly higher LNC, no difference in the number of positive nodes or survival was found. Similarly, a prospective single-center trial of Tentes et al. (2007) compared radical surgery including periaortic LN resection for left-sided CC with conventional surgery in 124 patients. Although the authors mention that patients were randomly selected to either approach, the manuscript does not conform to most of the CONSORT guidelines. Again, despite a significantly higher LNC after radical resection, no significant difference in a number of involved nodes or survival was noted although improved survival was reported in the subgroup of stage III patients after radical resection (70% vs 19%, $P = .04$).

9.3.5.3 Central (apical) lymph node resection

Hashiguchi et al. (2011) recently reported retrospective data on 914 T2–T4 CC patients in whom LNs were anatomically mapped and classified as horizontal nodes (epicolic/paracolic nodes), mesocolic nodes, and nodes at the origin of the main arterial trunk. They found that resection of main trunk nodes did improve neither staging accuracy nor survival compared to resection of pericolic and mesocolic nodes alone. Similarly, Ikeda et al. (2007) found no difference in survival of separate cohorts of stage II and stage III rectosigmoid cancer whether or not the main trunk (apical) nodes were prophylactically resected.

9.3.5.4 High versus low ligation of the inferior mesenteric artery

High ligation of the inferior mesenteric artery (IMA) may allow to maximally remove LNs at their origin from the aorta. However, several studies have shown that LN at the origin of the IMA, similarly to para-aortic LNs, are infrequently invaded, and when positive predict rapid systemic disease progression (Adachi et al., 1998; Huh et al., 2012; Kang et al., 2011; Lee et al., 2017; Rao et al., 2018). Comparative randomized and nonrandomized trials have failed to show the superiority of a "high tie" in terms of LN harvest and OS (Cirocchi et al., 2012; Fujii, Ishibe, Ota, Suwa, et al., 2018; Fujii, Ishibe, Ota, Watanabe, et al., 2018; Si et al., 2019; Yang et al., 2018). Moreover, proximal ligation of the IMA may compromise blood supply to the proximal colon, and increase the risk of anastomotic leakage (Zeng & Su, 2018). In addition, high ligation can damage the sympathetic nerve supply to the pelvic organs, resulting in functional urogenital complications (Campbell et al., 2018). A recent metaanalysis of high versus low ligation of the IMA showed that high ligation resulted in a significantly higher odds of anastomotic leakage and urethral dysfunction (OR 1.29; 95% CI 1.08–1.55 and OR 2.45; 95% 1.39–4.33, respectively), but failed to improve either the risk of LR, disease-free survival, or overall survival (Si et al., 2019).

9.3.5.5 Complete mesocolic excision and D3 dissection versus standard resection

CME encompasses ligation of mesenteric vessels at their origin, and complete excision of the intact mesenteric envelopes along embryological planes between mesentery and retroperitoneum (Hohenberger et al., 2009; Sehgal & Coffey, 2014). In single-center studies, routine application of the CME technique resulted in exceptionally good oncological outcomes (Sondenaa et al., 2014). Data from Erlangen, Germany suggest that routine application of CME resulted in an excellent oncological outcome with 5-year cancer-specific survival rates of 91.4% in stage II and 70.2% in stage III CC (Hohenberger et al., 2009; Sehgal & Coffey, 2014). However, these benefits remain untested in comparative prospective trials, and some have argued that CME represents a new nomenclature for a sound surgical approach for CC that many have since long implemented (Hogan & Winter, 2009). Japanese D3 dissection is generally considered similar to CME, except for the length of bowel and area of the mesentery that are resected, which are lower in the D3 technique.

A recent systematic review has aggregated the evidence from retrospective comparisons of CME and D3 surgery with standard colonic resection (Alhassan et al., 2019).

Survival data were equivocal: only nine studies reported long-term oncologic outcomes, only four of which reported improved overall or disease-free survival or decreased local or distant recurrence in favor of CME. The two studies reporting lower LR after CME had very high LR rates in the conventional group (20.6% and 14.8%), which questions the quality of surgery in these patients. Only one study reported intraoperative complications, and it demonstrated a significantly higher incidence of injury to the superior mesenteric vein and spleen, although overall morbidity was similar. The authors conclude that, based on the available evidence, the superiority of CME is not demonstrated. The metaanalysis by Wang et al. confirmed that CME results in a 23% higher risk of surgical complications (Hajibandeh et al., 2020; Wang et al., 2020). Pooled survival data suggested better outcomes after CME, but results were from a limited number of retrospective studies and heterogeneity was very high (I^2 ranging from 46% to 98%).

Clearly, results from randomized trials are needed to establish the value of CME. Several trials are ongoing. The Chinese LAparoscopic Right Colectomy for right-sided CC trial (RELARC, NCT02619942) will randomize patients between laparoscopic CME versus standard (D2) resection; the primary endpoint is disease-free survival at 3 years (Lu et al., 2017; Tan & Wu, 2018). Similar randomized trials were initiated in Egypt (NCT02526836), Russia (NCT04364373), and Ukraine (RICON trial, NCT03200834).

In conclusion, the dynamics of metastatic spread in CC is complex, heterogeneous, and incompletely understood. However, recent data suggest that the traditional paradigm of stepwise progression, with eventual establishment of distant metastases as the final step in time, does not hold in most CC patients. Phylogenetic and autopsy studies support the notion that metastatic spread occurs early in the course of the disease and that lymphatic and systemic spread occur independently from each other. Results from clinical studies suggest that efforts to remove all potentially cancer-invaded nodes do not reduce the risk of local or systemic recurrence.

9.3.5.6 *Effect of extended lymphadenectomy on isolated lymph node recurrence*

In rectal cancer, LR has been a frequent and dreaded manifestation of disease recurrence after surgery alone. This propensity to recur locally is explained by the anatomical confinement of the (meso)rectum by the bony pelvis, bladder, and genital organs and by the fact that the mesorectum may harbor tumor deposits several centimeters distally from the lower margin of the cancer in the bowel wall (Heald & Ryall, 1986). One of the arguments proposed in favor of extensive lymphadenectomy in CC is that it may prevent locally recurrent nodal disease (Alhassan et al., 2019). However, there is little evidence for this assumption. Indeed, published data suggest that in CC, local (anastomotic, nodal, or mesenteric) recurrence is far less common. In a recent population-based analysis of 2282 CC patients from the Netherlands, the LR rate was 6.6% (Elferink et al., 2012). Locoregional recurrence was defined as tumor regrowth in or nearby the primary site, irrespective of the presence of distant metastases. In multivariate analysis, advanced T stage (3 or 4), node-positive disease, left-sided cancer, and the absence of adjuvant therapy were independent predictors of locoregional recurrence. Strikingly similar findings were reported by Yun et al. (2008), who noted a 6.1% LR rate in 994 CC patients who underwent curative resection. In only approximately 10% of these patients was the LR situated at the regional LNs. The most powerful predictor of LR was TNM stage pN2. Liska et al. (2017)

found an LR rate of 4.4% in 1397 patients treated for CC over a 14-year period. In multivariate analysis, LR was associated with advanced stage, bowel obstruction or tumor invasion, margin involvement, and lymphovascular invasion.

We recently performed a systematic review on the incidence of locoregional LN recurrence after surgery with curative intent for CC (PROSPERO registration CRD42020203288). The results show that the overall risk of locoregional nodal recurrence is low (Table 9.1) (Adachi et al., 1998; Huh et al., 2012; Kang et al., 2011; Lee et al., 2017; Rao et al., 2018).

In addition, no significant difference was found between studies describing standard surgery versus those who used extensive (D3, CME) lymphadenectomy (Fig. 9.2).

TABLE 9.1 Overview of clinical studies describing the risk of isolated lymph node recurrence after surgery for colon cancer.

Extent of the lymphadenectomy	Study	Incidence of LLNR (%)
Standard/D2		
	Bruzzi et al. (2019)	1.2
	Ding et al. (2010)	0.7
	Leijssen et al. (2019)	1.5
	Liang et al. (2007)	1.1
	Liska et al. (2017)	0.8
	Moertel et al. (1990)	6.3
	Numata et al. (2019)	0.9
	Park et al. (2015)	1
	Peracchia et al. (1991)	1.5
	Willett et al. (1984)	7.3
CME/D3	Akgun et al. (2018)	0.3
	Cho et al. (2015)	0.6
	Kang et al. (2014)	1.6
	Kawamura et al. (2000)	2
	Kim et al. (2016)	1.8
	Lee et al. (2019)	0.8
	Nagasaki et al. (2019)	1.3
	Numata et al. (2019)	2.6
	Rosemurgy et al. (1988)	1.3
	Shin et al. (2014)	3
	Siani and Pulica (2014)	3.8

LLNR, Local lymph node recurrence.

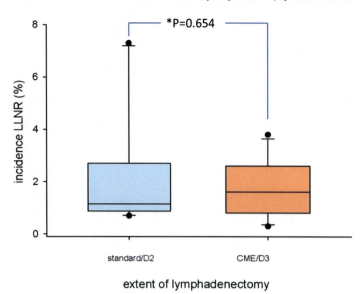

FIGURE 9.2 Overview of clinical studies reporting the risk of LLNR after surgery for colon cancer using either standard surgery or extended lymphadenectomy (complete mesocolic excision or D3 resection). Box plots represent median and interquartile range. $P = .654$, Mann–Whitney U test. *LLNR*, Local lymph node recurrence.

9.3.5.7 Lymph node counts and outcome

The presence of cancer-invaded nodes has important therapeutic and prognostic implications in CC. Therefore a minimal number of LN should be examined by the pathologist to lower the probability of false-negative findings. Most professional guidelines propose a minimal LNC of 12 (Horne et al., 2014). Over the past decade, numerous clinical studies have consistently reported a strong positive correlation between overall survival and the LNC in CC (Bernhoff et al., 2012; Chen & Bilchik, 2006; Le Voyer et al., 2003; Parsons et al., 2011; Swanson et al., 2003; Vather et al., 2009). It is tempting to imply, therefore, that a *causal* relation exists between the removal of mesenteric nodes and survival, and the observed association between nodal counts and survival is often cited by proponents of extensive lymphadenectomy. However, the relation between LNC and survival is confounded by a range of clinicopathological variables (Table 9.2), and neither a *therapeutic* effect nor stage migration adequately explains the observed correlation (Willaert et al., 2018). Probably, one of the main confounding factors is the patient's cancer-immune status: when a significant immune response is mobilized against the cancer, the prognosis will be better and at the same time, more and larger LN will be detectable in the mesentery (He et al., 2018; Märkl et al., 2016; Markl, 2015). The same argument may also explain the observation that the lymph node ratio (LNR), which represents the fraction of the LNC which is cancer invaded, is a more powerful prognosticator than the N stage in stage III disease (Ceelen et al., 2010). Indeed, the denominator of the LNR encompasses the prognostic information contained by the number of LNs examined.

CME, Complete mesocolic excision.

TABLE 9.2 Overview of clinical and pathological variables that confound the association between lymph node count (LNC) and survival in colon cancer.

Confounding variables	Effect on LNC	
Patient characteristics	↑ age, ↓ socioeconomic status, noncaucasian race	↓
	gender, body mass index	?
Tumor characteristics	↑ tumor diameter, ↑ lymph node size, T stage, overall cancer stage, lymphocytic infiltration, MSI-H phenotype, immune response, right colon cancer	↑
	↑ tumor grade	↓
	Mucinous differentiation, lymphovascular and perineural invasion	?
Treatment factors	Open versus minimally invasive resection	none
	Extensive lymphadenectomy (CME, D3)	
	Colorectal versus general surgeons, advanced fellowship training, extensive surgery (CME, D3)	↑
	Surgeon volume	?
	Neoadjuvant chemotherapy	↑
	More recent time period	↑
Institutional factors	High-volume centers, teaching hospitals, significant CC surgical practice, academic pathology laboratories	↑
Factors related to pathology examination	Xylene/alcohol fat clearance, embedding of the entire mesentery versus traditional dissection, *ex vivo* intraarterial methylene blue injection, tattooing of neoplasms during colonoscopy, pathologists interested in CRC, use of a standardized protocol to evaluate CC specimens	↑

9.4 Anatomical patterns of lymph node metastasis in colon cancer

In addition to the time course of lymphatic versus distant spread, anatomical patterns of lymphatic metastasis may provide important information on the underlying biology. In the current TNM staging system, the anatomical location of involved LNs is no longer a component of the N stage. In Western countries, pathologist reports the number of examined and metastatic nodes, but not their anatomical location. However, in the far East, the LNs from a colorectal cancer specimen are mapped and collected separately according to their anatomical location as paracolic nodes (L1), intermediate nodes (L2), and apical nodes (L3). Subsequently, the extent of mesenteric resection is classified as D1, D2, or D3, where the latter corresponds with the technique of CME. When LNs are mapped according to anatomical station in the mesentery, interesting information is available that allows to reconstruct the lymphatic spread pathways. Specifically, this allows to verify whether progression from the primary tumor to central nodes occurs orderly and stepwise from

T→L1→L2→L3, or whether one or more LN stations are "skipped" (when the distal LN is cancer invaded, but the more proximal one is not).

Several authors have reported the incidence of these "skipped" LN metastasis based on anatomical mapping (Table 9.3) (Lu et al., 2017; Tan & Wu, 2018). The reported incidence varies widely between 1% and 22%. In the largest series, recently reported by Kataoka et al. (2019), the overall incidence of skipped LN metastases was 8%, and higher among L3 nodes compared to L2 nodes. The same group analyzed the relationship between anatomical location of the primary cancer and the number of skipped metastases and found that the proportion of patients with a "skipped" pattern was higher in right compared to left CC (13.7% vs 9.0%; $P < 0.001$) (Kataoka et al., 2020).

Taken together, the data suggest that up to one in four CC patients harbors "skipped" LN metastases. This finding is based on studies using routine pathological analysis and H&E staining, and more detailed molecular or genomic analyses may shed a different light. Zhang (2020) performed whole-exome sequencing on a series of matched CRC samples consisting of primary tumors, metastatic LNs (paracolic, intermediate, and central), liver metastases, and normal tissue to reconstruct clonal evolutionary pathways. In 9 out of 10 patients (90%), a "skipped" metastatic pattern was observed (Fig. 9.3).

TABLE 9.3 Incidence of "skipped" lymph node metastases in clinical studies that have anatomically mapped invaded lymph nodes according to location (paracolic, intermediate, and central) after surgery for colon cancer.

Reference		N	%skipped
Shida et al. (1992)	Colon	148	8.1
Malassagne et al. (1993)	Colon	197	1.5
Tagliacozzo (1997)	Right colon	144	3.5
Yamamoto et al. (1998)	Colon	270	10.4
Merrie (2001)	Colon	54	18.5
Kanemitsu (2006)	Sigmoid	421	1
Shiozawa et al. (2007)	Colon	323	6.5
Eiholm (2010)	Right colon	11	9.1
Lan (2011)	Right colon	103	1.9
Tan et al. (2010)	Colon	93	18
Kanemitsu et al. (2013)	Right colon	142	4.2
Perrakis et al. (2014)	Transverse colon	45	6.7
Liang et al. (2015)	Colon	202	19.8
Benz et al. (2015)	Right colon	51	2
Bao et al. (2016)	Colon	184	22.3
Bao et al. (2018)	Colon	167	19.2
Kataoka (2019)	Colon	446	8

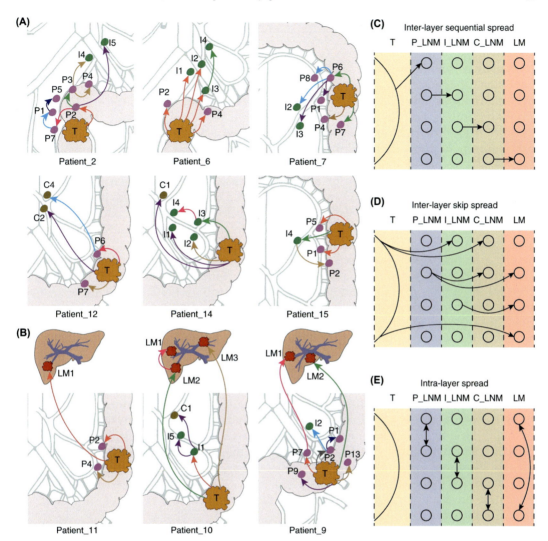

FIGURE 9.3 Parsimonious metastatic routes in nine colon cancer patients, based on exome sequencing of primary tumors and metastatic locations. Metastatic routes between lesions are denoted by arrows and colored according to the smallest involved subclone. For instance, in Patient_2, P2 is seeded by the red subclone in the primary tumor, thus this route is labeled in red. P3 was then seeded by the red and green subclones from P2, so this route is labeled in green, since the green subclone was smaller. Three metastatic modes were extracted from 61 metastatic events: c interlayer sequential spread, d interlayer skip spread, and e intralayer spread. The metastatic network is dissected into five layers, spanning from primary tumor (T) to pericolin, intermediate, or central lymph nodes (P, I, C), and LM. The arrows indicate metastatic routes. *LM*, Liver metastases. *Reprinted from Zhang, C. (2020). Mapping the spreading routes of lymphatic metastases in human colorectal cancer. Nature Communications, 11 (1993), 1–11. https://doi.org/10.1038/s41467-020-15886-6.*

9.5 Conclusion and implications for research

Knowledge of how and when CC gives rise to lymphatic metastases has immediate consequences for the surgical strategy. As it stands, the available evidence allows to draw several conclusions. First, in the majority of patients, metastatic spread occurs early, at random, and without obvious temporal or anatomical pattern. The risks of lymphatic and distant (systemic) metastasis seem to be independent, and the latter probably cannot be prevented by removing affected LNs before they seed distant disease. Second, multiple comparative trials have shown that extensive removal of the mesentery and/or potentially involved LNs increases the risk of complications, but does not affect the rate of local or systemic disease recurrence. There is, however, a consistent relation between LNC and survival, but a causal mechanism between both is highly unlikely. The observed association is probably explained by confounders such as experience and expertise, molecular status, and immune response. Similar observations were made in other cancer types: in esophageal, pancreatic, and ovarian cancer, there is a similar association between LN count and survival, but (extensive) lymphadenectomy does not result in a survival benefit (Huang & Chen, 2017). Obviously, these results do not negate the importance of an adequate LN count to allow accurate staging, and the appropriateness of a surgical technique that is guided by embryological planes (Gouvas et al., 2016).

Despite the width of the available evidence, several limitations should be mentioned. First, the (phylo)genetic data are based on a limited number of clinical samples, and given the significant heterogeneity of CC, these studies may not represent the full spectrum of cancer behavior. Second, the majority of clinical studies is either retrospective or included only a limited number of patients. Therefore the results of ongoing randomized trials comparing extensive lymphadenectomy (CME) with standard surgery are awaited. In addition, the molecular and (epi)genetic mechanisms that drive lymphatic metastasis in CC should be further elucidated in larger datasets, ideally in the context of translational studies that combine these molecular endpoints with novel intraoperative imaging techniques to identify LNs and lymphatic pathways (Ushijima et al., 2020). These efforts may ultimately result in a precision surgical approach that allows to tailor the extent of surgery to the biology of the cancer.

References

Adachi, Y., Inomata, M., Miyazaki, N., Sato, K., Shiraishi, N., & Kitano, S. (1998). Distribution of lymph node metastasis and level of inferior mesenteric artery ligation in colorectal cancer. *Journal of Clinical Gastroenterology, 26*(3), 179–182.

Akgun, E., Yoldas, T., Caliskan, C., Karabulut, B., Doganavsargil, B., & Akarca, U. S. (2018). Prognostic factors for isolated locoregional recurrences in colon cancer and survival after salvage surgery: A cohort study. *Indian Journal of Surgery, 80*(5), 428–434. Available from https://doi.org/10.1007/s12262-017-1623-1.

Alhassan, N., Yang, M., Wong-Chong, N., Liberman, A. S., Charlebois, P., Stein, B., Fried, G. M., & Lee, L. (2019). Comparison between conventional colectomy and complete mesocolic excision for colon cancer: A systematic review and pooled analysis: A review of CME versus conventional colectomies. *Surgical Endoscopy, 33*(1), 8–18. Available from https://doi.org/10.1007/s00464-018-6419-2.

Alves, J. M., Prado-López, S., Cameselle-Teijeiro, J. M., & Posada, D. (2019). Rapid evolution and biogeographic spread in a colorectal cancer. *Nature Communications, 10*(1), 5139. Available from https://doi.org/10.1038/s41467-019-12926-8.

Avanzini, S., & Antal, T. (2019). Cancer recurrence times from a branching process model. *PLoS Computational Biology, 15*(11), e1007423. Available from https://doi.org/10.1371/journal.pcbi.1007423.

Bao, F., Deng, Z. G., Wang, D., Xian-Yu, J. B., Li, G. Q., Xiang, C. H., Xiang, R. C., & Zhi, X. (2018). Factors influencing lymph node skip metastasis in colorectal cancer: A retrospective study. *ANZ Journal of Surgery, 88*(7−8), 770−774. Available from https://doi.org/10.1111/ans.14268.

Bao, F., Zhao, L. Y., Balde, A. I., Liu, H., Yan, J., Li, T. T., Chen, H., & Li, G. X. (2016). Prognostic impact of lymph node skip metastasis in Stage III colorectal cancer. *Colorectal Disease: The Official Journal of the Association of Coloproctology of Great Britain and Ireland, 18*(9), O322−O329. Available from https://doi.org/10.1111/codi.13465.

Benz, S. R., Tannapfel, A., Tam, Y., & Stricker, I. (2015). [Complete mesocolic excision for right-sided colon cancer—The role of central lymph nodes]. *Zentralblatt fur Chirurgie, 140*(4), 449−452. Available from https://doi.org/10.1055/s-0034-1383133.

Bernhoff, R., Holm, T., Sjovall, A., Granath, F., Ekbom, A., & Martling, A. (2012). Increased lymph node harvest in patients operated on for right-sided colon cancer: A population-based study. *Colorectal Disease, 14*(6). Available from https://doi.org/10.1111/j.1463-1318.2012.03020.x.

Bork, U., Rahbari, N. N., Schölch, S., Reissfelder, C., Kahlert, C., Büchler, M. W., Weitz, J., & Koch, M. (2015). Circulating tumour cells and outcome in non-metastatic colorectal cancer: A prospective study. *British Journal of Cancer, 112*(8), 1306−1313. Available from https://doi.org/10.1038/bjc.2015.88.

Bruzzi, M., Auclin, E., Lo Dico, R., Voron, T., Karoui, M., Espin, E., Cianchi, F., Weitz, J., Buggenhout, A., Malafosse, R., Denimal, F., Le Malicot, K., Vernerey, D., Douard, R., Emile, J. F., Lepage, C., Laurent-Puig, P., & Taieb, J. (2019). Influence of molecular status on recurrence site in patients treated for a stage III colon cancer: A post hoc analysis of the PETACC-8 trial. *Annals of Surgical Oncology, 26*(11), 3561−3567. Available from https://doi.org/10.1245/s10434-019-07513-6.

Campbell, A., Macdonald, A., Oliphant, R., Russell, D., & Fogg, Q. A. (2018). Neurovasculature of high and low tie ligation of the inferior mesenteric artery. *Surgical and Radiologic Anatomy*. Available from https://doi.org/10.1007/s00276-018-2092-3.

Ceelen, W., Van Nieuwenhove, Y., & Pattyn, P. (2010). Prognostic value of the lymph node ratio in stage III colorectal cancer: A systematic review. *Annals of Surgical Oncology, 17*(11), 2847−2855. Available from https://doi.org/10.1245/s10434-010-1158-1.

Chen, S. L., & Bilchik, A. J. (2006). More extensive nodal dissection improves survival for stages I to III of colon cancer—A population-based study. *Annals of Surgery, 244*(4), 602−610. Available from https://doi.org/10.1097/01.sla.0000237655.11717.50.

Cho, M. S., Baek, S. J., Hur, H., Min, B. S., Baik, S. H., & Kim, N. K. (2015). Modified complete mesocolic excision with central vascular ligation for the treatment of right-sided colon cancer: Long-term outcomes and prognostic factors. *Annals of Surgery, 261*(4), 708−715. Available from https://doi.org/10.1097/SLA.0000000000000831.

Cirocchi, R., Trastulli, S., Farinella, E., Desiderio, J., Vettoretto, N., Parisi, A., Boselli, C., & Noya, G. (2012). High tie versus low tie of the inferior mesenteric artery in colorectal cancer: A RCT is needed. *Surgical Oncology, 21*(3), e111−e123. Available from https://doi.org/10.1016/j.suronc.2012.04.004.

Dang, H. X., Krasnick, B. A., White, B. S., Grossman, J. G., Strand, M. S., Zhang, J., Cabanski, C. R., Miller, C. A., Fulton, R. S., Goedegebuure, S. P., Fronick, C. C., Griffith, M., Larson, D. E., Goetz, B. D., Walker, J. R., Hawkins, W. G., Strasberg, S. M., Linehan, D. C., Lim, K. H., ... Fields, R. C. (2020). The clonal evolution of metastatic colorectal cancer. *Science Advances, 6*(24), eaay9691. Available from https://doi.org/10.1126/sciadv.aay9691.

Degiuli, M., Reddavid, R., Ricceri, F., Di Candido, F., Ortenzi, M., Elmore, U., Belluco, C., Rosati, R., Guerrieri, M., Spinelli, A., De Nardi, P., Cianflocca, D., Borghi, F., Rega, D., Delrio, P., Milone, M., Domenico De Palma, G., Restivo, A., Deidda, S., ... Marsanic, P. (2020). Segmental colonic resection is a safe and effective treatment option for colon cancer of the splenic flexure: A nationwide retrospective study of the Italian society of surgical oncology-colorectal cancer network collaborative group. *Diseases of the Colon and Rectum, 63*(10), 1372−1382. Available from https://doi.org/10.1097/dcr.0000000000001743.

Ding, P. R., An, X., Zhang, R. X., Fang, Y. J., Li, L. R., Chen, G., Wu, X. J., Lu, Z. H., Lin, J. Z., Kong, L. H., Wan, D. S., & Pan, Z. Z. (2010). Elevated preoperative neutrophil to lymphocyte ratio predicts risk of recurrence following curative resection for stage IIA colon cancer. *International Journal of Colorectal Disease, 25*(12), 1427−1433. Available from https://doi.org/10.1007/s00384-010-1052-0.

Eiholm, S., & Ovesen, H. (2010). Total mesocolic excision versus traditional resection in right-sided colon cancer - method and increased lymph node harvest. *Danish Medical Bulletin, 57*(12), A4224.

Eiholm, S., & Ovesen, H. (2010). Total mesocolic excision versus traditional resection in right-sided colon cancer - method and increased lymph node harvest. *Danish Medical Bulletin, 57*(12), A4224.

Elferink, M. A. G., Visser, O., Wiggers, T., Otter, R., Tollenaar, R., Langendijk, J. A., & Siesling, S. (2012). Prognostic factors for locoregional recurrences in colon cancer. *Annals of Surgical Oncology, 19*(7), 2203−2211. Available from https://doi.org/10.1245/s10434-011-2183-4.

Ferlay, J., Colombet, M., Soerjomataram, I., Dyba, T., Randi, G., Bettio, M., Gavin, A., Visser, O., & Bray, F. (2018). Cancer incidence and mortality patterns in Europe: Estimates for 40 countries and 25 major cancers in 2018. *European Journal of Cancer.* Available from https://doi.org/10.1016/j.ejca.2018.07.005.

Finlay, I. G., Meek, D., Brunton, F., & McArdle, C. S. (1988). Growth rate of hepatic metastases in colorectal carcinoma. *The British Journal of Surgery, 75*(7), 641−644.

Fisher, B. (2008). Biological research in the evolution of cancer surgery: A personal perspective. *Cancer Research, 68*(24), 10007−10020. Available from https://doi.org/10.1158/0008-5472.can-08-0186.

Fujii, S., Ishibe, A., Ota, M., Suwa, H., Watanabe, J., Kunisaki, C., & Endo, I. (2018). Short-term and long-term results of a randomized study comparing high tie and low tie inferior mesenteric artery ligation in laparoscopic rectal anterior resection: Subanalysis of the HTLT (high tie vs. low tie) study. *Surgical Endoscopy, 33*(4), 1100−1110. Available from https://doi.org/10.1007/s00464-018-6363-1.

Fujii, S., Ishibe, A., Ota, M., Watanabe, K., Watanabe, J., Kunisaki, C., & Endo, I. (2018). Randomized clinical trial of high versus low inferior mesenteric artery ligation during anterior resection for rectal cancer. *BJS Open, 2*(4), 195−202. Available from https://doi.org/10.1002/bjs5.71.

Gilshtein, H., Ghuman, A., Dawoud, M., Yellinek, S., Kent, I., Sharp, S. P., Nagarajan, A., & Wexner, S. D. (2020). Total neoadjuvant treatment for rectal cancer: Preliminary experience. *The American Surgeon, 87*(5), 708−713, 3134820951499. Available from https://doi.org/10.1177/0003134820951499.

Gouvas, N., Agalianos, C., Papaparaskeva, K., Perrakis, A., Hohenberger, W., & Xynos, E. (2016). Surgery along the embryological planes for colon cancer: A systematic review of complete mesocolic excision. *International Journal of Colorectal Disease, 31*(9), 1577−1594. Available from https://doi.org/10.1007/s00384-016-2626-2.

Hajibandeh, S., Hajibandeh, S., Hussain, I., Zubairu, A., Akbar, F., & Maw, A. (2020). Comparison of extended right hemicolectomy, left hemicolectomy and segmental colectomy for splenic flexure colon cancer: A systematic review and meta-analysis. *Colorectal Disease: The Official Journal of the Association of Coloproctology of Great Britain and Ireland, 22*(12), 1885−1907. Available from https://doi.org/10.1111/codi.15292.

Hashiguchi, Y., Hase, K., Ueno, H., Mochizuki, H., Shinto, E., & Yamamoto, J. (2011). Optimal margins and lymphadenectomy in colonic cancer surgery. *British Journal of Surgery, 98*(8), 1171−1178. Available from https://doi.org/10.1002/bjs.7518.

He, W. Z., Xie, Q. K., Hu, W. M., Kong, P. F., Yang, L., Yang, Y. Z., Jiang, C., Yin, C. X., Qiu, H. J., Zhang, H. Z., Zhang, B., & Xia, L. P. (2018). An increased number of negative lymph nodes is associated with a higher immune response and longer survival in colon cancer patients. *Cancer Management and Research, 10*, 1597−1604. Available from https://doi.org/10.2147/cmar.s160100.

Heald, R. J., & Ryall, R. D. H. (1986). Recurrence and survival after total mesorectal excision for rectal-cancer. *Lancet, 1*(8496), 1479−1482. Available from https://doi.org/10.1016/s0140-6736(86)91510-2.

Hellman, S. (1994). Karnofsky memorial lecture. Natural history of small breast cancers. *Journal of Clinical Oncology: Official Journal of the American Society of Clinical Oncology, 12*(10), 2229−2234. Available from https://doi.org/10.1200/jco.1994.12.10.2229.

Hellman, S. (2005). Premise, promise, paradigm and prophesy. *Nature Clinical Practice. Oncology, 2*(7), 325. Available from https://doi.org/10.1038/ncponc0220.

Hogan, A. M., & Winter, D. C. (2009). Complete mesocolic excision (CME): A "novel" concept? *Journal of Surgical Oncology, 100*(3), 182−183. Available from https://doi.org/10.1002/jso.21310.

Hohenberger, W., Weber, K., Matzel, K., Papadopoulos, T., & Merkel, S. (2009). Standardized surgery for colonic cancer: Complete mesocolic excision and central ligation − Technical notes and outcome. *Colorectal Disease: The Official Journal of the Association of Coloproctology of Great Britain and Ireland, 11*(4), 354−364, discussion 364−5. Available from https://doi.org/10.1111/j.1463-1318.2008.01735.x.

Horne, J., Bateman, A. C., Carr, N. J., & Ryder, I. (2014). Lymph node revealing solutions in colorectal cancer: Should they be used routinely? *Journal of Clinical Pathology*, 67(5), 383–388. Available from https://doi.org/10.1136/jclinpath-2013-202146.

Hu, Z., & Curtis, C. (2020). Looking backward in time to define the chronology of metastasis. *Natural Communication*, 11(1), 3213. Available from https://doi.org/10.1038/s41467-020-16995-y.

Hu, Z., Ding, J., Ma, Z., Sun, R., Seoane, J. A., Scott Shaffer, J., Suarez, C. J., Berghoff, A. S., Cremolini, C., Falcone, A., Loupakis, F., Birner, P., Preusser, M., Lenz, H. J., & Curtis, C. (2019). Quantitative evidence for early metastatic seeding in colorectal cancer. *Nature Genetics*, 51(7), 1113–1122. Available from https://doi.org/10.1038/s41588-019-0423-x.

Huang, C., & Chen, Y. (2017). Lymphangiogenesis and colorectal cancer. *Saudi Medical Journal*, 38(3), 237–244. Available from https://doi.org/10.15537/smj.2017.3.16245.

Huh, J. W., Kim, Y. J., & Kim, H. R. (2012). Distribution of lymph node metastases is an independent predictor of survival for sigmoid colon and rectal cancer. *Annals of Surgery*, 255(1), 70–78. Available from https://doi.org/10.1097/SLA.0b013e31823785f6.

Ikeda, Y., Shimabukuro, R., Saitsu, H., Saku, M., & Maehara, Y. (2007). Influence of prophylactic apical node dissection of the inferior mesenteric artery on prognosis of colorectal cancer. *Hepato-Gastroenterology*, 54(79), 1985–1987, http://Go to ISI://WOS:000251892700021.

Jones, D., Pereira, E. R., & Padera, T. P. (2018). Growth and immune evasion of lymph node metastasis. *Frontiers in Oncology*, 8, 36. Available from https://doi.org/10.3389/fonc.2018.00036.

Kanemitsu, Y., Hirai, T., Komori, K., & Kato, T. (2006). Survival benefit of high ligation of the inferior mesenteric artery in sigmoid colon or rectal cancer surgery. *The British Journal of Surgery*, 93(5), 609–615. doi:10.1002/bjs.5327.

Kanemitsu, Y., Komori, K., Kimura, K., & Kato, T. (2013). D3 lymph node dissection in right hemicolectomy with a no-touch isolation technique in patients with colon cancer. *Diseases of the Colon and Rectum*, 56(7), 815–824. Available from https://doi.org/10.1097/DCR.0b013e3182919093.

Kang, J., Hur, H., Min, B. S., Kim, N. K., & Lee, K. Y. (2011). Prognostic impact of inferior mesenteric artery lymph node metastasis in colorectal cancer. *Annals of Surgical Oncology*, 18(3), 704–710. Available from https://doi.org/10.1245/s10434-010-1291-x.

Kang, J., Kim, I. K., Kang, S. I., Sohn, S. K., & Lee, K. Y. (2014). Laparoscopic right hemicolectomy with complete mesocolic excision. *Surgical Endoscopy*, 28(9), 2747–2751. Available from https://doi.org/10.1007/s00464-014-3521-y.

Kataoka, K., Beppu, N., Shiozawa, M., Ikeda, M., Tomita, N., Kobayashi, H., Sugihara, K., & Ceelen, W. (2020). Colorectal cancer treated by resection and extended lymphadenectomy: Patterns of spread in left- and right-sided tumours. *The British Journal of Surgery*, 107(8), 1070–1078. Available from https://doi.org/10.1002/bjs.11517.

Kataoka, K., Ysebaert, H., Shiozawa, M., Reynders, D., Ikeda, M., Tomita, N., Goetghebeur, E., & Ceelen, W. (2019). Prognostic significance of number versus location of positive mesenteric nodes in stage iii colon cancer. *European Journal of Surgical Oncology*, 45(10), 1862–1869. doi:10.1016/j.ejso.2019.05.022.

Kataoka, K., Ysebaert, H., Shiozawa, M., Reynders, D., Ikeda, M., Tomita, N., Goetghebeur, E., & Ceelen, W. (2019). Prognostic significance of number versus location of positive mesenteric nodes in stage III colon cancer. *European Journal of Surgical Oncology: The Journal of the European Society of Surgical Oncology and the British Association of Surgical Oncology*, 45(10), 1862–1869. Available from https://doi.org/10.1016/j.ejso.2019.05.022.

Kawamura, Y. J., Umetani, N., Sunami, E., Watanabe, T., Masaki, T., & Muto, T. (2000). Effect of high ligation on the long-term result of patients with operable colon cancer, particularly those with limited nodal involvement. *European Journal of Surgery*, 166(10), 803–807. Available from https://doi.org/10.1080/110241500447443.

Kim, J. W., Kim, J. Y., Kang, B. M., Lee, B. H., Kim, B. C., & Park, J. H. (2016). Short- and long-term outcomes of laparoscopic surgery vs open surgery for transverse colon cancer: A retrospective multicenter study. *OncoTargets and Therapy*, 9, 2203–2209. Available from https://doi.org/10.2147/OTT.S103763.

Klein, C. A. (2009). Parallel progression of primary tumours and metastases. *Nature Reviews. Cancer*, 9(4), 302–312. Available from https://doi.org/10.1038/nrc2627.

Knijn, N., van Erning, F. N., Overbeek, L. I., Punt, C. J., Lemmens, V. E., Hugen, N., & Nagtegaal, I. D. (2016). Limited effect of lymph node status on the metastatic pattern in colorectal cancer. *Oncotarget*, 7(22), 31699–31707. Available from https://doi.org/10.18632/oncotarget.9064.

La Torre, M., Nigri, G., Mazzuca, F., Ferri, M., Botticelli, A., Lorenzon, L., Pilozzi, E., & Ziparo, V. (2014). Standard versus limited colon resection for high risk T1 colon cancer. A matched case-control study. *Journal of Gastrointestinal and Liver Diseases*, *23*(3), 285−290. Available from https://doi.org/10.15403/jgld.2014.1121.233.mlt.

Lafitte, M., Lecointre, C., & Roche, S. (2019). Roles of exosomes in metastatic colorectal cancer. *American Journal of Physiology-Cell Physiology*, *317*(5), C869−C880. Available from https://doi.org/10.1152/ajpcell.00218.2019.

Lan, Y. T., Lin, J. K., Jiang, J. K., Chang, S. C., Liang, W. Y., & Yang, S. H. (2011). Significance of lymph node retrieval from the terminal ileum for patients with cecal and ascending colonic cancers. *Annals of Surgical Oncology*, *18*(1), 146−52. doi:10.1245/s10434-010-1270-2.

Le Voyer, T. E., Sigurdson, E. R., Hanlon, A. L., Mayer, R. J., Macdonald, J. S., Catalano, P. J., & Haller, D. G. (2003). Colon cancer survival is associated with increasing number of lymph nodes analyzed: A secondary survey of intergroup trial INT-0089. *Journal of Clinical Oncology*, *21*(15), 2912−2919. Available from https://doi.org/10.1200/Jco.2003.05.062.

Lee, S. H., Lee, J. L., Kim, C. W., Lee, H. I., Yu, C. S., & Kim, J. C. (2017). Oncologic significance of para-aortic lymph node and inferior mesenteric lymph node metastasis in sigmoid and rectal adenocarcinoma. *European Journal of Surgical Oncology: The Journal of the European Society of Surgical Oncology and the British Association of Surgical Oncology*, *43*(11), 2076−2083. Available from https://doi.org/10.1016/j.ejso.2017.08.014.

Lee, S. Y., Yeom, S. S., Kim, C. H., Kim, Y. J., & Kim, H. R. (2019). Distribution of lymph node metastasis and the extent of lymph node dissection in descending colon cancer patients. *ANZ Journal of Surgery*, *89*(9), E373−e378. Available from https://doi.org/10.1111/ans.15400.

Leijssen, L. G. J., Dinaux, A. M., Amri, R., Kunitake, H., Bordeianou, L. G., & Berger, D. L. (2019). The impact of a multivisceral resection and adjuvant therapy in locally advanced colon cancer. *Journal of Gastrointestinal Surgery: Official Journal of the Society for Surgery of the Alimentary Tract*, *23*(2), 357−366. Available from https://doi.org/10.1007/s11605-018-3962-z.

Liang, J. T., Huang, K. C., Lai, H. S., Lee, P. H., & Jeng, Y. M. (2007). Oncologic results of laparoscopic versus conventional open surgery for stage II or III left-sided colon cancers: A randomized controlled trial. *Annals of Surgical Oncology*, *14*(1), 109−117. Available from https://doi.org/10.1245/s10434-006-9135-4.

Liang, J. T., Lai, H. S., Huang, J., & Sun, C. T. (2015). Long-term oncologic results of laparoscopic D3 lymphadenectomy with complete mesocolic excision for right-sided colon cancer with clinically positive lymph nodes. *Surgical Endoscopy and Other Interventional Techniques*, *29*(8), 2394−2401. Available from https://doi.org/10.1007/s00464-014-3940-9.

Liska, D., Stocchi, L., Karagkounis, G., Elagili, F., Dietz, D. W., Kalady, M. F., Kessler, H., Remzi, F. H., & Church, J. (2017). Incidence, patterns, and predictors of locoregional recurrence in colon cancer. *Annals of Surgical Oncology*, *24*(4), 1093−1099. Available from https://doi.org/10.1245/s10434-016-5643-z.

Liu, Y., & Cao, X. (2016). Characteristics and significance of the pre-metastatic niche. *Cancer Cell*, *30*(5), 668−681. Available from https://doi.org/10.1016/j.ccell.2016.09.011.

Lu, Y. J., Wang, P., Peng, J., Wang, X., Zhu, Y. W., & Shen, N. (2017). Meta-analysis reveals the prognostic value of circulating tumour cells detected in the peripheral blood in patients with non-metastatic colorectal cancer. *Scientific Reports*, *7*(1), 905. Available from https://doi.org/10.1038/s41598-017-01066-y.

Malassagne, B., Valleur, P., Serra, J., Sarnacki, S., Galian, A., Hoang, C., & Hautefeuille, P. (1993). Relationship of apical lymph node involvement to survival in resected colon carcinoma. *Diseases of the Colon and Rectum*, *36*(7), 645−653. Available from https://doi.org/10.1007/bf02238591.

Markl, B. (2015). Stage migration vs immunology: The lymph node count story in colon cancer. *World Journal of Gastroenterology*, *21*(43), 12218−12233. Available from https://doi.org/10.3748/wjg.v21.i43.12218.

Märkl, B., Wieberneit, J., Kretsinger, H., Mayr, P., Anthuber, M., Arnholdt, H. M., & Schenkirsch, G. (2016). Number of intratumoral T lymphocytes is associated with lymph node size, lymph node harvest, and outcome in node-negative colon cancer. *American Journal of Clinical Pathology*, *145*(6), 826−836. Available from https://doi.org/10.1093/ajcp/aqw074.

Md Yusof, K., Rosli, R., Abdullah, M., & K A., A.-K. (2020). The roles of non-coding RNAs in tumor-associated lymphangiogenesis. *Cancers (Basel)*, *12*(11), 3290. Available from https://doi.org/10.3390/cancers12113290.

Merrie, A. E., Phillips, L. V., Yun, K., & McCall, J. L. (2001). Skip metastases in colon cancer: assessment by lymph node mapping using molecular detection. *Surgery*, *129*(6), 684−691. doi:10.1067/msy.2001.113887.

Moertel, C. G., Fleming, T. R., Macdonald, J. S., Haller, D. G., Laurie, J. A., Goodman, P. J., Ungerleider, J. S., Emerson, W. A., Tormey, D. C., Glick, J. H., et al. (1990). Levamisole and fluorouracil for adjuvant therapy of resected colon carcinoma. *The New England Journal of Medicine*, 322(6), 352−358. Available from https://doi.org/10.1056/nejm199002083220602.

Nagasaki, T., Akiyoshi, T., Fukunaga, Y., Tominaga, T., Yamaguchi, T., Konishi, T., Fujimoto, Y., Nagayama, S., & Ueno, M. (2019). The short- and long-term feasibility of laparoscopic surgery in colon cancer patients with bulky tumors. *Journal of Gastrointestinal Surgery: Official Journal of the Society for Surgery of the Alimentary Tract*, 23(9), 1893−1899. Available from https://doi.org/10.1007/s11605-019-04114-2.

Naxerova, K., Reiter, J. G., Brachtel, E., Lennerz, J. K., van de Wetering, M., Rowan, A., Cai, T., Clevers, H., Swanton, C., Nowak, M. A., Elledge, S. J., & Jain, R. K. (2017). Origins of lymphatic and distant metastases in human colorectal cancer. *Science (New York, N.Y.)*, 357(6346), 55−60. Available from https://doi.org/10.1126/science.aai8515.

Numata, M., Sawazaki, S., Aoyama, T., Tamagawa, H., Sato, T., Saeki, H., Saigusa, Y., Taguri, M., Mushiake, H., Oshima, T., Yukawa, N., Shiozawa, M., Rino, Y., & Masuda, M. (2019). D3 lymph node dissection reduces recurrence after primary resection for elderly patients with colon cancer. *International Journal of Colorectal Disease*, 34(4), 621−628. Available from https://doi.org/10.1007/s00384-018-03233-7.

Park, J. H., Kim, M. J., Park, S. C., Hong, C. W., Sohn, D. K., Han, K. S., & Oh, J. H. (2015). Difference in time to locoregional recurrence between patients with right-sided and left-sided colon cancers. *Diseases of the Colon and Rectum*, 58(9), 831−837. Available from https://doi.org/10.1097/DCR.0000000000000426.

Parsons, H. M., Tuttle, T. M., Kuntz, K. M., Begun, J. W., McGovern, P. M., & Virnig, B. A. (2011). Association between lymph node evaluation for colon cancer and node positivity over the past 20 years. *Journal of the American Medical Association*, 306(10), 1089−1097. Available from https://doi.org/10.1001/jama.2011.1285.

Peracchia, A., Sarli, L., Carreras, F., Pietra, N., Longinotti, E., & Gafà, M. (1991). Locoregional recurrences following curative surgery for colon cancer. *Annali Italiani di Chirurgia*, 62(1), 37−42, discussion 43−44. Available from http://www.embase.com/search/results?subaction = viewrecord&from = export&id = L21884611.

Perrakis, A., Weber, K., Merkel, S., Matzel, K., Agaimy, A., Gebbert, C., & Hohenberger, W. (2014). Lymph node metastasis of carcinomas of transverse colon including flexures. Consideration of the extramesocolic lymph node stations. *International Journal of Colorectal Disease*, 29(10), 1223−1229. Available from https://doi.org/10.1007/s00384-014-1971-2.

Rahbari, N. N., Bork, U., Motschall, E., Thorlund, K., Büchler, M. W., Koch, M., & Weitz, J. (2012). Molecular detection of tumor cells in regional lymph nodes is associated with disease recurrence and poor survival in node-negative colorectal cancer: A systematic review and meta-analysis. *Journal of Clinical Oncology: Official Journal of the American Society of Clinical Oncology*, 30(1), 60−70. Available from https://doi.org/10.1200/jco.2011.36.9504.

Rao, X., Zhang, J., Liu, T., Wu, Y., Jiang, Y., Wang, P., Chen, G., Pan, Y., Wu, T., Liu, Y., Wan, Y., Huang, S., & Wang, X. (2018). Prognostic value of inferior mesenteric artery lymph node metastasis in cancer of the descending colon, sigmoid colon and rectum. *Colorectal Disease: The Official Journal of the Association of Coloproctology of Great Britain and Ireland*, 20(6), O135−O142. Available from https://doi.org/10.1111/codi.14105.

Reinert, T., Henriksen, T. V., Christensen, E., Sharma, S., Salari, R., Sethi, H., Knudsen, M., Nordentoft, I., Wu, H. T., Tin, A. S., Heilskov Rasmussen, M., Vang, S., Shchegrova, S., Frydendahl Boll Johansen, A., Srinivasan, R., Assaf, Z., Balcioglu, M., Olson, A., Dashner, S., . . . Lindbjerg Andersen, C. (2019). Analysis of plasma cell-free DNA by ultradeep sequencing in patients with stages I to III colorectal cancer. *JAMA Oncology*, 5(8), 1124−1131. Available from https://doi.org/10.1001/jamaoncol.2019.0528.

Reiter, J. G., Hung, W. T., Lee, I. H., Nagpal, S., Giunta, P., Degner, S., Liu, G., Wassenaar, E. C. E., Jeck, W. R., Taylor, M. S., Farahani, A. A., Marble, H. D., Knott, S., Kranenburg, O., Lennerz, J. K., & Naxerova, K. (2020). Lymph node metastases develop through a wider evolutionary bottleneck than distant metastases. *Nature Genetics*, 52(7), 692−700. Available from https://doi.org/10.1038/s41588-020-0633-2.

Rosemurgy, A. S., Block, G. E., & Shihab, F. (1988). Surgical treatment of carcinoma of the abdominal colon. *Surgery Gynecology and Obstetrics*, 167(5), 399−406. Available from http://www.embase.com/search/results?subaction = viewrecord&from = export&id = L18279574.

Rouffet, F., Hay, J. M., Vacher, B., Fingerhut, A., Elhadad, A., Flamant, Y., Mathon, C., Gainant, A., Benhamida, F., Bernard, J. L., Breil, P., Chipponi, J., Cour, J. C., Cras, C., Dazza, F., Delalande, J. P., Descottes, B., Pouget, X., Desmaizieres, F., . . . Timmermans, M. (1994). Curative resection for left colonic-carcinoma—Hemicolectomy vs segmental colectomy—A prospective, controlled, multicenter trial. *Diseases of the Colon & Rectum*, 37(7), 651−659. Available from https://doi.org/10.1007/bf02054407.

Sehgal, R., & Coffey, J. C. (2014). Historical development of mesenteric anatomy provides a universally applicable anatomic paradigm for complete/total mesocolic excision. *Gastroenterol Report (Oxford)*, 2(4), 245–250. Available from https://doi.org/10.1093/gastro/gou046.

Sharma, R. (2020). An examination of colorectal cancer burden by socioeconomic status: Evidence from globocan 2018. *EPMA Journal*, 11(1), 95–117. Available from https://doi.org/10.1007/s13167-019-00185-y.

Shida, H., Ban, K., Matsumoto, M., Masuda, K., Imanari, T., Machida, T., & Yamamoto, T. (1992). Prognostic-significance of location of lymph-node metastases in colorectal-cancer. *Diseases of the Colon & Rectum*, 35(11), 1046–1050. Available from https://doi.org/10.1007/bf02252994.

Shin, J. W., Amar, A. H. Y., Kim, S. H., Kwak, J. M., Baek, S. J., Cho, J. S., & Kim, J. (2014). Complete mesocolic excision with D3 lymph node dissection in laparoscopic colectomy for stages II and III colon cancer: Long-term oncologic outcomes in 168 patients. *Techniques in Coloproctology*, 18(9), 795–803. Available from https://doi.org/10.1007/s10151-014-1134-z.

Shiozawa, M., Akaike, M., Yamada, R., Godai, T., Yamamoto, N., Saito, H., Sugimasa, Y., Takemiya, S., Rino, Y., & Imada, T. (2007). Clinicopathological features of skip metastasis in colorectal cancer. *Hepato-Gastroenterology*, 54(73), 81–84.

Si, M. B., Yan, P. J., Du, Z. Y., Li, L. Y., Tian, H. W., Jiang, W. J., Jing, W. T., Yang, J., Han, C. W., Shi, X. E., Yang, K. H., & Guo, T. K. (2019). Lymph node yield, survival benefit, and safety of high and low ligation of the inferior mesenteric artery in colorectal cancer surgery: A systematic review and meta-analysis. *International Journal of Colorectal Disease*, 34(6), 947–962. Available from https://doi.org/10.1007/s00384-019-03291-5.

Siani, L. M., & Pulica, C. (2014). Stage I-IIIC right colonic cancer treated with complete mesocolic excision and central vascular ligation: Quality of surgical specimen and long term oncologic outcome according to the plane of surgery. *Minerva Chirurgica*, 69(4), 199–208.

Sloothaak, D. A., Sahami, S., van der Zaag-Loonen, H. J., van der Zaag, E. S., Tanis, P. J., Bemelman, W. A., & Buskens, C. J. (2014). The prognostic value of micrometastases and isolated tumour cells in histologically negative lymph nodes of patients with colorectal cancer: A systematic review and meta-analysis. *European Journal of Surgical Oncology: The Journal of the European Society of Surgical Oncology and the British Association of Surgical Oncology*, 40(3), 263–269. Available from https://doi.org/10.1016/j.ejso.2013.12.002.

Sondenaa, K., Quirke, P., Hohenberger, W., Sugihara, K., Kobayashi, H., Kessler, H., Brown, G., Tudyka, V., D'Hoore, A., Kennedy, R. H., West, N. P., Kim, S. H., Heald, R., Storli, K. E., Nesbakken, A., & Moran, B. (2014). The rationale behind complete mesocolic excision (CME) and a central vascular ligation for colon cancer in open and laparoscopic surgery: Proceedings of a consensus conference. *International Journal of Colorectal Disease*, 29(4), 419–428. Available from https://doi.org/10.1007/s00384-013-1818-2.

Sun, B., Zhou, Y. M., Fang, Y. T., Li, Z. Y., Gu, X. D., & Xiang, J. B. (2019). Colorectal cancer exosomes induce lymphatic network remodeling in lymph nodes. *International Journal of Cancer*, 145(6), 1648–1659. Available from https://doi.org/10.1002/ijc.32196.

Swanson, R. S., Compton, C. C., Stewart, A. K., & Bland, K. I. (2003). The prognosis of T3N0 colon cancer is dependent on the number of lymph nodes examined. *Annals of Surgical Oncology*, 10(1), 65–71. Available from https://doi.org/10.1245/aso.2003.03.058.

Tagliacozzo, S., & Tocchi, A. (1997). Extended mesenteric excision in right hemicolectomy for carcinoma of the colon. *International Journal of Colorectal Disease*, 12(5), 272–275. Available from https://doi.org/10.1007/s003840050104.

Tagliacozzo, S., & Tocchi, A. (1997). Extended mesenteric excision in right hemicolectomy for carcinoma of the colon. *International Journal of Colorectal Disease*, 12(5), 272–275.

Tan, K. Y., Kawamura, Y. J., Mizokami, K., Sasaki, J., Tsujinaka, S., Maeda, T., Nobuki, M., & Konishi, F. (2010). Distribution of the first metastatic lymph node in colon cancer and its clinical significance. *Colorectal Disease: The Official Journal of the Association of Coloproctology of Great Britain and Ireland*, 12(1), 44–47. Available from https://doi.org/10.1111/j.1463-1318.2009.01924.x.

Tan, Y., & Wu, H. (2018). The significant prognostic value of circulating tumor cells in colorectal cancer: A systematic review and meta-analysis. *Current Problems in Cancer*, 42(1), 95–106. Available from https://doi.org/10.1016/j.currproblcancer.2017.11.002.

Tentes, A.-A. K., Mirelis, C., Karanikiotis, C., & Korakianitis, O. (2007). Radical lymph node resection of the retroperitoneal area for left-sided colon cancer. *Langenbecks Archives of Surgery*, 392(2), 155–160. Available from https://doi.org/10.1007/s00423-006-0143-4.

Torino, F., Bonmassar, E., Bonmassar, L., Vecchis, L. D., Barnabei, A., Zuppi, C., Capoluongo, E., & Aquino, A. (2013). Circulating tumor cells in colorectal cancer patients. *Cancer Treatment Reviews, 39*(7), 759–772. Available from https://doi.org/10.1016/j.ctrv.2012.12.007, [Epub ahead of print].

Ulintz, P. J., Greenson, J. K., Wu, R., Fearon, E. R., & Hardiman, K. M. (2018). Lymph node metastases in colon cancer are polyclonal. *Clinical Cancer Research: An Official Journal of the American Association for Cancer Research, 24*(9), 2214–2224. Available from https://doi.org/10.1158/1078-0432.ccr-17-1425.

Ushijima, H., Kawamura, J., Ueda, K., Yane, Y., Yoshioka, Y., Daito, K., Tokoro, T., Hida, J. I., & Okuno, K. (2020). Visualization of lymphatic flow in laparoscopic colon cancer surgery using indocyanine green fluorescence imaging. *Scientific Report, 10*(1), 14274. Available from https://doi.org/10.1038/s41598-020-71215-3.

Vather, R., Sammour, T., Kahokehr, A., Connolly, A. B., & Hill, A. G. (2009). Lymph node evaluation and long-term survival in stage II and stage III colon cancer: A national study. *Annals of Surgical Oncology, 16*(3), 585–593. Available from https://doi.org/10.1245/s10434-008-0265-8.

Wang, X., Zheng, Z., Chen, M., Lu, X., Huang, S., Huang, Y., & Chi, P. (2020). Subtotal colectomy, extended right hemicolectomy, left hemicolectomy, or splenic flexure colectomy for splenic flexure tumors: A network meta-analysis. *International Journal of Colorectal Disease, 36*(2), 311–322. Available from https://doi.org/10.1007/s00384-020-03763-z.

Wei, Q., Ye, Z., Zhong, X., Li, L., Wang, C., Myers, R. E., Palazzo, J. P., Fortuna, D., Yan, A., Waldman, S. A., Chen, X., Posey, J. A., Basu-Mallick, A., Jiang, B. H., Hou, L., Shu, J., Sun, Y., Xing, J., Li, B., & Yang, H. (2017). Multiregion whole-exome sequencing of matched primary and metastatic tumors revealed genomic heterogeneity and suggested polyclonal seeding in colorectal cancer metastasis. *Annals of Oncology: Official Journal of the European Society for Medical Oncology/ESMO, 28*(9), 2135–2141. Available from https://doi.org/10.1093/annonc/mdx278.

Willaert, W., Cosyns, S., & Ceelen, W. (2018). Biology-based surgery: The extent of lymphadenectomy in cancer of the colon. *European Surgical Research. Europaische Chirurgische Forschung. Recherches Chirurgicales Europeennes, 59*(5–6), 371–379. Available from https://doi.org/10.1159/000494831.

Willett, C. G., Tepper, J. E., Cohen, A. M., Orlow, E., & Welch, C. E. (1984). Failure patterns following curative resection of colonic carcinoma. *Annals of Surgery, 200*(6), 685–690. Available from https://doi.org/10.1097/00000658-198412000-00001.

Yamamoto, Y., Takahashi, K., Yasuno, M., Sakoma, T., & Mori, T. (1998). Clinicopathological characteristics of skipping lymph node metastases in patients with colorectal cancer. *Japanese Journal of Clinical Oncology, 28*(6), 378–382. Available from https://doi.org/10.1093/jjco/28.6.378.

Yang, Y., Wang, G., He, J., Zhang, J., Xi, J., & Wang, F. (2018). High tie versus low tie of the inferior mesenteric artery in colorectal cancer: A meta-analysis. *International Journal of Surgery (London, England), 52*, 20–24. Available from https://doi.org/10.1016/j.ijsu.2017.12.030.

Yun, H. R., Lee, L. J., Park, J. H., Cho, Y. K., Cho, Y. B., Lee, W. Y., Kim, H. C., Chun, H. K., & Yun, S. H. (2008). Local recurrence after curative resection in patients with colon and rectal cancers. *International Journal of Colorectal Disease, 23*(11), 1081–1087. Available from https://doi.org/10.1007/s00384-008-0530-0.

Zeng, J., & Su, G. (2018). High ligation of the inferior mesenteric artery during sigmoid colon and rectal cancer surgery increases the risk of anastomotic leakage: A meta-analysis. *World Journal of Surgical Oncology, 16*(1), 157. Available from https://doi.org/10.1186/s12957-018-1458-7.

Zhang, C. (2020). Mapping the spreading routes of lymphatic metastases in human colorectal cancer. *Nature Communications, 11*(1993), 1–11. Available from https://doi.org/10.1038/s41467-020-15886-6.

Zhao, Z. M., Zhao, B., Bai, Y., Iamarino, A., Gaffney, S. G., Schlessinger, J., Lifton, R. P., Rimm, D. L., & Townsend, J. P. (2016). Early and multiple origins of metastatic lineages within primary tumors. *Proceedings of the National Academy of Sciences of the United States of America, 113*(8), 2140–2145. Available from https://doi.org/10.1073/pnas.1525677113.

Treatment

10

Neoadjuvant treatment and lymph node metastasis in rectal cancer

Jesse P. Wright, Alexandra Elias and John R.T. Monson

Center for Colon and Rectal Surgery, Digestive Health and Surgery Institute, AdventHealth, Orlando, FL, United states

10.1 Introduction

With over 44,000 new diagnoses of rectal cancer annually, colorectal cancer was the third most common cancer diagnosis in 2019, as well as the third most common cause of cancer-related death in the United States (American Cancer Society, 2019). Surgical resection of rectal cancer remains the mainstay of treatment with or without neoadjuvant and/or adjuvant therapy, depending on clinical or pathologic stage. Lymph node assessment for metastasis is critical in determining oncologic prognosis and treatment, including therapy timing and options. Lymph node involvement is the most important prognostic factor and varies by the number and site of involved nodes (Compton et al., 1999). Clinical and pathologic staging algorithms have placed increasingly more importance on lymph node evaluation and yield during subsequent guideline editions. Presently, the College of American Pathologists and National Comprehensive Cancer Network (NCCN) recommend using the American Joint Committee on Cancer (AJCC) tumor, lymph node, metastasis (TNM) staging system (now in its eighth iteration) for the staging of colorectal cancers (AJCC, 2017; Burgart et al., 2020; NCCN, n.d.).

While varying numbers of harvested lymph nodes have been reported as a recommended minimum, it has been shown that 12–15 negative lymph nodes predict for regional node negativity (Compton et al., 1999). A retrospective study of 538 patients from 1980 to 1989 with stage III colorectal cancer suggested that at least 10 lymph nodes need to be examined to avoid understaging (Tang et al., 1995). Another 1989 study of 103 colorectal specimens demonstrated that if at least 13 lymph nodes were examined, more than 90% of specimens with metastases would be identified, thus suggesting 13 as the "magic number" (Scott, Grace & Scott, 1989). Multivariate survival analysis in a 2019 study demonstrated that at least 12 lymph nodes examined were an independent prognostic feature of good overall survival (OS), disease-free survival (DFS), and distant metastasis-free survival, although not local recurrence-free survival (Wang et al., 2019). Supported by these and countless other trials, the AJCC currently recommends retrieval of 12 lymph nodes following resection for accurate pathologic staging (AJCC, 2017).

10.2 Importance of lymph node yield

As noted, lymph node metastasis is one of the most powerful prognostic factors for recurrence and OS in colon and rectal cancer. Understanding the details of these relationships and how to both efficiently and accurately quantify surgical lymph node yield, especially in the era of neoadjuvant therapy, has prompted changes in practice and treatment paradigms.

In the era before routine neoadjuvant therapy for locally advanced rectal cancer, it was seen that higher lymph node yields resulted in improved survival. Heald (1988) pioneered sharp dissection of the "holy plane," citing the importance of proper total mesorectal excision (TME) to completely remove the rectum, together with the surrounding mesorectum lymphovascular fatty tissue. In 2001 Tepper et al. were one of the first groups to investigate outcomes by both tumor and nodal stages in rectal cancer. They assessed 527 pathologically node-negative patients (T3 or T4, N0) treated with upfront TME followed by adjuvant chemoradiotherapy (CRT) as part of a previously reported Intergroup 0114 trial cohort (Tepper et al., 1997, 2001). The authors reported a significant improvement in OS ($P = .003$), with a mean follow-up time of 7.5 years, as well as improvement in time to disease recurrence ($P = .02$) with the increasing number (by quartiles) of harvested lymph nodes. Cut-off for improvement was found to be 14 nodes (Tepper et al., 2001).

Kim et al. aimed to investigate the prognostic effect of retrieved lymph nodes in patients with locally advanced rectal cancer (clinical stages II and III) who underwent upfront TME followed by adjuvant CRT. The authors reviewed 900 patients between 1989 and 2006 and confirmed that an absolute higher lymph node yield alone, regardless of tumor metastasis, resulted in improved cancer-specific and recurrence-free survival, particularly in stage II patients (Kim et al., 2009). This reinforced the concept of optimal surgical resection and pathologic evaluation of rectal cancer specimens.

There remained, however, significant heterogeneity and lack of consensus in recommendations for optimal lymph node yields. Thus, in one of the largest studies of lymph node analyses to date, Chou et al. interrogated the surveillance, epidemiology, and end results (SEER) database (sponsored by the National Cancer Institute) on patients who underwent resection for stages I–III colorectal cancer. They identified over 153,000 patients between 1994 and 2005 who had undergone total mesocolorectal excision. Interestingly, the authors found that the mean number of lymph nodes harvested for the entire population was 12 (± 9.3), which indicates that approximately half (49%) of patients had less than 12 nodes harvested. Of the approximately 25,500 rectal cancer specimens, the mean number of lymph nodes harvested was 10.2 (± 8.8), and the median number was 8. The mean number of lymph nodes increased between stage II and stage III disease from 10.2 to 12.9, respectively. There was a significant decrease in nodal harvest with every 10-year incremental increase in age. Additionally, and most relevant to future studies, this analysis showed that 25% of rectal cancer patients received neoadjuvant (chemo) radiation with a significant decrease in lymph node yield from 10.1 to 8.3 ($P < .001$) (Chou et al., 2010).

10.3 Lymph node yield

As the importance of lymph node yield to the improvement of long-term outcomes became more evident, surgeons, pathologists, and other invested parties sought to investigate how to maximize effective lymph node yield. There are several interconnected clinical, pathologic, and patient demographic variables that seem to affect lymph node harvest.

The anatomic distribution of mesorectal lymph nodes within a TME specimen is known to be unevenly spaced (Canessa et al., 2001; Langman et al., 2015; Mekenkamp et al., 2009). In a prospective study of 244 TME specimens reviewed by one histopathologist over a 6-year period, 40% of lymph nodes were found in the mesorectum (vs sigmoid mesentery or vascular pedicle), and 68% were above the peritoneal reflection, with a progressive decrease in lymph node number and density as one moves distally. 89% of identified nodes were found adjacent to or above the primary tumor, and only 7% of nodes were found anteriorly (Canessa et al., 2001; Langman et al., 2015). This may explain the Dutch trial data, which revealed a higher yield of lymph nodes in low anterior resection specimens compared to abdominoperineal resection specimens (Kapiteijn et al., 2001).

While the American College of Surgeons Commission on Cancer has declared a near-complete or complete TME of rectal cancer, a specific quality metric for surgical technique, the pathologic assessment of the specimen, and enclosed lymph nodes within the mesorectum are equally as imperative (Commission on Cancer, n.d.). The numbers of lymph nodes identified have been shown to vary widely, based mainly on pathologic method used in specimen examination (Compton et al., 1999). The College of American Pathologists publishes protocols (now version 4.1.0.0, 2/2020) for the evaluation of resected colon and rectal specimen and poignantly notes, "surgical technique, surgery volume, and patient factors (e.g., age and anatomic variation) alter the actual number of nodes in a resection specimen, but the diligence and skill of the pathologist in identifying and harvesting lymph nodes in the resection specimen also are major factors" (AJCC, 2017; Burgart et al., 2020; NCCN, n.d.). Importantly, the evolution of pathologic protocols has paralleled our continuing understanding of the importance of lymph nodes, especially following neoadjuvant therapy.

In a retrospective study of 985 patients with colorectal cancer (361 of whom had rectal cancer) between 1981 and 1996, Ratto et al. (1999) demonstrated that changes in pathologic assessment protocols of mesocolorectal lymph node examination resulted in significant increases in numbers of retrieved lymph nodes when controlling for surgical technique. Subsequently, the identification of more nodes resulted in a significant increase in upstaging. The authors note, "simply preparing the specimen more accurately and patiently viewing and palpating the excised tissues allows an adequate pathologic examination in most patients." When more diligent methods were trialed, the authors report at least 13 nodes were detectable in approximately 90% of cases (Compton et al., 1999). Wang et al. (2019) confirmed that utilizing a specific fat clearance method in a cohort of 237 patients undergoing TME, achieved by submerging mesorectal adipose tissue into a clearing solution for 24 hours, resulted in a higher lymph node yield compared to their traditional (palpation and visualization) method, bringing their average yield from 6.9 to 22.6 nodes ($P < .001$).

However, even when the examination process was standardized, Mekenkamp et al. found there were significant variations in lymph node retrieval between the 49 pathology laboratories and individual pathologists they analyzed. Their analysis confirmed that regardless of whether patients received radiotherapy (RT), fewer collected lymph nodes in node-negative patients resulted in decreased interval-free survival, while node-positive disease did not demonstrate any improvements based on number of nodes collected (Mekenkamp et al., 2009).

The current NCCN guidelines for rectal cancer report that if less than 12 lymph nodes are initially identified at the time of preparation, the pathologist is recommended to resubmit more tissue, and if there are still inadequate nodes, a comment should be made in the final report that an extensive search for adequate lymph nodes was undertaken (NCCN, n.d.).

Of note, increased awareness of the importance of diligent lymph node identification during pathologic assessment seems to be leading to improved lymph node retrieval over time. Chou et al. noted a 2.5% annual increase in lymph node retrieval in rectal cancer over in their SEER analysis from 1994 to 2005. They concluded that this is secondary to multiple factors, including increased awareness of the importance of accurate lymph node identification and increased colorectal cancer screening leading to younger age of patients at diagnosis, which was shown to predict higher lymph node yield (Chou et al., 2010).

10.4 Neoadjuvant therapy and lymph node yield

Following the outcomes of several landmark studies from the turn of the century, neoadjuvant therapy, specifically neoadjuvant CRT, has become the mainstay treatment for the management of locally advanced rectal cancer (clinical stages II and III), followed by TME and adjuvant chemotherapy (Kapiteijn et al., 2001; NCCN, n.d.; Påhlman, 1997; Rolf et al., 2004). Neoadjuvant therapy followed by resection has been shown to improve local recurrence rates when compared to adjuvant radiation (or CRT), but it does not improve OS. Globally, neoadjuvant therapy allows for assessment of a tumor's general response to therapy, facilitates potential "downstaging" of disease, and may even permit organ preservation.

Understanding the treatment effects of CRT on nodal tissue is essential in interpreting the pathologic sequelae of neoadjuvant therapy. Not only can the size and dimensions of the primary tumor be affected, but there can also be an equally profound impact on the lymph nodes containing disease (Mekenkamp et al., 2009). Histologically, CRT results in tumor regression within lymph nodes and subsequent lymphocyte depletion, stromal atrophy, and replacement with increasing fibrosis and adipocytes (Lindebjerg et al., 2011; MacGregor et al., 2012). Although desirable as a clinical response, these morphologic and structural changes can make lymph nodes difficult to identify or isolate during pathologic assessment and have resulted in new challenges for pathologists, oncologists, and surgeons when establishing treatment plans based on final pathologic findings.

There is a growing body of evidence that neoadjuvant therapy not only results in the aforementioned structural changes in harvested nodes but a concomitant decrease in the total number of nodes harvested at the time of resection (Amajoyi et al., 2013; Miller et al., 2012; Mechera et al., 2017; Rullier et al., 2008). In fact, due to the significant decrease in the number of lymph nodes harvested after neoadjuvant therapy, there are questions if 12 harvested lymph nodes should be the required yield for proper staging.

In data analysis from patients included within the Dutch rectal cancer trial, a total of 1227 patients who had received neoadjuvant RT (short course 5×5 Gy) were evaluated to assess factors that affect lymph node yield. They found that the number of retrieved lymph nodes was based on multiple factors. Importantly, they discovered that the number of lymph nodes in patients receiving neoadjuvant RT was decreased compared to surgery alone (8.5 vs 6.9, $P < .0001$). They also noted that retrieved lymph nodes were smaller, especially positive nodes, in the neoadjuvant RT cohort when compared to the surgical cohort (Kapiteijn et al., 2001).

In one of the early investigations into this phenomenon, Amajoyi et al. (2013) reported a 30% decrease in the mean lymph node yield (13 to 9, $P = .001$) in patients who received neoadjuvant therapy over a 20-year period. Similarly, Rullier et al. (2008) reported a 24% decrease in the mean number of lymph nodes harvested (17 to 13, $P < .001$). Other studies demonstrated similar results, which decreases from 19 to 13 ($P < .05$) and 10 to 7 ($P < .001$) (Baxter et al., 2005; Wichmann et al., 2002). In 2012, Johnstone et al. expanded upon these findings by performing the first systematic review of seven trials assessing lymph node yield following neoadjuvant CRT in locally advanced rectal. The authors report that six of the seven trials showed a significant, 7%–53% (mean 28%), decrease in lymph node yield in the neoadjuvant treatment arm, and only 64% of patients who received neoadjuvant therapy had adequate (12 or more) lymph nodes harvested. The authors express concern with the current guideline of 12 lymph nodes in patients receiving neoadjuvant therapy (Miller et al., 2012).

In 2017 Mechera et al. performed a formal metaanalysis of 34 studies assessing lymph node yield following neoadjuvant therapy (radiation or CRT) compared to those who did not receive any neoadjuvant therapy (control) from 1980 to 2015. Their analysis confirmed the use of neoadjuvant CRT ($n = 5784$) led to a mean reduction of 3.9 total lymph nodes harvested compared to the control arm ($n = 10,406$), and neoadjuvant radiation ($n = 1628$) led to mean reduction of 2.1 lymph nodes. There was an average decrease in positive lymph nodes in the CRT cohort by 0.7 compared to the control arm. The authors report confidently, "We present now clear evidence that neoadjuvant CRT and RT decrease the number of harvested lymph nodes." Interestingly, 9 of the studies included in the trial failed to demonstrate a significant association with identifying at least 12 lymph nodes on OS or DFS, thus continuing to call into question the importance of these metrics following neoadjuvant therapy (Miller et al., 2012).

To further investigate prognostic oncologic impact of lymph node yield following neoadjuvant CRT in rectal cancer, Xu et al. queried the National Cancer Database (NCDB) over 5-year period from 2006 to 2011. Over 25,500 patients were identified who underwent proctectomy for stages I–III rectal cancer, 62% of whom received neoadjuvant CRT.

On final pathology, 68% of the entire cohort had 12 or more lymph nodes within their mesorectal resection specimen. The median lymph node yield for the neoadjuvant therapy cohort was 13 (IQR 9−18), with 63% having 12 lymph nodes, while the surgery alone cohort yielded a median 15 (IQR 12−21) nodes, with 76% having 12. Of most interest, in those patients who received neoadjuvant therapy, the 5-year OS for a lymph node yield <12 was 77.2% compared to 81.2% ($P < .0001$) for a yield 12. In the propensity-adjusted analysis, patients who received neoadjuvant therapy with <12 lymph node yield were independently associated with a 20% increased hazard of death. The authors conclude that lymph node yield <12 was independently associated with worse 5-year OS. Multivariable analysis found that inadequate (<12) lymph node harvest was associated with decreased OS, independent of neoadjuvant therapy, margin status, clinical and pathological stage, patient and hospital factors, and adjuvant therapy (Xu et al., 2017).

With the general trend of decrease in nodal yield following therapy, what is the prognostic impact of finding only negative nodes on final pathologic evaluation (yN0)? Tsai et al. evaluated 372 patients with stages II and III rectal cancer who received neoadjuvant CRT and were found to have yN0 disease on final pathology. 66% of the cohort had clinically positive disease before therapy and were downstaged following treatment on final pathologic assessment, and the median number of lymph nodes harvested was 7, with only 24% of the cohort having a harvest of more than 12 nodes. Disease relapse, cancer-specific survival, and OS were improved with increased nodal yield, even when all nodes were negative (Tsai et al., 2011).

It is increasingly clear that the introduction of neoadjuvant therapy into the treatment algorithm for locally advanced rectal cancer has not only changed the oncologic outcomes for stage-specific disease but has also raised more questions as to how to continue to interpret pathologic findings.

10.5 Lymph node ratio

Despite the demonstrated decrease in total lymph node harvest following neoadjuvant therapy, there has been a growing interest in alternative prognostic metric as it relates to lymph node tumor involvement and total yield. Berger et al. were one of the first groups to investigate the role of assessing a positive lymph node ratio—the ratio of positive lymph nodes to total harvested lymph nodes—in colon cancer. Although the concept had been popularized in gastric cancer, the authors identified a clear negative correlation with a higher lymph node ratio and OS in their population of nonmetastatic colon cancer patients from the INT-0089 study dataset (Bando et al., 2002; Berger et al., 2005). Interestingly, the authors found that the 5-year OS for a patient with N1 disease with a lymph node ratio >40% was worse than patients with N2 disease with a lymph node ratio <20% (60% vs 73%, respectively), inferring the ratio of positive nodes is more prognostic than the absolute number of harvested positive nodes. The authors caution, however, that a ratio-based system of analysis should not, as of yet, become a substitute for an adequate oncologic lymph node dissection (Berger et al., 2005).

Rosenberg et al. confirmed Berger's findings in their analysis of 3026 patients with colon and rectal cancer (1763 colon and 1263 rectal) over a 25-year period. When specifically assessing the rectal cancer patients, 45% had locally advanced disease and received neoadjuvant CRT, and the mean number of lymph nodes resected was 16.6. Although there was not a subgroup analysis comparing those who received neoadjuvant therapy to those who did not, the rectal cancer cohort, as a whole, demonstrated significantly improved 5-year OS with lower lymph node ratio (grouped in quartiles) compared to those with a higher ratio (Rosenberg et al., 2008).

In 2016 Chang et al. first assessed the utility of the lymph node ratio specifically in rectal cancer following neoadjuvant therapy compared to surgery alone. The neoadjuvant cohort had significantly lower lymph node retrieval compared to the surgery alone cohort (6.68 vs 11.54, $P < .01$), consistent with previous, aforementioned findings (Chang et al., 2016). Interestingly, there was no significance difference in lymph node ratio between the two cohorts (0.122 vs 0.161, $P = .361$), even when using the lymph node ratio quartiles established by Rosenberg et al. (2008). The authors concluded that there are no predicted improvements in DFS or OS based on lymph node ratios (Chang et al., 2016).

Xu et al. also found no significant difference in positive lymph node ratio between those who received neoadjuvant therapy and those who did not (0.812 and 0.813, $P = .97$) in their evaluation of 25,500 rectal cancer patients from an NCDB analysis. The authors report that in patients who received neoadjuvant treatment, the lymph node ratio was independently associated with an 81% increase in hazard of death ($P < .0001$), while in those that did not receive neoadjuvant treatment, the ratio was associated with a 154% increase in hazard of death ($P < .0001$) (Xu et al., 2017).

10.6 Impact of nodal involvement with complete clinical response after neoadjuvant therapy

Neoadjuvant therapy has led to significant changes in patient care plans over the years, as organ and sphincter preservation become options, and even complete clinical responses with watch-and-wait protocols become a reality (Habr-Gama et al., 2013). However, while one of the goals of neoadjuvant therapy is to downstage tumors, thus allowing organ preservation, this prevents definitive nodal staging. In the setting of complete clinical tumor response, residual nodal involvement has been reported to be between 6% and 17% (Erkan et al., 2019; Hughes et al., 2006; Tulchinsky et al., 2006). While tumor response can be assessed digitally and by endoscopy with or without local excision, lymph node status cannot be clinically reliably determined. Advocates for radical surgery despite complete clinical response cite the inaccuracy of nodal restaging with current locoregional imaging modalities and the potential negative effects of residual nodal disease (Memon et al., 2015). Residual positive mesorectal lymph nodes can lead to local recurrence, distant metastases, and ultimately worse oncologic outcomes, particularly if patients do not get systemic chemotherapy secondary to misstaging (Newton et al., 2016). Erkan et al. (2019), in a review of NCDB data from 2004 to 2014 for 5156 patients with ypT0N − and 527 with ypT0N disease, found that residual nodal involvement despite complete tumor regression was associated with worse 5-year OS compared to complete pathologic response. Thus the authors suggested considering additional chemotherapy, perhaps with a different regimen, rather than waiting alone for a watch-and-wait approach.

10.7 Effects of total neoadjuvant therapy

With the growing body of evidence evaluating nodal response to neoadjuvant CRT, there are now new questions and variables to consider as total neoadjuvant therapy (TNT), the administration of the full extent of CRT and chemotherapy preoperatively, is becoming more mainstream in the treatment of locally advanced rectal cancer. The RAPIDO randomized control trial was the first landmark study to assess outcomes following TNT compared to the traditional neoadjuvant CRT followed by TME and then adjuvant chemotherapy pathway. Although full results are to be published, the rate of pathological complete response was 27.7% compared to 13.8% ($P < .001$) in the experimental (TNT) and standard arms, respectively (Nilsson et al., 2013). The PRODIGE 23 trial, the second landmark study comparing TNT to traditional therapy, also demonstrated increased pathological complete response rates of 27.7% versus 13.8% ($P < .001$) between TNT and control arms, respectively (Conroy et al., 2019). Although there was no mention of nodal morphologic changes or comparison of nodal status between cohorts, with an increase in complete response rates, there is clearly a more robust histologic/pathologic response within nodal tissues that will have to be further investigated in the future. However, as reported earlier in the Tsai study, yN0 disease has been found to have improved outcomes with increased nodal harvest (Tsai et al., 2011).

Although no data has been specifically reported on the effects of neoadjuvant chemotherapy alone on rectal cancer lymph node yield, the preliminary data from the FOXTROT trial assessed oncologic outcomes in a feasibility study of neoadjuvant and adjuvant chemotherapy versus adjuvant chemotherapy alone. The authors found, histologically, that preoperative chemotherapy resulted in significant TNM downstaging compared with the control cohort ($P = .04$) and significant decreases in apical node (node closest to high inferior mesenteric artery [IMA] ligation) involvement (1% vs 20% $P < .0001$), although there were no differences in median lymph nodes harvested (21 vs 22) between the two cohorts (Agbamu et al., 2012).

Mullen et al. investigated the role of neoadjuvant chemotherapy alone on lymph node yield in colon cancer after querying the NCDB. The authors identified 9077 patients who received a colectomy for colon cancer in 2014, of whom 963 received neoadjuvant chemotherapy. There was a significantly higher percentage of patients with inadequate nodal yield (<12) than adequate yield (≥12), and neoadjuvant chemotherapy was found on multivariable analysis to be a negative predictor of adequate yield (OR 0.44, $P < .0001$) (Mullen et al., 2018).

10.8 Conclusion

In a study examining T3 colorectal specimens over 45 years, Goldberg concludes, "The predictive probability increased as the number of recovered lymph nodes increased, suggesting there is no minimum number that reliably or accurately stages all patients. Thus all palpable lymph nodes should be recovered, including those that are 1 or 2 mm" (Goldstein, 2002). But Wichmann et al. (2002) demonstrated that if 12 lymph nodes are considered, the number needed to accurately stage II tumors, following neoadjuvant therapy, only 20% of patients had adequate lymph node sampling. This has helped lead to 2020 NCCN rectal cancer guidelines which state, "To date, the number of lymph nodes needed to accurately stage neoadjuvant-treated cases is unknown. However, it is not known what is the clinical significance of this in the neoadjuvant setting, as postoperative therapy is indicated in all patients who receive preoperative therapy regardless of the surgical pathology results." Clearly, as our treatment paradigms evolve, lymph nodes will continue to play a pivotal role, and much is yet to be understood, particularly in the setting of neoadjuvant therapy.

References

Agbamu, D. A., Day, N., Walsh, C. J., Hendrickse, C. W., Langman, G., Pallan, A., Lowe, A., Ostrowski, J., Steward, M., Callaway, M., Falk, S., Thomas, M. G., Wong, N., Hartley, J., MacDonald, A. W., Blunt, D., Cohen, P., Dawson, P., Lowdell, C. P., & Lees, N. (2012). Feasibility of preoperative chemotherapy for locally advanced, operable colon cancer: The pilot phase of a randomised controlled trial. *The Lancet Oncology, 13*(11), 1152−1160. Available from https://doi.org/10.1016/S1470-2045(12)70348-0.

AJCC. (2017). *AJCC cancer staging manual. Cancer staging man*. American Joint Committee on Cancer.

Amajoyi, R., Lee, Y., Recio, P. J., & Kondylis, P. D. (2013). Neoadjuvant therapy for rectal cancer decreases the number of lymph nodes harvested in operative specimens. *American Journal of Surgery, 205*(3), 289−292. Available from https://doi.org/10.1016/j.amjsurg.2012.10.020.

American Cancer Society. (2019). *Cancer facts & figures*. American Cancer Society.

Bando, E., Yonemura, Y., Taniguchi, K., Fushida, S., Fujimura, T., & Miwa, K. (2002). Outcome of ratio of lymph node metastasis in gastric carcinoma. *Annals of Surgical Oncology, 9*(8), 775−784. Available from https://doi.org/10.1245/ASO.2002.10.011.

Baxter, N. N., Morris, A. M., Rothenberger, D. A., & Tepper, J. E. (2005). Impact of preoperative radiation for rectal cancer on subsequent lymph node evaluation: A population-based analysis. *International Journal of Radiation Oncology Biology Physics, 61*(2), 426−431. Available from https://doi.org/10.1016/j.ijrobp.2004.06.259.

Berger, A. C., Sigurdson, E. R., LeVoyer, T., Hanlon, A., Mayer, R. J., Macdonald, J. S., Catalano, P. J., & Haller, D. G. (2005). Colon cancer survival is associated with decreasing ratio of metastatic to examined lymph nodes. *Journal of Clinical Oncology, 23*(34), 8706−8712. Available from https://doi.org/10.1200/JCO.2005.02.8852.

Bugard, L.J., Kakar, S., Shi, C., Berho, M.E., Driman, D.K., Fitzgibbons, P., Frankel, W.L., Hill, K.A., Jessup, J., Krasinskas, A.M., Washington, M.K. (2020). College of American pathologists guidelines. *Protocol for the Examination of Resection Specimens from Patients with Primary Carcinoma of the Colon and Rectum*, 4.1.0.0. Availbale at https://documents.cap.org/protocols/cp-gilower-colonrectum-resection-20-4100.pdf.

Canessa, C. E., Badía, F., Fierro, S., Fiol, V., & Háyek, G. (2001). Anatomic study of the lymph nodes of the mesorectum. *Diseases of the Colon and Rectum, 44*(9), 1333−1336. Available from https://doi.org/10.1007/BF02234794.

Chang, K. H., Kelly, N. P., Duff, G. P., Condon, E. T., Waldron, D., & Coffey, J. C. (2016). Neoadjuvant therapy does not affect lymph node ratio in rectal cancer. *Surgeon, 14*(5), 270−273. Available from https://doi.org/10.1016/j.surge.2015.06.002.

Chou, J. F., Row, D., Gonen, M., Liu, Y. H., Schrag, D., & Weiser, M. R. (2010). Clinical and pathologic factors that predict lymph node yield from surgical specimens in colorectal cancer: A population-based study. *Cancer, 116*(11), 2560−2570. Available from https://doi.org/10.1002/cncr.25032.

Commission on Cancer. (n.d.). *Optimal resources for cancer care 2020 standards*. American College of Surgeons.

Compton, C., Fielding, & Burgart, L. (1999). *Prognostic factors in colorectal cancer*. College of American Pathologists Consensus Statement.

Conroy, T., Lamfichekh, N., Etienne, P., Rio, E., Francois, E., Mesgouez-Nebout, N., Vendrely, V., Artignan, X., Bouché, O., Gargot, D., Boige, V., Bonichon-Lamichhane, N., Louvet, C., Morand, C., De La Fouchardiere, C., Juzyna, B., Rullier, E., Marchal, F., Castan, F., & Borg, C. (2019). Total neoadjuvant therapy with mFOLFIRINOX versus preoperative chemoradiation in patients with locally advanced rectal cancer: Final results of PRODIGE 23 phase III trial, a UNICANCER GI trial. *Journal of Clinical Oncology, 38*.

Erkan, A., Mendez, A., Trepanier, M., Kelly, J., Nassif, G., Albert, M. R., Lee, L., & Monson, J. R. T. (2019). Impact of residual nodal involvement after complete tumor response in patients undergoing neoadjuvant (chemo) radiotherapy for rectal cancer. *Surgery (United States)*, *166*(4), 648–654. Available from https://doi.org/10.1016/j.surg.2019.03.026.

Goldstein, N. S. (2002). Lymph node recoveries from 2427 pT3 colorectal resection specimens spanning 45 years: Recommendations for a minimum number of recovered lymph nodes based on predictive probabilities. *American Journal of Surgical Pathology*, *26*(2), 179–189. Available from https://doi.org/10.1097/00000478-200202000-00004.

Habr-Gama, A., Sabbaga, J., Gama-Rodrigues, J., Julião, G. P. S., Proscurshim, I., Aguilar, P. B., Nadalin, W., & Perez, R. O. (2013). Watch and wait approach following extended neoadjuvant chemoradiation for distal rectal cancer: Are we getting closer to anal cancer management? *Diseases of the Colon and Rectum*, *56*(10), 1109–1117. Available from https://doi.org/10.1097/DCR.0b013e3182a25c4e.

Heald, R. J. (1988). The "Holy Plane" of rectal surgery. *Journal of the Royal Society of Medicine*, *81*(9), 503–508. Available from https://doi.org/10.1177/014107688808100904.

Hughes, R., Glynne-Jones, R., Grainger, J., Richman, P., Makris, A., Harrison, M., Ashford, R., Harrison, R. A., Livingstone, J. I., McDonald, P. J., Meyrick Thomas, J., Mitchell, I. C., Northover, J. M. A., Phillips, R., Wallace, M., Windsor, A., & Novell, J. R. (2006). Can pathological complete response in the primary tumour following pre-operative pelvic chemoradiotherapy for T3–T4 rectal cancer predict for sterilisation of pelvic lymph nodes, a low risk of local recurrence and the appropriateness of local excision? *International Journal of Colorectal Disease*, *21*(1), 11–17. Available from https://doi.org/10.1007/s00384-005-0749-y.

Kapiteijn, E., Marijnen, C. A. M., Nagtegaal, I. D., Putter, H., Steup, W. H., Wiggers, T., Rutten, H. J. T., Pahlman, L., Glimelius, B., Van Krieken, J. H. J. M., Leer, J. W. H., & Van De Velde, C. J. H. (2001). Preoperative radiotherapy combined with total mesorectal excision for resectable rectal cancer. *New England Journal of Medicine*, *345*(9), 638–646. Available from https://doi.org/10.1056/NEJMoa010580.

Kim, Y. W., Kim, N. K., Min, B. S., Lee, K. Y., Sohn, S. K., & Cho, C. H. (2009). The influence of the number of retrieved lymph nodes on staging and survival in patients with stage II and III rectal cancer undergoing tumor-specific mesorectal excision. *Annals of Surgery*, *249*(6), 965–972. Available from https://doi.org/10.1097/SLA.0b013e3181a6cc25.

Langman, G., Patel, A., & Bowley, D. M. (2015). Size and distribution of lymph nodes in rectal cancer resection specimens. *Diseases of the Colon and Rectum*, *58*(4), 406–414. Available from https://doi.org/10.1097/DCR.0000000000000321.

Lindebjerg, J., Hansborg, N., Ploen, J., Rafaelsen, S., Jorgensen, J. C. R., & Jakobsen, A. (2011). Factors influencing reproducibility of tumour regression grading after high-dose chemoradiation of locally advanced rectal cancer. *Histopathology*, *59*(1), 18–21. Available from https://doi.org/10.1111/j.1365-2559.2011.03888.x.

MacGregor, T. P., Maughan, T. S., & Sharma, R. A. (2012). Pathological grading of regression following neoadjuvant chemoradiation therapy: The clinical need is now. *Journal of Clinical Pathology*, *65*(10), 867–871. Available from https://doi.org/10.1136/jclinpath-2012-200958.

Mechera, R., Schuster, T., Rosenberg, R., & Speich, B. (2017). Lymph node yield after rectal resection in patients treated with neoadjuvant radiation for rectal cancer: A systematic review and meta-analysis. *European Journal of Cancer*, *72*, 84–94. Available from https://doi.org/10.1016/j.ejca.2016.10.031.

Mekenkamp, L. J. M., Van Krieken, J. H. J. M., Marijnen, C. A. M., Van De Velde, C. J. H., & Nagtegaal, I. D. (2009). Lymph node retrieval in rectal cancer is dependent on many factors-the role of the tumor, the patient, the surgeon, the radiotherapist, and the pathologist. *American Journal of Surgical Pathology*, *33*(10), 1547–1553. Available from https://doi.org/10.1097/PAS.0b013e3181b2e01f.

Memon, S., Lynch, A. C., Bressel, M., Wise, A. G., & Heriot, A. G. (2015). Systematic review and meta-analysis of the accuracy of MRI and endorectal ultrasound in the restaging and response assessment of rectal cancer following neoadjuvant therapy. *Colorectal Disease*, *17*(9), 748–761. Available from https://doi.org/10.1111/codi.12976.

Miller, E. D., Robb, B. W., Cummings, O. W., & Johnstone, P. A. S. (2012). The effects of preoperative chemoradiotherapy on lymph node sampling in rectal cancer. *Diseases of the Colon and Rectum*, *55*(9), 1002–1007. Available from https://doi.org/10.1097/DCR.0b013e3182536d70.

Mullen, M. G., Shah, P. M., Michaels, A. D., Hassinger, T. E., Turrentine, F. E., Hedrick, T. L., & Friel, C. M. (2018). Neoadjuvant chemotherapy is associated with lower lymph node counts in colon cancer. *American Surgeon*, *84*(6), 996–1002. Available from https://doi.org/10.1177/000313481808400655.

NCCN. (n.d.). *NCCN clinical practice guidelines in oncology rectal cancer version 6. 2020*. National Comprehensive Cancer Network.

Newton, A. D., Li, J., Jeganathan, A. N., Mahmoud, N. N., Epstein, A. J., & Paulson, E. C. (2016). A nomogram to predict lymph node positivity following neoadjuvant chemoradiation in locally advanced rectal cancer. *Diseases of the Colon and Rectum, 59*(8), 710–717. Available from https://doi.org/10.1097/DCR.0000000000000638.

Nilsson, P. J., van Etten, B., Hospers, G. A. P., Påhlman, L., van de Velde, C. J. H., Beets-Tan, R. G. H., Blomqvist, L., Beukema, J. C., Kapiteijn, E., Marijnen, C. A. M., Nagtegaal, I. D., Wiggers, T., & Glimelius, B. (2013). Short-course radiotherapy followed by neo-adjuvant chemotherapy in locally advanced rectal cancer—The RAPIDO trial. *BMC Cancer, 13*, 279. Available from https://doi.org/10.1186/1471-2407-13-279.

Påhlman, L. (1997). Improved survival with preoperative radiotherapy in resectable rectal cancer. *New England Journal of Medicine, 336*(14), 980–987. Available from https://doi.org/10.1056/NEJM199704033361402.

Ratto, C. M. D., Sofo, L. M. D., Ippoliti, M. M. D., Merico, M. M. D., Bossola, M. M. D., Vecchio, F. M. M. D., Doglietto, G. B. M. D., & Crucitti, F. M. D. (1999). Accurate lymph-node detection in colorectal specimens resected for cancer is of prognostic significance. *Diseases of the Colon & Rectum, 42*(2), 143–154. Available from https://doi.org/10.1007/bf02237119.

Rolf, S., Heinz, B., Werner, H., Claus, R., Christian, W., Rainer, F., Peter, M., Jörg, T., Eva, H., Hess, C. F., Johann-H., K., Torsten, L., Heinz, S., & Rudolf, R. (2004). Preoperative versus postoperative chemoradiotherapy for rectal cancer. *New England Journal of Medicine, 351*(17), 1731–1740. Available from https://doi.org/10.1056/nejmoa040694.

Rosenberg, R., Friederichs, J., Schuster, T., Gertler, R., Maak, M., Becker, K., Grebner, A., Ulm, K., Höer, H., Nekarda, H., & Siewert, J. R. (2008). Prognosis of patients with colorectal cancer is associated with lymph node ratio a single-center analysis of 3026 patients over a 25-year time period. *Annals of Surgery, 248*(6), 968–977. Available from https://doi.org/10.1097/SLA.0b013e318190eddc.

Rullier, A., Laurent, C., Capdepont, M., Vendrely, V., Belleannée, G., Bioulac-Sage, P., & Rullier, E. (2008). Lymph nodes after preoperative chemoradiotherapy for rectal carcinoma: Number, status, and impact on survival. *American Journal of Surgical Pathology, 32*(1), 45–50. Available from https://doi.org/10.1097/PAS.0b013e3180dc92ab.

Scott, K. W. M., Grace, R. H., & Scott, W. M. (1989). Detection of lymph node metastases in colorectal carcinoma before and after fat clearance. *British Journal of Surgery, 76*, 1165–1167.

Tang, R., Wang, J. Y., Chen, J. S., Chang-Chien, C. R., Tang, S., Lin, S. E., You, Y. T., Hsu, K. C., Ho, Y. S., & Fan, H. A. (1995). Survival impact of lymph node metastasis in TNM stage III carcinoma of the colon and rectum. *Journal of the American College of Surgeons, 180*(6), 705–712.

Tepper, J. E., O'Connell, M. J., Niedzwiecki, D., Hollis, D., Compton, C., Benson, A. B., Cummings, B., Gunderson, L., Macdonald, J. S., & Mayer, R. J. (2001). Impact of number of nodes retrieved on outcome in patients with rectal cancer. *Journal of Clinical Oncology, 19*(1), 157–163. Available from https://doi.org/10.1200/JCO.2001.19.1.157.

Tepper, J. E., O'Connell, M. J., Petroni, G. R., Hollis, D., Cooke, E., Benson, A. B., Cummings, B., Gunderson, L. L., Macdonald, J. S., & Martenson, J. A. (1997). Adjuvant postoperative fluorouracil-modulated chemotherapy combined with pelvic radiation therapy for rectal cancer: Initial results of intergroup 0114. *Journal of Clinical Oncology, 15*(5), 2030–2039. Available from https://doi.org/10.1200/JCO.1997.15.5.2030.

Tsai, C. J., Crane, C. H., Skibber, J. M., Rodriguez-Bigas, M. A., Chang, G. J., Feig, B. W., Eng, C., Krishnan, S., Maru, D. M., & Das, P. (2011). Number of lymph nodes examined and prognosis among pathologically lymph node-negative patients after preoperative chemoradiation therapy for rectal adenocarcinoma. *Cancer, 117*(16), 3713–3722. Available from https://doi.org/10.1002/cncr.25973.

Tulchinsky, H., Rabau, M., Shacham-Shemueli, E., Goldman, G., Geva, R., Inbar, M., Klausner, J. M., & Figer, A. (2006). Can rectal cancers with pathologic T0 after neoadjuvant chemoradiation (ypT0) be treated by transanal excision alone? *Annals of Surgical Oncology, 13*(3), 347–352. Available from https://doi.org/10.1245/ASO.2006.03.029.

Wang, Y., Zhou, M., Yang, J., Sun, X., Zou, W., Zhang, Z., Zhang, J., Shen, L., Yang, L., & Zhang, Z. (2019). Increased lymph node yield indicates improved survival in locally advanced rectal cancer treated with neoadjuvant chemoradiotherapy. *Cancer Medicine, 8*(10), 4615–4625. Available from https://doi.org/10.1002/cam4.2372.

Wichmann, M. W., Müller, C., Meyer, G., Strauss, T., Hornung, H. M., Lau-Werner, U., Angele, M. K., & Schildberg, F. W. (2002). Effect of preoperative radiochemotherapy on lymph node retrieval after resection of rectal cancer. *Archives of Surgery, 137*(2), 206–210. Available from https://doi.org/10.1001/archsurg.137.2.206.

Xu, Z., Berho, M. E., Becerra, A. Z., Aquina, C. T., Hensley, B. J., Arsalanizadeh, R., Noyes, K., Monson, J. R. T., & Fleming, F. J. (2017). Lymph node yield is an independent predictor of survival in rectal cancer regardless of receipt of neoadjuvant therapy. *Journal of Clinical Pathology, 70*(7), 584–592. Available from https://doi.org/10.1136/jclinpath-2016-203995.

Further reading

Heald, R., & Ryall, R. D. (1986). Recurrence and survival after total mesorectal excision for rectal cancer. *Lancet, 1*(8496), 91510–91512. Available from https://doi.org/10.1016/s0140-6736.

Heald, R., Moran, B., Ryall, R., Sexton, R., & MacFarlane, J. K. (1978). Rectal cancer: The basingstoke experience of total mesorectal excision. *The Archives of Surgery, 133*(8), 894–898.

Scott, K. W. M., & Grace, R. H. (1989). Detection of lymph node metastases in colorectal carcinoma before and after fat clearance. *British Journal of Surgery, 76*(11), 1165–1167. Available from https://doi.org/10.1002/bjs.1800761118.

Wang, H., Safar, B., Wexner, S., Zhao, R. H., Cruz-Correa, M., & Berho, M. (2009). Lymph node harvest after proctectomy for invasive rectal adenocarcinoma following neoadjuvant therapy: Does the same standard apply? *Diseases of the Colon and Rectum, 52*(4), 549–557. Available from https://doi.org/10.1007/DCR.0b013e31819eb872.

11

Complete mesocolic excision in colon cancer

Alice C. Westwood[1,2], Jim P. Tiernan[3] and Nicholas P. West[1,2]

[1]Pathology & Data Analytics, Leeds Institute of Medical Research at St. James's, University of Leeds, Leeds, United Kingdom [2]Department of Histopathology, Leeds Teaching Hospitals NHS Trust, Leeds, United Kingdom [3]John Goligher Colorectal Unit, Leeds Teaching Hospitals NHS Trust, Leeds, United Kingdom

11.1 Introduction

Colorectal cancer is the third most common malignant disease in men and women worldwide with over 1.8 million new cases diagnosed each year (World Cancer Research Fund, 2020). Approximately two-thirds of these cases occur in the colon, with the remaining one-third occurring in the rectum. Since the description of total mesorectal excision (TME) for rectal cancer by Professor Bill Heald in Basingstoke 40 years ago and subsequent international roll out of this technique, the quality of rectal cancer surgery has radically improved around the world, which has led to significantly improved oncological outcomes including a marked reduction in local recurrence and improvement in survival (Heald et al., 1982; Kapiteijn et al., 2002; Martling et al., 2000; Quirke et al., 2009; West et al., 2008; Wibe et al., 2002). Similar international standardization of the surgical approach for colon cancer has not yet been achieved to date.

Following similar principles to TME, Professor Werner Hohenberger and colleagues introduced complete mesocolic excision (CME) with central vascular ligation (CVL) in Erlangen during the 1980s before publishing their long-term experience of this standardized surgical technique in 2009 (Hohenberger et al., 2009). Hohenberger described sharp dissection along the embryological tissue planes, with resection of the mesocolon in an intact package covered by peritoneum and fascia including the associated lymphatics, lymph nodes, and blood vessels, to prevent tumor spillage or transcoelomic spread during surgery (Hohenberger et al., 2009). Hohenberger reported impressive improvements in cancer-specific survival, which have since been replicated in more recent studies

(Bertelsen et al., 2015; Hohenberger et al., 2009; Storli et al., 2014). In centers performing CME with CVL, improvement in the oncological quality of the surgical specimen has been documented with higher rates of mesocolic plane resection, greater distance between the tumor and the central vascular tie, and increased lymph node yield (Heald et al., 1982; Kapiteijn et al., 2002; Martling et al., 2000; Quirke et al., 2009; West et al., 2008; Wibe et al., 2002).

Despite the reported benefits and an international consensus meeting recommending that CME with CVL should be the standard surgical procedure for colon cancer, this has not yet materialized to date (Søndenaa et al., 2014). There remain many controversies regarding the benefits of CME, potential complications, and suitability of the technique for all colon cancer patients. This chapter describes the oncological and surgical principles of CME, the evidence and controversies surrounding the technique, and the importance of independent pathological assessment of surgical specimen quality.

11.2 Outline of the key anatomy

Anatomically, distal to the duodenojejunal flexure, the mesentery is a continuous structure with the mesocolon in direct continuity with the mesorectum (Byrnes et al., 2019; Coffey et al., 2020; Culligan et al., 2012). Similarly, the embryological tissue planes that are utilized in TME are not unique to the rectum. The mesocolon, to a variable degree, is invested in a layer of visceral fascia which makes the principles of TME surgery easily transferrable to colon cancer surgery (Heald et al., 1982; Hohenberger et al., 2009). CME involves sharp dissection in the embryologically defined mesocolic plane to remove an intact envelope of mesentery lined by peritoneum and visceral fascia (where present) along with its corresponding lymphatic and venous drainage (Hohenberger et al., 2009). Lymphatic spread in colon cancer, similar to that of rectal cancer, primarily follows the path of the supplying arteries. A comprehensive understanding of the colorectal vasculature and its variations is therefore essential to understand the lymphatic drainage of the colon and to appreciate the technical aspects of CME surgery (Hohenberger et al., 2009; Jamieson & Dobson, 1909).

The blood supply to the colon is received via branches of the superior and inferior mesenteric arteries. Importantly, there can be significant anatomical variation between individuals and this may influence the type of surgical resection performed. The right colon and transverse colon are supplied by the ileocolic and the middle colic arteries; these are both branches of the superior mesenteric artery. The right colic artery shows significant anatomical variation; it originates from the superior mesenteric artery in only 10%−15% of patients and more commonly arises from the middle colic artery (Hohenberger et al., 2009; Vandamme & Bonte, 1990). The remainder of the colon is supplied by branches of the inferior mesenteric artery. The Arc of Riolan, also known as the central anastomotic mesenteric artery, is an inconstant artery that connects the proximal superior mesenteric artery, or one of its primary branches, to the inferior mesenteric artery, or one of its primary branches; classically it is described as connecting the middle colic branch of the superior mesenteric artery with the left colic branch of the inferior mesenteric artery (Xie et al., 2015).

Due to the significant anatomical variation between individuals, it has been proposed that three-dimensional computed tomography (CT) angiography could be used preoperatively to analyze the vascular anatomy and demonstrate the distribution of arteries feeding the cancer (Kijima et al., 2014). This could be achieved through modification of routine preoperative CT staging and may be particularly useful in laparoscopic surgery for right-sided colon cancer, where the reduced view makes identifying variations in the mesenteric vessels difficult and may help to reduce the risk of intraoperative bleeding (Hirai et al., 2013; Kijima et al., 2014).

The lymphatic drainage of the colon and the surgical principle of removing colon cancer as an intact package was first described by Jamieson and Dobson in Leeds back in 1909 (Jamieson & Dobson, 1909). The Japanese Society for Cancer of the Colon and Rectum (JSCCR) classify colonic lymph nodes into three broad groups based on their proximity to the bowel wall (Hashiguchi et al., 2020). The initial lymphatic drainage from a cancer is usually to the paracolic (D1) lymph nodes, a series of nodes close to the bowel wall that run along the marginal arteries. The next level of drainage is to the intermediate (D2) lymph nodes; these nodes run along the branches of the supplying superior and inferior mesenteric arteries. The final level of drainage that is routinely surgically resected en bloc is to the central (D3) lymph nodes. On the right side, the central nodes lie at the origin of the ileocolic, right colic, and middle colic arteries; on the left side, the central nodes run along the inferior mesenteric artery from its origin at the aorta to the branching of the left colic artery (Hashiguchi et al., 2020). The concept of CME as described by Hohenberger includes CVL or "high-tie" of the supplying arteries at or very close to their roots to facilitate complete removal of the central (D3) lymph nodes (Hohenberger et al., 2009).

11.3 Principles of CME surgery

Hohenberger's description of CME includes three key components. Firstly, sharp dissection along the embryological tissue planes to separate the visceral and retroperitoneal (parietal) fascia, leading to a surgical specimen containing the entire regional mesocolon in an intact envelope lined by peritoneum and fascia (where relevant). This aims to avoid any significant breach of the visceral fascia and peritoneal layers thereby preventing the spread of tumor cells into the peritoneal cavity. Secondly, perform CVL of the supplying colonic arteries to remove the central (D3) lymph nodes and facilitate a maximal lymph node harvest. Lastly, resection of an adequate length of bowel to remove any longitudinal pericolic lymph nodes that may be involved (Heald et al., 1982; Kapiteijn et al., 2002; Martling et al., 2000; Quirke et al., 2009; West et al., 2008; Wibe et al., 2002). The relative importance of each of these components is not fully understood due to the lack of randomized controlled trial evidence; however, it is believed that preservation of the mesocolic planes is the most important variable with a small amount of additional benefit from the high vascular ligation. It is important to note that surgical technique and complexity differ depending on the location of the tumor.

Techniques and terminology for colon cancer surgery, unlike rectal cancer surgery, are not standardized and differ worldwide (Culligan et al., 2013). In Japan, similar to CME, the JSCCR advocate careful dissection along the embryological tissue planes. However, unlike CME, the extent of lymph node dissection is dependent on the intraoperative assessment of depth of tumor invasion and the extent of lymph node metastases (Hashiguchi et al., 2020). Clinically T1 tumors undergo D2 resection (standard intermediate-level vascular ligation) and T2—T4 tumors undergo a D3 resection (high-tie), although some T2 tumors may be deemed suitable for D2 resection (Hashiguchi et al., 2020; West et al., 2012). Several Japanese studies have demonstrated that longitudinal lymphatic spread beyond the immediate pericolic area is very rare: 0% for left-sided tumors and 1%—4% for right-sided tumors (Morikawa et al., 1994; Toyota et al., 1995). However, a study by Tan et al. on the location of the first metastatic lymph node in colon cancer found that in 16% of cases (15/93), the first node was a pericolic node more than 5 cm from the tumor (Tan et al., 2010; West et al., 2012). Despite this, Japanese longitudinal resection margins are rarely more than 10 cm from the tumor, and significantly shorter than resections performed in Europe (Kobayashi et al., 2014; West et al., 2012). Although the length of colon has been shown to differ, a comparison by West et al. (2012) found the rate of mesocolic plane excision was high in cases from Japan, comparable to those performed by Hohenberger's group in Erlangen, with equally impressive long-term outcomes.

11.3.1 Role of minimally invasive CME surgery

CME surgery as described by Hohenberger was performed using an open approach; however, slight modifications to the technique have been made to adapt it for both laparoscopic and robotic-assisted surgery, while maintaining the same key principles (Cho et al., 2015). Minimally invasive surgery for colon cancer has significant benefits compared to an open technique, including shorter hospital stay, reduction in postoperative pain, and recovery time with similar oncological outcomes (Athanasiou et al., 2017; Guillou et al., 2005). CME is acknowledged to be a technically challenging procedure, and there has been some debate as to whether laparoscopic surgery should be performed in all cases, especially for tumors in more anatomically complex areas such as the transverse colon.

There have been several retrospective comparative studies and one randomized controlled trial evaluating the efficacy of laparoscopic CME versus open CME, predominantly in right hemicolectomies (Bertelsen et al., 2015; Hohenberger et al., 2009; Storli et al., 2014). The majority of studies found laparoscopic CME to have similar operative times, oncological benefit, safety profile, and specimen quality when compared to open surgery (Bae et al., 2014; Han et al., 2013; Munkedal et al., 2014; Shin et al., 2018; Siani et al., 2017; Storli et al., 2013; West et al., 2014). The only randomized controlled trial to date comparing laparoscopic and open D3 surgery in Japan showed equivalent oncological outcomes between the two groups, but a lower complication rate and shorter length of hospital stay in the laparoscopic group (Kitano et al., 2017; Yamamoto et al., 2014); these findings have been reflected in the other retrospective comparative studies (Bae et al., 2014; Cho et al., 2015; Shin et al., 2018; Storli et al., 2013).

The main focus for robotic CME to date has been in right-sided colon cancer, and studies have shown robotic surgery to have the same oncological benefit and safety profile as that of both open and laparoscopic CME surgery (Petz et al., 2017; Spinoglio et al., 2018). Robotic CME was associated with an increase in lymph node yield and low conversion rates, but longer operation times (Petz et al., 2017; Spinoglio et al., 2018; Widmar et al., 2017; Yozgatli et al., 2019). The results for robotic surgery are promising, but further studies looking at long-term outcomes are required.

11.4 Oncological benefits of CME

CME has not been universally adopted due to concerns over potential morbidity with many centers internationally still performing "conventional" colon cancer surgery (hemicolectomy with intermediate or low-level vascular tie) (Abdelkhalek et al., 2018; Ng et al., 2020). There has been similar international debate about the potential oncological benefits and limitations of CME.

The 2009 paper by Hohenberger et al. describes the stepwise change in surgical technique and development of a standardized CME approach in Erlangen between 1978 and 2002; the primary outcome, 5-year cancer-related survival, was analyzed during three different time periods: 1978–84 (pre-CME), 1985–94 (development of CME), and 1995–2002 (implementation of CME). They reported improvements in 5-year cancer-related survival after surgery from 82.1% in 1978–84 to 89.1% in 1995–2002, and a reduction in local 5-year recurrence rates from 6.5% in 1978–84 to 3.6% in 1995–2002 (Hohenberger et al., 2009). It is important to note the limitations of Hohenberger's series; firstly, the study was carried out over 24 years and therefore has a range of potential confounders that may have influenced outcome; and secondly the study is nonrandomized and we do not fully understand the influence of patient selection. Although several studies have shown CME to be associated with significantly greater disease-free or disease-specific survival (Bertelsen et al., 2015; Storli et al., 2014) and lower rates of local and distant recurrence (Galizia et al., 2014; Merkel et al., 2016) compared to conventional colon resection, these findings are not reflected universally; four studies reported no significant difference in overall survival (Bertelsen et al., 2015; Kotake et al., 2015; Olofsson et al., 2016; Tagliacozzo & Tocchi, 1997). It is important to note that all these studies are nonrandomized and the oncological benefits of CME are still debated (Olofsson et al., 2016). Confounding factors, such as the use of adjuvant chemotherapy in patients with stage III disease, may lead to underestimation of the impact of CME.

Many studies comparing CME versus conventional surgery do not take into account the importance of independent pathological quality assessment. It is essential that CME is not simply seen as a reflection of vascular ligation height but also takes into account mesocolic integrity. It is likely that the benefits of CVL/D3 surgery will be negated if there are substantial defects in the mesocolon, leading to local and peritoneal recurrence. All studies to date have been retrospective in nature and as of yet, there are no published randomized controlled trials comparing CME to non-CME surgery with independent pathological assessment of the plane of surgery. However, two randomized controlled trials are currently ongoing: the RELARC trial (NCT02619942) comparing D2 dissection versus CME in laparoscopic right hemicolectomy for right-sided colon cancer (Lu et al., 2016); and the COLD trial (NCT03009227) comparing oncological outcomes for D2 versus D3 lymph node dissection in colonic cancer (Karachun et al., 2019). The results of both are eagerly anticipated.

We believe that the key benefit of CME surgery is improved quality of the surgical specimen, and most importantly removal of an intact mesocolon with its peritoneal and fascial linings. Prior to the description of CME there was marked variability in the plane of surgery achieved in many centers (West et al., 2008). Two studies have shown that surgeons regularly performing CME were more likely to produce an intact specimen in the mesocolic plane than those performing conventional non-CME surgery, independent of the height of the vascular ligation (Kobayashi et al., 2014; West, Hohenberger, et al., 2010; West, Sutton, et al., 2010). One study comparing surgical techniques in colon cancer found that 88.5% of specimens following CME were resected in the mesocolic plane compared to only 47.4% following conventional surgery (Kobayashi et al., 2014). A further recent study by Ng et al. (2020) at a quaternary referral center where some surgeons undertake CME in specific circumstances found that two-thirds of all specimens were resected in the mesocolic plane. West et al. (2008) demonstrated that surgery performed in the mesocolic plane confers a 15% improvement in 5-year overall survival compared to surgery performed in the muscularis propria plane; this was further increased to a 27% 5-year survival benefit in stage III disease and remained significant on multivariate analysis.

Another proposed benefit of the widespread adoption of CME is the standardization of surgical technique which should in turn lead to an improvement in specimen quality (Emmanuel & Haji, 2016; West, Hohenberger, et al., 2010; West, Sutton, et al., 2010) and patient outcomes (Bokey et al., 2003; Enker et al., 1979). A study conducted in Denmark assessed the effect of a surgical education program on the quality of the surgical specimens by comparing one hospital in which surgeons had undergone a CME education program to the other five hospitals in the region that utilized conventional non-CME techniques (West, Hohenberger, et al., 2010; West, Sutton, et al., 2010). Surgeons trained in CME produced specimens with a higher rate of mesocolic plane surgery (75% vs 48%), a greater distance between the tumor and vascular tie (105 vs 84 mm), a superior lymph node yield (28 vs 18), and improved 4-year disease-free survival [85.8% (95% confidence interval (CI) 81.4%−90.1%) vs 73.4% (66.2%−80.6%)] (Bertelsen et al., 2015; West, Hohenberger, et al., 2010; West, Sutton, et al., 2010). A subsequent extension of this educational program across the other surgical units led to an improvement in the overall rate of mesocolic plane resection from 58% to 77% ($P < .001$) (Sheehan-Dare et al., 2018). There is also evidence that CME techniques can be learned and adopted over a relatively short period of time through such educational programs (Guo et al., 2012; West, Hohenberger, et al., 2010; West, Sutton, et al., 2010).

11.4.1 Which patients benefit from CME?

There remains ambiguity over which patients benefit from CME. Logically, CME should be most beneficial to patients with stage III disease as they are most at risk of cancer dissemination at the time of surgery, although this is highly dependent on the location of the involved lymph nodes. If, for example, a patient only has paracolic lymph node metastases, then CME may confer no additional benefit to that of conventional intermediate-ligation surgery. In addition, if a patient has D3 lymph node metastases with micrometastatic disease higher in the lymphatic chain, then even CME surgery will not remove all of the disease. The patients with stage III disease who may be most likely to benefit are those with involved D2 nodes with no disease in the D3 nodes; if these patients undergo conventional surgery they would previously have been classified as Dukes' C2 (involved apical node), however with CME they would be "downstaged" to Dukes' C1, which is known to have a better outcome (Dukes, 1932; Gabriel et al., 1935).

Data from studies assessing oncological outcomes in stage III colon cancer are not entirely clear about the degree of benefit with CME. One study found no significant improvement in 4-year disease-free survival between CME and non-CME in stage III (Bertelsen et al., 2015). However, a meta-analysis by Wang et al. found CME to be associated with improved 3-year survival for stage III (hazard ratio (HR) 0.69, 95% CI 0.60−0.80), findings similar to other studies (Bernhoff et al., 2015; Galizia et al., 2014; Merkel et al., 2016; Wang et al., 2017). The use of adjuvant chemotherapy in most patients with stage III might act as a confounding factor here as it may reduce the micrometastatic burden in lymph nodes not removed in non-CME surgery; this may contribute to underestimation of the effect of CME on survival in stage III colon cancer (Wang et al., 2017).

Such radical surgery for patients with stage I−II disease has been questioned by some (Hashiguchi et al., 2020); however, a significant proportion of patients with stage I−II colorectal cancer will die with recurrent disease (5-year survival is 84% in stage II and 92% stage I in England) (Office for National Statistics, 2019). One possible explanation for locoregional recurrence is the presence of lymph node skip metastases where metastatic tumor is present in the apical nodes, but has not been identified in the paracolic or intermediate nodes (Wang et al., 2017). It is understood that colonic lymphatic spread generally follows the supplying arteries, from the paracolic nodes to the intermediate and finally the central nodes; however, lymph node metastases do not always follow this stepwise degree of spread. The true incidence of skip metastases is not known: one study found 1.8% of cases identified by histopathology, but occult tumor cells identified by molecular analysis in 18.0% of cases (Merrie et al., 2001), and a further two studies reporting that 18% of patients with lymph node involvement had skip metastasis (Liang et al., 2007; Tan et al., 2010). CME increases the chances of identifying skip metastases due to the wider lymphadenectomy, which may lead to upstaging, adjuvant chemotherapy and thereby improving prognosis and survival (Liang et al., 2007; Tan et al., 2010).

A further explanation for disease recurrence in stage I and II diseases is the presence of other high-risk features associated with locoregional and distant spread. Extramural venous invasion is a well-established independent poor prognostic indicator with a high risk of distant metastasis, and there is some evidence that intramural venous spread is also prognostically important (Betge et al., 2012; Santos et al., 2013). In addition, there is evidence that invasion of small vessels (including lymphatics, capillaries, and postcapillary venules) (Lim et al., 2010; Santos et al., 2013) and perineural invasion (Huh et al., 2010; Knijn et al., 2016; Ueno et al., 2013) are of significant prognostic value. It is now a requirement according to the eighth edition of tumour nodes metastasis (TNM) staging that the presence or absence of venous, lymphatic, and perineural invasion should all be assessed and reported independently, and the Royal College of Pathologists recommends that the deepest level of invasion should also be recorded (Brierley et al., 2017; Loughrey et al., 2018). CME may confer additional benefit to these "high-risk" stage I and II patients as removal of an intact mesocolon would be more likely to completely remove intravascular or perineural disease, whereas conventional surgery could potentially leave residual disease behind.

Two multicenter studies showed that CME was a significant predictive factor for 3- and 4-year disease-free survival in patients with stage I and stage II disease (Bertelsen et al., 2015; Storli et al., 2014). Bertlesen et al. found that 4-year disease-free survival in stage I was 100% in the CME group compared to 89.8% (95% CI 83.1−96.6) in the non-CME group ($P = .046$); in patients with stage II, 4-year disease-free survival was 91.9% (95% CI 87.2−96.6) in the CME group versus 77.9% (95% CI 71.6−81.4) in the non-CME group ($P = .0033$) (Bertelsen et al., 2015). However, other studies have shown no significant improvement in overall survival in stage I or II disease (Merkel et al., 2016).

11.5 Potential limitations of CME and associated controversies

11.5.1 Complications

Technically, CME is considered more challenging than conventional colon cancer surgery, at least during the learning phase. A recent meta-analysis showed that CME is associated with an increase in morbidity such as intraoperative blood loss when compared to non-CME surgery (Galizia et al., 2014; Wang et al., 2017). CME in the right colon is technically more difficult than the conventional counterpart as it requires dissection up to the root of the right branch of the middle colic artery and vein and subsequent central ligation; it is at this point that critical structures such as the superior mesenteric vein (SMV) have a higher tendency to be damaged. Freund et al. (2016) reported injury to the SMV in 1.6% of cases when performing central ligation for right-sided colon cancer; possible mechanisms for these injuries include greater anatomical variability in the right colon and avulsion of the middle colic vein due to excessive traction. The incidence of other intraoperative injuries has also been reported to be higher in CME compared with non-CME; a study by Bertelsen et al. reported a higher rate of intraoperative injuries including SMV injury, splenic injury, and injury to other segments of the colon in CME with CVL (9.1% vs 3.6%, $P < .001$) (Bertelsen et al., 2015; Hohenberger et al., 2009; Storli et al., 2014).

A recent meta-analysis found CME to be associated with more postoperative surgical complications; these included anastomotic leakage, intraabdominal abscess, wound infection, and postoperative obstruction (Wang et al., 2017). However, some studies comparing CME and conventional surgery found no significant differences in short-term outcomes (Bertelsen et al., 2016, 2011). Detrimental complications that can occur following CME include damage to retroperitoneal structures such as the ureter, tail of pancreas, and the duodenum as well as genitourinary dysfunction, sexual dysfunction, chyle leak, and refractory diarrhea (Gouvas et al., 2012; Koh & Tan, 2019). An important consideration is that any increase in postoperative complications may increase hospital stay and lead to a delay in patients receiving adjuvant treatment. Crucially, there appears to be no significant difference in postoperative mortality after CME (Bertelsen et al., 2016; Kim et al., 2016; Merkel et al., 2016; Wang et al., 2017).

11.5.2 Technical difficulties

It is likely that not all patients will significantly benefit from a high vascular ligation, and it has been postulated that radical lymphadenectomy is more challenging and complex in patients with a higher body mass index (BMI) (Alhassan et al., 2019). Many studies of D3 lymphadenectomy have been performed in Asian patients, where a lower BMI is well known compared to Western populations (Kobayashi et al., 2014; West et al., 2012). Therefore some Western surgeons have questioned whether the increase in complexity, operation time, and postoperative complications are worthwhile for more obese Western patients (Chow & Kim, 2014). However, Erlangen and other European units have shown that variability in BMI does not affect the feasibility of CME, and other centers have demonstrated that laparoscopic CME is still possible even in obese patients (Chow & Kim, 2014; Gouvas et al., 2012; Hohenberger et al., 2009; Storli et al., 2013).

11.5.3 CME in transverse colon cancer

Transverse colon and flexure cancers represent only a small proportion (10%) of colorectal cancers; however, CME for these cases is very challenging (Lê et al., 2006; Mori et al., 2017). A recent multicenter study demonstrated that extended right hemicolectomy for transverse colon cancer is safer and more oncologically sound than transverse colectomy (Milone et al., 2020). Typically for transverse colon and flexure cancers either an extended right or subtotal colectomy can be performed. An extended right hemicolectomy is performed with ligation of the roots of the middle colic artery and vein, along with the ileocolic pedicle (Kim et al., 2016; Milone et al., 2020). A subtotal colectomy includes ligation of the ileocolic, right colic, middle colic, and inferior mesenteric pedicle (Martínez-Pérez et al., 2019).

As previously discussed, lymphatic spread is generally considered to occur in a stepwise approach. However, some studies have suggested there are aberrant routes of spread in the transverse colon specifically (Saha et al., 2013) and that this may be a key factor in the higher local recurrence rates noted for these cancers (Sjövall et al., 2007). In cases of advanced transverse colon cancer, including both flexures, lymph node metastases can be found within the gastroepiploic arcade along the greater curvature of the stomach, along the lower edge of the left pancreas, and in front of the pancreatic head (Perrakis et al., 2014; Weber et al., 2013). A study by Perrakis et al. (2014) assessing cancer of the transverse colon and flexures found metastases in the infrapancreatic lymph node region and gastroepiploic arcade in 20% and 12.5%, respectively, of cases with stage III disease. This aberrant lymphatic spread can be explained by the identification of vascular and lymphatic vessels running between the transverse colon and the greater omentum as well as the uncinate process of the pancreas (Bertelsen et al., 2015; Hohenberger et al., 2009; Storli et al., 2014). For this reason, in CME for transverse colon cancer, the corresponding right gastroepiploic artery arcade and the lymph nodes over the pancreatic head and inferior aspect of the left pancreas should also ideally be removed to maximize the chances of complete tumor removal (Merkel et al., 2016; Perrakis et al., 2014). There are inherent difficulties with this procedure resulting from the anatomical peculiarities and complex anatomy of the transverse colon; therefore not all surgeons undertaking CME for transverse colon cancer will perform such a radical lymphadenectomy as described by Hohenberger. In addition, due to the limited numbers of transverse colon cancer resections, individual surgeons may not gain adequate experience in the procedure; subsequently there has been suggestion that all transverse colon cancer surgery should be performed at specialist centers, although this is controversial and not internationally supported.

Although outcomes following laparoscopic CME appear similar to that for open CME in right-sided colon cancer, relatively few studies include patients with transverse colon cancer. All randomized trials comparing open colon surgery to laparoscopic, including Kitano et al., have excluded tumors of the transverse colon (Kitano et al., 2017; Yamamoto et al., 2014). This is likely due to the complexity of performing lymph node dissection laparoscopically around the middle colic artery and vein as well the technical difficulties associated with ligation of the root of the middle colic vessels (Athanasiou et al., 2017; Ozben et al., 2020). A recent meta-analysis of studies including transverse colon cancer found that laparoscopic surgery was safe and had similar oncological outcomes to that of open surgery when performed by experienced surgeons (Athanasiou et al., 2017); however, future high-quality randomized controlled trials are required to further investigate this.

The reluctance to perform laparoscopic CME on patients with transverse colon cancer due to technical difficulties could be overcome by the use of robotic surgery; this aims to reduce the limitations of laparoscopy by offering an improved view and greater range of movement. The efficacy of robotic surgery at reducing difficulties associated with laparoscopic surgery has already been demonstrated in some studies with reduction in conversion to open surgery (Dolejs et al., 2017) and increased use of intracorporeal anastomosis (Trastulli et al., 2015). A retrospective observational study comparing short-term outcomes of robotic CME for transverse colon cancer with conventional laparoscopic colectomy found that robotic CME could be performed with similar morbidity and increased lymph node yield, albeit with longer operating times (Ozben et al., 2020).

11.6 Importance of pathological quality control in CME surgery

Pathologists are the ideal candidates to assess the quality of CME specimens; they are independent of the surgical procedure, have experience of seeing a range of specimens from multiple surgeons, and have an opportunity to provide direct feedback through demonstrating photographic records during postoperative multidisciplinary team meetings (MDTs). The Medical Research Council (MRC) CR07 trial assessed the effect of the plane of mesorectal surgery on local recurrence in rectal cancer and demonstrated an improvement in the quality of the specimens resected over the duration of the trial; these improvements included an increased frequency of mesorectal plane excision (from less than 50% in 1998 to over 60% in 2005), a reduced frequency of muscularis propria plane excision (from over 20% in 1998 to approximately 10% in 2005), and a reduction in circumferential resection margin involvement (from 21% in 1998 to 10% in 2005; $P < .001$) (Quirke et al., 2009). This is highly likely to have been influenced through direct feedback of specimen quality through the MDTs. The sustained improvement in long-term rectal cancer outcomes over the past 20–30 years has been partly attributed to the corresponding improvement in the quality of the resection specimen (Koh & Tan, 2019), although other factors such as the introduction of MRI for operative planning and neoadjuvant treatment have also played a major role. When assessing the quality of colon cancer resection specimens, there are several aspects to consider and these will be discussed in turn.

11.6.1 Integrity of the mesocolon (plane of surgery)

An optimal colon cancer resection specimen should include the primary tumor and surrounding mesocolon containing all of the structures associated with locoregional tumor spread in an intact peritoneal and fascial-lined package (where appropriate). An intact mesocolon containing the relevant lymphatics, lymph nodes, and blood vessels is a key principle of CME surgery, as described by Hohenberger (Hohenberger et al., 2009; Jamieson & Dobson, 1909). Several studies have shown improved patient outcomes following CME or D3 surgery (Hohenberger et al., 2009; Kitano et al., 2017), and the majority of this benefit is thought to be due to preservation of the mesocolic plane thereby preventing tumor cells' shedding into the peritoneal cavity or parts of the mesocolon being left behind resulting in intraabdominal recurrence (West et al., 2008). Keeping the mesocolon intact with no or only very minor surface defects was shown to translate into a 15% overall survival advantage at 5 years compared with surgery showing defects in the mesocolon down to the muscularis propria in one large single-center retrospective study (West et al., 2008). This survival advantage increased further to 27% when looking at patients with stage III disease, resulting in twice as many patients alive at 5 years, likely because these patients have mesocolic disease beyond the bowel wall and are thereby at highest risk of tumor dissemination following disruption of the mesocolon. This study was performed at a time when CME was not undertaken in the institution and only 32% of specimens were removed in the mesocolic plane with 24% showing defects down onto or into the muscularis propria (West et al., 2008). Unfortunately, data were not available for recurrence patterns and distant metastases in this study.

Grading the plane of surgery in colon cancer specimens is therefore a critical component when assessing the quality of the surgical specimen, providing immediate feedback on likely prognosis. A three-tier grading system was initially developed for the MRC CLASICC trial (Guillou et al., 2005), which was largely based on the mesorectal grading system used in the MRC CR07 trial (Heald et al., 1982; Kapiteijn et al., 2002; Martling et al., 2000; Quirke et al., 2009; West et al., 2008; Wibe et al., 2002). Assessment of the plane of surgery should be performed initially on the intact fresh specimen with confirmation of the presence and depth of any defects on the intact formalin-fixed specimen and subsequent cross-sectional slices. Grading should be performed according to the "worst" area within the lymphatic drainage field of the tumor. The grade categories consist of: the *mesocolic plane* (the optimal plane) where the specimen has an intact peritoneal and fascial surface with only very minor surface defects allowed (no greater than 5 mm in depth); the *intramesocolic plane* (intermediate quality) where the specimen has a moderate amount of mesocolon with significant defects in the mesocolon to a depth of more than 5 mm, but not extending down to the muscularis propria; and the *muscularis propria plane* (poor quality) where the specimen has at least one defect exposing the muscularis propria or beyond into the submucosa or mucosa, to include iatrogenic perforations (Fig. 11.1) (West et al., 2008).

The grading of colon cancer resections not only depends on the plane of surgical dissection but also on the training of the pathologist and ensuring excellent communication across the MDT such that defects caused either during specimen extraction or transport are not incorrectly interpreted. As with any subjective assessment, the grading system demonstrates an element of both intraobserver and interobserver variations. One study

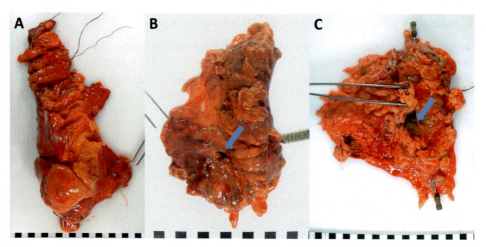

FIGURE 11.1 Grading the plane of surgery according to the presence and depth of mesocolic disruptions, including the (A) mesocolic plane, (B) intramesocolic plane, and (C) muscularis propria plane. Mesocolic defects are indicated by arrows.

found that when grading colon cancer specimens from specimen photographs, the interobserver concordance was poor (κ value < 0.40); this low concordance was particularly notable for specimens resected in the muscularis propria plane, possibility representing the difficulty in assessing the depth of defects from photographs (Bae et al., 2014; Kitano et al., 2017; Koh & Tan, 2019; Munkedal et al., 2014; Shin et al., 2018; Storli et al., 2013; Yamamoto et al., 2014). In the same study, intraobserver variability was fair to good (κ value 0.4−0.7). Following the results of this study, refinements to the mesocolic grading system were proposed to reduce some of the subjectivity and interobserver variability (Munkedal et al., 2016). These included clarification that only the mesocolon in the tumor lymphatic drainage field should be assessed (defined using the Japanese classification of lymphatic drainage), the distal 10 mm of the specimen should be ignored due to the irregularity that can be expected at the distal margin, and isolated disruptions of the mesocolic windows (areas of fused peritoneum + /−fascia with no intervening fat) should not be regarded as significant defects and downgraded. It was also clarified that mesocolon integrity should be assessed independent of the extent of surgery, with central radicality, that is, height of the vascular ligation, being assessed through a different measure described below (Munkedal et al., 2016). Despite some subjectivity in mesocolic grading between centers, grading within a single center by a cohesive pathology team can be very useful to document variability between surgeons and over time. Recognition of potential interobserver variability is vital and highlights the importance of central review of the specimen in the clinical trial setting to remove any bias. The international FOxTROT trial (NCT00647530), evaluating the use of neoadjuvant chemotherapy for advanced colon cancer, has built in prospective assessment of mesocolic grading by local pathologists across more than 80 centers in three countries; these local gradings will be compared with that determined following central review of specimen photographs to understand the level of interobserver variation across a range of centers in the clinical trial setting (Group, 2012). The local and central gradings will also be compared to short- and long-term patient outcomes. When mesorectal grading was performed prospectively by local pathologists in the MRC CR07 trial, a clear independent association with long-term outcomes was demonstrated leading to its widespread introduction in pathology reporting guidance internationally (Quirke et al., 2006). The FOxTROT trial has completed recruitment and the results of the local and central grading and relationship to outcomes are eagerly awaited.

11.6.2 Distance between the tumor and the central arterial ligation point

CVL, or high vascular tie, is an essential component of CME as described by the Erlangen group. Central ligation of the supplying colonic arteries and draining veins allows for a maximal regional lymph node harvest including the central D3 lymph nodes. The linear distance between the tumor edge and the central arterial ligation point is a recognized quality measure, although on an individual patient basis is limited by its use may be flawed as in most cases the length of the original vessel is not known (West, Hohenberger, et al., 2010; West, Sutton, et al., 2010). A study comparing conventional intermediate-level ligation surgery in the United Kingdom with CME in Germany found a significantly greater distance between the tumor and high-tie in patients following CME for both right- and left-sided tumors when measured on the fresh ex vivo specimen (81.4 mm vs 128.7 mm in right-sided tumors; $P < .0001$, 97.0 mm vs 145.0 mm in left-sided tumors; $P < .001$); this was also significantly correlated with an increased lymph node yield (West, Hohenberger, et al., 2010; West, Sutton, et al., 2010). These findings have been reflected in other studies comparing CME and conventional surgery (Galizia et al., 2014; Kobayashi et al., 2014; West, Hohenberger, et al., 2010; West, Sutton, et al., 2010). Specimens graded in the mesocolic plane have been shown to be associated with a greater distance between the tumor and high-tie as well as an increased lymph node yield (Bae et al., 2014; Han et al., 2013; Munkedal et al., 2014; Shin et al., 2018; Siani et al., 2017; Storli et al., 2013; West et al., 2014).

Although central ligation height is an important marker of surgical radicality, there are inherent difficulties in its measurement that need to be considered. Firstly, the distance from the tumor to the central ligation point should ideally be performed on the fresh ex vivo specimen to prevent the shrinkage that is associated with formalin fixation; however in practice, this is not always possible if specimens are transferred to the laboratory already in fixative (Fig. 11.2) (Munkedal et al., 2017).

Transfer of fresh specimens may not be possible if the pathological analysis will take place in a different hospital to the surgery, if there is likely to be a delay in specimen handling, for example, over a weekend, or more recently during the coronavirus disease (COVID-19) pandemic when the risk of virus shedding from feces has led to immediate and elongated fixation in many centers. One way around this is to either perform the measurement in theater or to obtain fresh specimen photographs alongside a metric scale in theater and transfer along with the specimen for independent analysis by a pathologist. Secondly, the length of the main supplying vessel can show significant individual variation across a population; several studies have shown the positive correlation between height and weight of individuals and increased central ligation height (Kobayashi et al.,

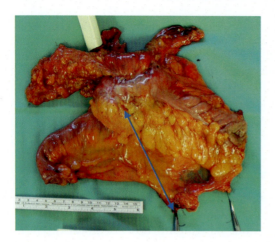

FIGURE 11.2 Measuring the central radicality by assessing the distance between the tumor (measured from the bowel wall) and the closest arterial tie. Note that measurements are best performed on the fresh specimen prior to shrinkage and distortion caused by formalin fixation.

2014; Munkedal et al., 2017). For this reason, on an individual patient basis the vessel length is of limited importance, however, across a population it does accurately reflect the radicality of an individual surgeon or surgical technique and can be used to audit practice and demonstrate improvement after training. Next, the distance between the tumor and the central tie is clearly related to the position of the tumor and individual vascular anatomy and does not always directly reflect the radicality of the surgery. For this reason, the distance between the central tie and the nearest bowel wall may be a more accurate measure of radicality by removing the variation in tumor position, although the measurement between the tumor and central tie may be more likely to be associated with the risk of dissemination and prognosis. Lastly, the measurement of the central ligation height on a resected specimen may not accurately reflect the length of the residual arterial stump. Pathologists can only comment on what tissue has been removed and not what has been left behind in the patient. A recent study in Denmark, measuring the residual arterial stump 2 days postoperatively by CT, found that the mean length of the residual arterial stump was 38 mm (95% CI 33–43 mm), significantly longer than the expectation of 10 mm; 10 mm was chosen as this represents a reasonable estimate of what would be left behind after CME and CVL recognizing that any less may lead to nerve damage (Munkedal et al., 2017). This corroborated previous findings from studies in Sweden and Leeds (Kaye et al., 2016; Spasojevic et al., 2011). In addition, the Danish study found no significant correlation between the distance between the tumor and high-tie and the residual arterial stump measurement (Munkedal et al., 2017). A follow-up study assessed the residual arterial stump length by CT 1 year following surgery in the same Danish cohort, and they found they were able to identify the residual arterial stump in 81% of patients and there was no significant difference in the length of the residual arterial stump measured 2 days after surgery and at 1 year (mean difference −1.7 mm; 95% CI (−3.8 to 0.5 mm), $P = .53$) (Munkedal et al., 2019). Immediate postoperative CT assessment of the residual arterial stump allows immediate feedback to the surgical team; however, this places an additional radiation and time burden on patients who otherwise would not have undergone this scan, and therefore is unlikely to be realistic in routine clinical practice.

However, a CT scan at 1 year is routinely performed to assess for disease recurrence, so this may be a feasible method to measure the residual arterial stump length, however by waiting for 1 year, the opportunity to give immediate real-time feedback to the surgical team is lost.

11.6.3 Length of bowel resected

Resection of a sufficient length of bowel is a key principle of CME as described by Hohenberger; the main purpose of this is to prevent longitudinal spread and harvest longitudinal lymph nodes that may potentially harbor metastases. However, this is a very different approach to that used in Japan where the "10 cm rule" dominates. A multicenter study by Kobayashi et al. compared different techniques for stage III colon cancer including conventional surgery (hemicolectomy with low vascular tie), CME with CVL, and D3 resection carried out in England, Erlangen, and Japan, respectively (Kobayashi et al., 2014). This study demonstrated that CME surgery yielded significantly longer lengths of colon than Japanese D3 surgery (median 355 mm vs 184 mm in right-sided tumors; $P = .0003$, 355 mm vs 146 mm in left-sided tumors; $P < .0001$); however, there was no significant difference in central ligation height (median 115 mm vs 103 mm in right-sided tumors, 128 mm vs 120 mm in left-sided tumors). In comparison, conventional surgery also yielded significantly longer lengths of large bowel when compared to D3 resections, but the central ligation height was significantly smaller (median central ligation height 81 mm vs 103 mm in right-sided tumors; $P = .037$, 100 mm vs 120 mm in left-sided tumors; $P = .034$) and the rate of mesocolic plane surgery was lower (mesocolic plane rate 47% vs 72%) (Kobayashi et al., 2014). This demonstrates that the length of bowel resected may not necessarily reflect the overall quality of the specimen (as judged by mesocolic plane and central radicality) and therefore on its own, length of bowel cannot be used as a surrogate marker for quality of surgery. The oncological value of removing a significant length of colon is not fully understood, but it is not believed to be significant for most patients except perhaps for a very small number showing longitudinal lymphatic spread beyond 10 cm from the tumor, but many of these are likely to be incurable with significant lymph node disease leading to blockage of the central lymphatic flow. Kobayashi et al. also demonstrated that the length of resected large bowel was an independent factor affecting the number of retrieved lymph nodes ($P < .0001$), a finding that has been reflected in other studies (Kobayashi et al., 2014; West et al., 2008; West, Hohenberger, et al., 2010, 2012; West, Sutton, et al., 2010). However, the increase in lymph node yield did not lead to an increase in involved lymph nodes (Kobayashi et al., 2014; West et al., 2012). As mentioned previously, the longitudinal spread of tumor is very rare (Morikawa et al., 1994; Toyota et al., 1995), raising the question as to whether resection of these additional lymph nodes are important. This appears to be validated by comparison of the long-term outcomes in Erlangen and Japan, which are both excellent despite the marked differences in operative approach (Hohenberger et al., 2009; Kitano et al., 2017). The oncological importance of the length of colon resected is being prospectively investigated in the international T-REX study (NCT02938481) (Shiozawa et al., 2020). T-REX has a target sample size of more than 4000 patients with preoperative stage I, II, or III colon cancer and aims to assess the metastatic lymph node distribution and identify the optimal length of bowel resection and central radicality in the surgical management of colon cancer (Shiozawa et al., 2020).

11.6.4 Lymph node yield

Lymph node yield has been found to be consistently higher in CME versus non-CME surgery (Alhassan et al., 2019) and has been associated with an increase from a median of 18 nodes to 30 nodes in one series (West, Hohenberger, et al., 2010; West, Sutton, et al., 2010); similarly, another independent series showed an increase in median nodes from 18 to 28 (West, Hohenberger, et al., 2010; West, Sutton, et al., 2010). The majority of these additional lymph nodes are likely to be paracolic (D1) and intermediate (D2) due to the additional length of bowel and mesentery removed with CME, as indicated in CME versus non-CME studies (West, Hohenberger, et al., 2010; West, Sutton et al., 2010). A sufficient lymph node harvest and examination is essential to allow accurate staging and ensure that patients who may benefit from adjuvant chemotherapy are identified. Low nodal yields have been shown to reduce the proportion of patients diagnosed with stage III disease and thereby restricting access to adjuvant chemotherapy (Morris et al., 2007). There has been debate internationally regarding the optimum number of lymph nodes required to accurately stage colorectal cancer; guidelines from the Royal College of Pathologists suggest that all lymph nodes within the specimen should be identified and examined, with the setting of a standard median number of at least 12 lymph nodes across a series of cases (Loughrey et al., 2018). Studies have shown that the proportion of cases in stage III continues to rise if more than 12 nodes are identified (Morris et al., 2007). However, in centers with excellent surgery and pathology, median yields in excess of 20 are anticipated, and with CME and the use of ancillary methods to identify small nodes, for example, methylene blue, the median number of nodes harvested can approach 40 (Jepsen et al., 2012). This approach has been shown to increase the number of involved nodes in early-stage (pT1) disease, especially tiny nodes containing metastatic tumor that are easily missed with standard palpation (Jepsen et al., 2012).

A strong emphasis has been placed on lymph node yield by some as a key marker of both the radicality and quality of surgery; however, this is fraught with difficulties and we do not recommend this to be used as the main quality measure without knowledge of the associated plane of surgery, specimen measurements (as described above), and knowledge of both host factors and tumor biology that are known to affect nodal yield. A good lymph node yield relies on meticulous pathological dissection and the numbers may remain low even in cases of radical surgery (Søndenaa et al., 2014). Pathologists should dedicate adequate time to performing a thorough lymph node search to ensure that all of the nodes in a specimen are identified and examined. They should make sure not to double count any nodes that are bisected, and where low yields are identified or there are significant time constraints, the use of ancillary techniques, for example, methylene blue injection is recommended.

Numbers of lymph nodes identified in resection specimens are affected by diverse clinicopathological factors, such as the stage of disease, with higher stage tumors associated with increasing lymph node yield (Morris et al., 2007); patient age, with number of nodes declining with increasing age (Morris et al., 2007); tumor diameter, with more nodes retrieved in larger tumors (Bertelsen et al., 2016; Kim et al., 2016; Merkel et al., 2016; Wang et al., 2017); and mismatch repair deficiency/microsatellite instability, which is associated with a greater number of nodes (Kim et al., 2013). Morris et al. (2007) also found that females were more likely to have an adequate lymphadenectomy (at least 12 nodes) when compared to males (odds ratio (OR) 1.19; 95% CI 1.07−1.33), similarly larger tumors (maximum tumor diameter, per cm OR 1.05; 95% CI 1.03−1.06) and more advanced tumors (pT4 tumors OR 3.03; 95% CI 2.20−4.76) were more likely to undergo adequate lymphadenectomy. When looking at whether lymph node yield is influenced by the pathologist undertaking the lymph node dissection, Morris et al. found that specialist MDT pathologists, defined as those who report on the majority of colorectal cancer specimens in their hospital and regularly attend MDTs, were much more likely than nonspecialist pathologists to achieve an adequate lymphadenectomy (OR 2.16; 95% CI 1.93−2.41) (Morris et al., 2007). Similarly, specialist surgeons who attend the colorectal MDT were more likely to produce an adequate lymphadenectomy (OR 1.40; 95% CI 1.24−1.58) (Morris et al., 2007). The number of lymph nodes also varies according to the location of the tumor; for example, more lymph nodes are present in right-sided and transverse tumor specimens compared to left-sided tumors (Kobayashi et al., 2014). This probably partly reflects the biological difference (more deficient mismatch repair in the right colon) but also differences in the anatomy with right hemicolectomy often containing more than one vascular arcade. The use of neoadjuvant therapy has also been shown to reduce lymph node yield in rectal cancer (Chetty & McCarthy, 2019; Scott et al., 2004), which may become more relevant for colon cancer as the use of neoadjuvant chemotherapy increases. Setting a minimum quota of lymph nodes for pathologists to identify in all colorectal specimens can often be unhelpful. We believe that it is crucial to remove and examine all of the key lymph nodes: the central lymph nodes, paracolic, and intermediate lymph nodes less than 10 cm from the tumor, rather than simply achieving a pre-specified lymph node yield without any consideration of tumor/host biology, surgical radicality, and pathological diligence.

Additional survival benefits following CME surgery are thought to be partly related to the increased lymph node yield and thereby more accurate lymph node staging. When Hohenberger first described CME, he demonstrated that a yield of more than 28 lymph nodes was independently associated with an improved 5-year cancer related survival (96.3% vs 90.7%, $P = .018$) in node-negative patients (Hohenberger et al., 2009). In addition, a Danish multicenter study found CME to be associated with significantly improved disease-free survival in patients with stage I and stage II disease compared to conventional colon cancer surgery (Bertelsen et al., 2015). The concept of lymph node ratio, the quotient of involved to examined lymph nodes, has also been linked to prognosis (Sjo et al., 2012). Other studies have also shown that the number of resected lymph nodes has a positive impact on survival even in patients with node-negative disease (Chang et al., 2007; Chen & Bilchik, 2006; Le Voyer et al., 2003; Sjo et al., 2012), although this improvement in outcome may be related to the higher quality of pathology in these centers and increased reporting of other high-risk features (Morris et al., 2007).

11.6.5 Specimen photography

Photography of the resection specimen is crucial to improving colon cancer outcomes by capturing and providing a permanent record of the quality of the specimen, which can be fed back to the surgical team through demonstration and discussion in the MDT meetings. The benefits of specimen photography have been proven in several studies and clinical trials across both colon and rectal cancer to assess the quality of the specimen and relationship to outcomes (Munkedal et al., 2014; Nakajima et al., 2014). Munkedal et al. (2014) demonstrated that the rate of mesocolic plane excision could be significantly increased by holding MDT meetings with review of the specimen photography, in which a pathologist gave feedback to the surgeons regarding the plane of surgery (52%—76%, $P = .02$). Specimen photography is also key in the clinical trial setting as it allows central review of specimens from multiple centers, thereby reducing the interobserver variation associated with mesocolic grading, as described earlier (Kobayashi et al., 2014; West, Hohenberger, et al., 2010; West et al., 2012; West, Sutton, et al., 2010). It is crucial that specimen photography is performed in a methodical and standardized way; below we have set out a simple method to achieve consistent, high-quality photographs that can easily be performed through routine clinical care.

Photographs should be taken with a high-resolution digital camera, ideally mounted on a fixed stand where possible to minimize any movement artifacts, and always taken from directly above the specimen to reduce distortion artifact. A metric scale should be included in all photographs to allow calibration with image analysis software, for example, ruler. Photographs should firstly be taken of the whole specimen (front and back), preferably before the specimen is opened and ideally in the fresh unfixed state. The mesentery should be laid out flat without being stretched under tension, with the proximal and distal aspects of the specimen clearly labeled (if not obvious) and the site of the tumor and vascular ties easily visible or indicated (Fig. 11.3). Close-ups of any mesocolic defects, perforations, or other features of interest can also be taken.

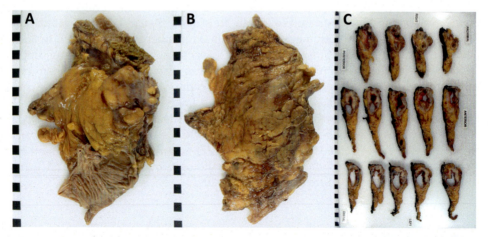

FIGURE 11.3 Specimen photography protocol including the (A) front and (B) back of the intact specimen, and also the (C) serial cross-sectional slices. Note that a metric scale is present in all images. The tumor, high-tie, proximal and distal aspects should be indicated, if not obvious, for example, by the use of preprinted labels.

Following adequate fixation of the specimen, the tumor segment should be sliced at 3–4 mm intervals and additional photographs of the cross-sectional slices should be taken; these should be laid out sequentially with the proximal and distal slice clearly labeled (Fig. 11.3).

11.7 Conclusion

The sustained improvement in rectal cancer outcomes over the past 20–30 years has partly been attributed to the corresponding improvement in the quality of surgery following the widespread introduction of TME. We believe that there are similar significant gains to be made by improving and standardizing colon cancer surgery, in particular concentrating on improving the proportion of tumors resected in the mesocolic plane. Studies that include an independent assessment of the quality of the surgical specimen, specifically through grading of the mesocolic plane, have consistently shown that CME surgery is more likely to produce an intact specimen in the mesocolic plane than conventional non-CME surgery as well as an improvement in the other proposed markers of surgical quality including an increased lymph node yield and central ligation height (Kobayashi et al., 2014; West et al., 2008; West, Hohenberger, et al., 2010; West et al., 2012). An improvement in the rate of mesocolic plane surgery has been associated with an improvement in 5-year survival rates particularly in patients with stage III colon cancer where survival benefit remained significant on multivariate analysis. Despite this, many studies published to date comparing CME to conventional colon cancer surgery have failed to appreciate the importance of independent pathological assessment of the quality of the surgical specimen to validate the surgical technique used in each group. Given the lack of international standardization in the precise definition of CME, we believe that this should be an essential component of all CME studies going forward.

A recent international consensus meeting recommended that extended lymphadenectomy techniques should be the standard procedure for colon cancer surgery (Søndenaa et al., 2014). However, the survival benefit of CME versus non-CME is still controversial and many authors question its suitability for use in all patients, especially those with early-stage disease or significant comorbidity. Despite many studies showing improvement in oncological outcomes when comparing CME or D3 lymphadenectomy, and conventional or D2 surgery, these findings are not reflected in all studies and it is important to note their limitations. To date, there are no results published from randomized controlled trials comparing CME to non-CME surgery with standardization and independent confirmation of resection in the mesocolic plane. It is likely that not all patients will significantly benefit from a high ligation and some series have reported an increase in postoperative morbidity. The concept of identifying involved lymph nodes intraoperatively has been explored in trials such as the multicenter GLiSten (Next-Generation intraoperative lymph node staging for Stratified colon cancer surgery) trial, which used 5-aminolevulinic acid (5-ALA) as a fluorescent probe (Andrew et al., 2016). Unfortunately, the use of 5-ALA to detect lymph node metastases was unsuccessful; however, the concept of using a more sensitive fluorescent probe to detect involved lymph nodes in colon cancer remains and could allow a stratified approach to colon cancer surgery in the future.

References

Abdelkhalek, M., Setit, A., Bianco, F., Belli, A., Denewer, A., Youssef, T. F., Falato, A., & Romano, G. M. (2018). Complete mesocolic excision with central vascular ligation in comparison with conventional surgery for patients with colon cancer - The experiences at two centers. *Annals of Coloproctology, 34*(4), 180−186. Available from https://doi.org/10.3393/ac.2017.08.05.

Alhassan, N., Yang, M., Wong-Chong, N., Liberman, A. S., Charlebois, P., Stein, B., Fried, G. M., & Lee, L. (2019). Comparison between conventional colectomy and complete mesocolic excision for colon cancer: A systematic review and pooled analysis. *Surgical Endoscopy, 33*(1), 8−18.

Andrew, H., Gossedge, G., Croft, J., Corrigan, N., Brown, J. M., West, N., Quirke, P., Tolan, D., Cahill, R., & Jayne, D. G. (2016). Next generation intraoperative lymph node staging for stratified colon cancer surgery (glisten): A multicentre, multinational feasibility study of fluorescence in predicting lymph node-positive disease.

Athanasiou, C. D., Robinson, J., Yiasemidou, M., Lockwood, S., & Markides, G. A. (2017). Laparoscopic vs open approach for transverse colon cancer. A systematic review and meta-analysis of short and long term outcomes. *International Journal of Surgery, 41*, 78−85.

Bae, S. U., Saklani, A. P., Lim, D. R., Kim, D. W., Hur, H., Min, B. S., Baik, S. H., Lee, K. Y., & Kim, N. K. (2014). Laparoscopic-assisted versus open complete mesocolic excision and central vascular ligation for right-sided colon cancer. *Annals of Surgical Oncology, 21*(7), 2288−2294.

Bernhoff, R., Martling, A., Sjövall, A., Granath, F., Hohenberger, W., & Holm, T. (2015). Improved survival after an educational project on colon cancer management in the county of Stockholm—A population based cohort study. *European Journal of Surgical Oncology (EJSO), 41*(11), 1479−1484.

Bertelsen, C., Neuenschwander, A., Jansen, J., Kirkegaard-Klitbo, A., Tenma, J., Wilhelmsen, M., Rasmussen, L., Jepsen, L., Kristensen, B., & Gögenur, I. (2016). Short-term outcomes after complete mesocolic excision compared with 'conventional' colonic cancer surgery. *British Journal of Surgery, 103*(5), 581−589.

Bertelsen, C. A., Bols, B., Ingeholm, P., Jansen, J. E., Neuenschwander, A. U., & Vilandt, J. (2011). Can the quality of colonic surgery be improved by standardization of surgical technique with complete mesocolic excision? *Colorectal Disease, 13*(10), 1123−1129.

Bertelsen, C. A., Neuenschwander, A. U., Jansen, J. E., Wilhelmsen, M., Kirkegaard-Klitbo, A., Tenma, J. R., Bols, B., Ingeholm, P., Rasmussen, L. A., & Jepsen, L. V. (2015). Disease-free survival after complete mesocolic excision compared with conventional colon cancer surgery: A retrospective, population-based study. *The Lancet Oncology, 16*(2), 161−168.

Betge, J., Pollheimer, M. J., Lindtner, R. A., Kornprat, P., Schlemmer, A., Rehak, P., Vieth, M., Hoefler, G., & Langner, C. (2012). Intramural and extramural vascular invasion in colorectal cancer: Prognostic significance and quality of pathology reporting. *Cancer, 118*(3), 628−638.

Bokey, E., Chapuis, P., Dent, O., Mander, B., Bissett, I., & Newland, R. (2003). Surgical technique and survival in patients having a curative resection for colon cancer. *Diseases of the Colon & Rectum, 46*(7), 860−866.

Brierley, J. D., Gospodarowicz, M. K., & Wittekind, C. (2017). *TNM classification of malignant tumours*. John Wiley & Sons.

Byrnes, K. G., Walsh, D., Lewton-Brain, P., McDermott, K., & Coffey, J. C. (2019). Anatomy of the mesentery: Historical development and recent advances. *Seminars in Cell & Developmental Biology, 92*, 4−11. Available from https://doi.org/10.1016/j.semcdb.2018.10.003.

Chang, G. J., Rodriguez-Bigas, M. A., Skibber, J. M., & Moyer, V. A. (2007). Lymph node evaluation and survival after curative resection of colon cancer: Systematic review. *Journal of the National Cancer Institute, 99*(6), 433−441.

Chen, S. L., & Bilchik, A. J. (2006). More extensive nodal dissection improves survival for stages I to III of colon cancer: A population-based study. *Annals of Surgery, 244*(4), 602.

Chetty, R., & McCarthy, A. J. (2019). Neoadjuvant chemoradiation and rectal cancer. *Journal of Clinical Pathology, 72*(2), 97−101.

Cho, M. S., Baek, S. J., Hur, H., Min, B. S., Baik, S. H., & Kim, N. K. (2015). Modified complete mesocolic excision with central vascular ligation for the treatment of right-sided colon cancer: Long-term outcomes and prognostic factors. *Annals of Surgery, 261*(4), 708−715.

Chow, C. F., & Kim, S. H. (2014). Laparoscopic complete mesocolic excision: West meets East. *World Journal of Gastroenterology: WJG, 20*(39), 14301.

References

Coffey, J. C., Walsh, D., Byrnes, K. G., Hohenberger, W., & Heald, R. J. (2020). Mesentery - A 'New' organ. *Emerging Topics in Life Sciences*, 4(2), 191–206. Available from https://doi.org/10.1042/etls20200006.

Culligan, K., Coffey, J. C., Kiran, R. P., Kalady, M., Lavery, I. C., & Remzi, F. H. (2012). The mesocolon: A prospective observational study. *Colorectal Disease*, 14(4), 421–428. Available from https://doi.org/10.1111/j.1463-1318.2012.02935.x, discussion 428–430.

Culligan, K., Remzi, F. H., Soop, M., & Coffey, J. C. (2013). Review of nomenclature in colonic surgery—proposal of a standardised nomenclature based on mesocolic anatomy. *The Surgeon: Journal of the Royal Colleges of Surgeons of Edinburgh and Ireland*, 11(1), 1–5. Available from https://doi.org/10.1016/j.surge.2012.01.006.

Dolejs, S. C., Waters, J. A., Ceppa, E. P., & Zarzaur, B. L. (2017). Laparoscopic versus robotic colectomy: A national surgical quality improvement project analysis. *Surgical Endoscopy*, 31(6), 2387–2396.

Dukes, C. E. (1932). The classification of cancer of the rectum. *The Journal of Pathology and Bacteriology*, 35(3), 323–332.

Emmanuel, A., & Haji, A. (2016). Complete mesocolic excision and extended (D3) lymphadenectomy for colonic cancer: is it worth that extra effort? A review of the literature. *International Journal of Colorectal Disease*, 31(4), 797–804.

Enker, W. E., Laffer, U. T., & Block, G. E. (1979). Enhanced survival of patients with colon and rectal cancer is based upon wide anatomic resection. *Annals of Surgery*, 190(3), 350.

Freund, M., Edden, Y., Reissman, P., & Dagan, A. (2016). Iatrogenic superior mesenteric vein injury: The perils of high ligation. *International Journal of Colorectal Disease*, 31(9), 1649–1651.

Gabriel, W., Dukes, C., & Bussey, H. (1935). Lymphatic spread in cancer of the rectum. *British Journal of Surgery*, 23(90), 395–413.

Galizia, G., Lieto, E., De Vita, F., Ferraraccio, F., Zamboli, A., Mabilia, A., Auricchio, A., Castellano, P., Napolitano, V., & Orditura, M. (2014). Is complete mesocolic excision with central vascular ligation safe and effective in the surgical treatment of right-sided colon cancers? A prospective study. *International Journal of Colorectal Disease*, 29(1), 89–97.

Gouvas, N., Pechlivanides, G., Zervakis, N., Kafousi, M., & Xynos, E. (2012). Complete mesocolic excision in colon cancer surgery: A comparison between open and laparoscopic approach. *Colorectal Disease*, 14(11), 1357–1364.

Group, Fo. C. (2012). Feasibility of preoperative chemotherapy for locally advanced, operable colon cancer: The pilot phase of a randomised controlled trial. *The Lancet Oncology*, 13(11), 1152–1160.

Guillou, P. J., Quirke, P., Thorpe, H., Walker, J., Jayne, D. G., Smith, A. M., Heath, R. M., & Brown, J. M. (2005). Short-term endpoints of conventional versus laparoscopic-assisted surgery in patients with colorectal cancer (MRC CLASICC trial): Multicentre, randomised controlled trial. *Lancet*, 365(9472), 1718–1726. Available from https://doi.org/10.1016/s0140-6736(05)66545-2.

Guo, P., Ye, Y., Jiang, K., Gao, Z., Wang, T., Yin, M., Wang, Y., Xie, Q., Yang, X., & Qu, J. (2012). Learning curve of complete mesocolic excision for colon cancer. *Zhonghua Wei Chang Wai Ke Za Zhi = Chinese Journal of Gastrointestinal Surgery*, 15(1), 28.

Han, D.-P., Lu, A.-G., Feng, H., Cao, Q.-F., Zong, Y.-P., Feng, B., & Zheng, M.-H. (2013). Long-term results of laparoscopy-assisted radical right hemicolectomy with D3 lymphadenectomy: Clinical analysis with 177 cases. *International Journal of Colorectal Disease*, 28(5), 623–629.

Hashiguchi, Y., Muro, K., Saito, Y., Ito, Y., Ajioka, Y., Hamaguchi, T., Hasegawa, K., Hotta, K., Ishida, H., & Ishiguro, M. (2020). Japanese Society for Cancer of the Colon and Rectum (JSCCR) guidelines 2019 for the treatment of colorectal cancer. *International Journal of Clinical Oncology*, 25(1), 1–42.

Heald, R., Husband, E., & Ryall, R. (1982). The mesorectum in rectal cancer surgery—The clue to pelvic recurrence? *British Journal of Surgery*, 69(10), 613–616.

Hirai, K., Yoshinari, D., Ogawa, H., Nakazawa, S., Takase, Y., Tanaka, K., Miyamae, Y., Takahashi, N., Tsukagoshi, H., & Toya, H. (2013). Three-dimensional computed tomography for analyzing the vascular anatomy in laparoscopic surgery for right-sided colon cancer. *Surgical Laparoscopy, Endoscopy & Percutaneous Techniques*, 23(6), 536–539.

Hohenberger, W., Weber, K., Matzel, K., Papadopoulos, T., & Merkel, S. (2009). Standardized surgery for colonic cancer: Complete mesocolic excision and central ligation—Technical notes and outcome. *Colorectal Disease: the Official Journal of the Association of Coloproctology of Great Britain and Ireland*, 11(4), 354–364. Available from https://doi.org/10.1111/j.1463-1318.2008.01735.x, discussion 364-5.

Huh, J. W., Kim, H. R., & Kim, Y. J. (2010). Prognostic value of perineural invasion in patients with stage II colorectal cancer. *Annals of Surgical Oncology, 17*(8), 2066−2072.

Jamieson, J. K., & Dobson, J. F. (1909). The lymphatics of the colon. *Proceedings of the Royal Society of Medicine, 2*, 149−174. (Surg Sect).

Jepsen, R. K., Ingeholm, P., & Lund, E. L. (2012). Upstaging of early colorectal cancers following improved lymph node yield after methylene blue injection. *Histopathology, 61*(5), 788−794.

Kapiteijn, E., Putter, H., & Van de Velde, C. (2002). Impact of the introduction and training of total mesorectal excision on recurrence and survival in rectal cancer in The Netherlands. *British Journal of Surgery, 89*(9), 1142−1149.

Karachun, A., Petrov, A., Panaiotti, L., Voschinin, Y., & Ovchinnikova, T. (2019). Protocol for a multicentre randomized clinical trial comparing oncological outcomes of D2 versus D3 lymph node dissection in colonic cancer (COLD trial). *BJS Open, 3*(3), 288.

Kaye, T. L., West, N. P., Jayne, D. G., & Tolan, D. J. (2016). CT assessment of right colonic arterial anatomy pre and post cancer resection−A potential marker for quality and extent of surgery? *Acta Radiologica, 57*(4), 394−400.

Kijima, S., Sasaki, T., Nagata, K., Utano, K., Lefor, A. T., & Sugimoto, H. (2014). Preoperative evaluation of colorectal cancer using CT colonography, MRI, and PET/CT. *World Journal of Gastroenterology: WJG, 20*(45), 16964.

Kim, N. K., Kim, Y. W., Han, Y. D., Cho, M. S., Hur, H., Min, B. S., & Lee, K. Y. (2016). Complete mesocolic excision and central vascular ligation for colon cancer: Principle, anatomy, surgical technique, and outcomes. *Surgical Oncology, 25*(3), 252−262.

Kim, Y. W., Jan, K. M., Jung, D. H., Cho, M. Y., & Kim, N. K. (2013). Histological inflammatory cell infiltration is associated with the number of lymph nodes retrieved in colorectal cancer. *Anticancer Research, 33*(11), 5143−5150.

Kitano, S., Inomata, M., Mizusawa, J., Katayama, H., Watanabe, M., Yamamoto, S., Ito, M., Saito, S., Fujii, S., Konishi, F., Saida, Y., Hasegawa, H., Akagi, T., Sugihara, K., Yamaguchi, T., Masaki, T., Fukunaga, Y., Murata, K., Okajima, M., & Shimada, Y. (2017). Survival outcomes following laparoscopic versus open D3 dissection for stage II or III colon cancer (JCOG0404): a phase 3, randomised controlled trial. *Lancet Gastroenterologia y Hepatologia, 2*(4), 261−268. Available from https://doi.org/10.1016/s2468-1253(16)30207-2.

Knijn, N., Mogk, S. C., Teerenstra, S., Simmer, F., & Nagtegaal, I. D. (2016). Perineural invasion is a strong prognostic factor in colorectal cancer. *The American Journal of Surgical Pathology, 40*(1), 103−112.

Kobayashi, H., West, N. P., Takahashi, K., Perrakis, A., Weber, K., Hohenberger, W., Quirke, P., & Sugihara, K. (2014). Quality of surgery for stage III colon cancer: Comparison between England, Germany, and Japan. *Annals of Surgical Oncology, 21*(Suppl. 3), S398−S404. Available from https://doi.org/10.1245/s10434-014-3578-9.

Koh, F. H., & Tan, K.-K. (2019). Complete mesocolic excision for colon cancer: is it worth it? *Journal of Gastrointestinal Oncology, 10*(6), 1215.

Kotake, K., Kobayashi, H., Asano, M., Ozawa, H., & Sugihara, K. (2015). Influence of extent of lymph node dissection on survival for patients with pT2 colon cancer. *International Journal of Colorectal Disease, 30*(6), 813−820.

Le Voyer, T., Sigurdson, E., Hanlon, A., Mayer, R., Macdonald, J., Catalano, P., & Haller, D. (2003). Colon cancer survival is associated with increasing number of lymph nodes analyzed: A secondary survey of intergroup trial INT-0089. *Journal of Clinical Oncology, 21*(15), 2912−2919.

Lê, P., Mehtari, L., & Billey, C. (2006). Carcinoma of the transverse colon. *Journal de Chirurgie, 143*(5), 285−293.

Liang, J.-T., Huang, K.-C., Lai, H.-S., Lee, P.-H., & Sun, C.-T. (2007). Oncologic results of laparoscopic D3 lymphadenectomy for male sigmoid and upper rectal cancer with clinically positive lymph nodes. *Annals of Surgical Oncology, 14*(7), 1980−1990.

Lim, S.-B., Yu, C. S., Jang, S. J., Kim, T. W., Kim, J. H., & Kim, J. C. (2010). Prognostic significance of lymphovascular invasion in sporadic colorectal cancer. *Diseases of the Colon & Rectum, 53*(4), 377−384.

Loughrey, M. B., Quirke, P., & Shepherd, N. A. (2018). *Standards and datasets for reporting cancers: Dataset for histopathological reporting of colorectal cancer* (V4, December 2017). The Royal College of Pathologists website. Available from https://www.rcpath.org/uploads/assets/c8b61ba0-ae3f-43f1-85ffd3ab9f17cfe6/G049-Dataset-for-histopathological-reporting-of-colorectal-cancer.pdf.

Lu, J.-Y., Xu, L., Xue, H.-D., Zhou, W.-X., Xu, T., Qiu, H.-Z., Wu, B., Lin, G.-L., & Xiao, Y. (2016). The Radical Extent of lymphadenectomy—D2 dissection versus complete mesocolic excision of LAparoscopic Right

Colectomy for right-sided colon cancer (RELARC) trial: Study protocol for a randomized controlled trial. *Trials*, 17(1), 582.

Martínez-Pérez, A., Reitano, E., Gavriilidis, P., Genova, P., Moroni, P., Memeo, R., Brunetti, F., & de'Angelis, N. (2019). What is the best surgical option for the resection of transverse colon cancer? *Annals of Laparoscopic and Endoscopic Surgery*, 4. Available from http://ales.amegroups.com/article/view/5305.

Martling, A. L., Holm, T., Rutqvist, L., Moran, B., Heald, R., & Cedermark, B. (2000). Effect of a surgical training programme on outcome of rectal cancer in the County of Stockholm. *The Lancet*, 356(9224), 93–96.

Merkel, S., Weber, K., Matzel, K., Agaimy, A., Göhl, J., & Hohenberger, W. (2016). Prognosis of patients with colonic carcinoma before, during and after implementation of complete mesocolic excision. *British Journal of Surgery*, 103(9), 1220–1229.

Merrie, A. E., Phillips, L. V., Yun, K., & McCall, J. L. (2001). Skip metastases in colon cancer: assessment by lymph node mapping using molecular detection. *Surgery*, 129(6), 684–691. Available from https://doi.org/10.1067/msy.2001.113887.

Milone, M., Degiuli, M., Allaix, M., Ammirati, C., Anania, G., Barberis, A., Belli, A., Bianchi, P., Bianco, F., & Bombardini, C. (2020). Mid-transverse colon cancer and extended versus transverse colectomy: Results of the Italian Society of Surgical Oncology Colorectal Cancer Network (SICO CCN) Multicenter Collaborative Study. *European Journal of Surgical Oncology*, 46(9), 1683–1688.

Mori, S., Kita, Y., Baba, K., Yanagi, M., Tanabe, K., Uchikado, Y., Kurahara, H., Arigami, T., Uenosono, Y., & Mataki, Y. (2017). Laparoscopic complete mesocolic excision via combined medial and cranial approaches for transverse colon cancer. *Surgery Today*, 47(5), 643–649.

Morikawa, E., Yasutomi, M., Shindou, K., Matsuda, T., Mori, N., Hida, J., Kubo, R., Kitaoka, M., Nakamura, M., Fujimoto, K., et al. (1994). Distribution of metastatic lymph nodes in colorectal cancer by the modified clearing method. *Diseases of the Colon and Rectum*, 37(3), 219–223. Available from https://doi.org/10.1007/bf02048158.

Morris, E. J. A., Maughan, N. J., Forman, D., & Quirke, P. (2007). Identifying stage III colorectal cancer patients: The influence of the patient, surgeon, and pathologist. *Journal of Clinical Oncology*, 25(18), 2573–2579.

Munkedal, D. L. E., Rosenkilde, M., West, N. P., & Laurberg, S. (2019). Routine CT scan one year after surgery can be used to estimate the level of central ligation in colon cancer surgery. *Acta Oncologica*, 58(4), 469–471.

Munkedal, D. L., Laurberg, S., Hagemann-Madsen, R., Stribolt, K. J., Krag, S. R., Quirke, P., & West, N. P. (2016). Significant individual variation between pathologists in the evaluation of colon cancer specimens after complete mesocolic excision. *Diseases of the Colon and Rectum*, 59(10), 953–961. Available from https://doi.org/10.1097/dcr.0000000000000671.

Munkedal, D., Rosenkilde, M., Nielsen, D., Sommer, T., West, N., & Laurberg, S. (2017). Radiological and pathological evaluation of the level of arterial division after colon cancer surgery. *Colorectal Disease*, 19(7), O238–O245.

Munkedal, D., West, N., Iversen, L., Hagemann-Madsen, R., Quirke, P., & Laurberg, S. (2014). Implementation of complete mesocolic excision at a university hospital in Denmark: An audit of consecutive, prospectively collected colon cancer specimens. *European Journal of Surgical Oncology (EJSO)*, 40(11), 1494–1501.

Nakajima, K., Inomata, M., Akagi, T., Etoh, T., Sugihara, K., Watanabe, M., Yamamoto, S., Katayama, H., Moriya, Y., & Kitano, S. (2014). Quality control by photo documentation for evaluation of laparoscopic and open colectomy with D3 resection for stage II/III colorectal cancer: Japan Clinical Oncology Group Study JCOG 0404. *Japanese Journal of Clinical Oncology*, 44(9), 799–806.

Ng, K.-S., West, N. P., Scott, N., Holzgang, M., Quirke, P., & Jayne, D. G. (2020). What factors determine specimen quality in colon cancer surgery? A cohort study. *International Journal of Colorectal Disease*, 35(5), 869–880.

Office for National Statistics. (2019). *Cancer survival by stage at diagnosis for England*.

Olofsson, F., Buchwald, P., Elmståhl, S., & Syk, I. (2016). No benefit of extended mesenteric resection with central vascular ligation in right-sided colon cancer. *Colorectal Disease*, 18(8), 773–778.

Ozben, V., de Muijnck, C., Sengun, B., Zenger, S., Agcaoglu, O., Balik, E., Aytac, E., Bilgin, I., Baca, B., & Hamzaoglu, I. (2020). Robotic complete mesocolic excision for transverse colon cancer can be performed with a morbidity profile similar to that of conventional laparoscopic colectomy. *Techniques in Coloproctology*, 24(10), 1035–1042.

Perrakis, A., Weber, K., Merkel, S., Matzel, K., Agaimy, A., Gebbert, C., & Hohenberger, W. (2014). Lymph node metastasis of carcinomas of transverse colon including flexures. Consideration of the extramesocolic lymph node stations. *International Journal of Colorectal Disease*, 29(10), 1223–1229.

Petz, W., Ribero, D., Bertani, E., Borin, S., Formisano, G., Esposito, S., Spinoglio, G., & Bianchi, P. (2017). Suprapubic approach for robotic complete mesocolic excision in right colectomy: Oncologic safety and short-term outcomes of an original technique. *European Journal of Surgical Oncology*, 43(11), 2060−2066.

Quirke, P., Sebag-Montefiore, D., Steele, R., Khanna, S., Monson, J., Holliday, A., Thompson, L., Griffiths, G., & Stephens, R. (2006). Local recurrence after rectal cancer resection is strongly related to the plane of surgical dissection and is further reduced by pre-operative short course radiotherapy. Preliminary results of the Medical Research Council (MRC) CR07 Trial. *Journal of Clinical Oncology*, 24(18 Suppl.), 3512, −3512. Available from https://doi.org/10.1200/jco.2006.24.18_suppl.3512.

Quirke, P., Steele, R., Monson, J., Grieve, R., Khanna, S., Couture, J., O'Callaghan, C., Myint, A. S., Bessell, E., & Thompson, L. C. (2009). Effect of the plane of surgery achieved on local recurrence in patients with operable rectal cancer: A prospective study using data from the MRC CR07 and NCIC-CTG CO16 randomised clinical trial. *The Lancet*, 373(9666), 821−828.

Saha, S., Johnston, G., Korant, A., Shaik, M., Kanaan, M., Johnston, R., Ganatra, B., Kaushal, S., Desai, D., & Mannam, S. (2013). Aberrant drainage of sentinel lymph nodes in colon cancer and its impact on staging and extent of operation. *The American Journal of Surgery*, 205(3), 302−306.

Santos, C., López-Doriga, A., Navarro, M., Mateo, J., Biondo, S., Martínez Villacampa, M., Soler, G., Sanjuan, X., Paules, M., & Laquente, B. (2013). Clinicopathological risk factors of stage II colon cancer: Results of a prospective study. *Colorectal Disease*, 15(4), 414−422.

Scott, N., Thorne, C., & Jayne, D. (2004). Lymph node retrieval after neoadjuvant radiotherapy for rectal adenocarcinoma. *Journal of Clinical Pathology*, 57(3), 335−336.

Sheehan-Dare, G. E., Marks, K. M., Tinkler-Hundal, E., Ingeholm, P., Bertelsen, C. A., Quirke, P., & West, N. P. (2018). The effect of a multidisciplinary regional educational programme on the quality of colon cancer resection. *Colorectal Disease*, 20(2), 105−115.

Shin, J. K., Kim, H. C., Lee, W. Y., Yun, S. H., Cho, Y. B., Huh, J. W., Park, Y. A., & Chun, H.-K. (2018). Laparoscopic modified mesocolic excision with central vascular ligation in right-sided colon cancer shows better short-and long-term outcomes compared with the open approach in propensity score analysis. *Surgical Endoscopy*, 32(6), 2721−2731.

Shiozawa, M., Ueno, H., Shiomo, A., Kim, N., Kim, J., Tsarkov, P., Grützmann, R., Dulskas, A., Liang, J., & Samalavičius, N. (2020). Study protocol for an International Prospective Observational Cohort Study for Optimal Bowel Resection Extent and Central Radicality for Colon Cancer (T-REX Study). *Japanese Journal of Clinical Oncology*, 51(1), 145−155.

Siani, L. M., Lucchi, A., Berti, P., & Garulli, G. (2017). Laparoscopic complete mesocolic excision with central vascular ligation in 600 right total mesocolectomies: Safety, prognostic factors and oncologic outcome. *The American Journal of Surgery*, 214(2), 222−227.

Sjo, O. H., Merok, M. A., Svindland, A., & Nesbakken, A. (2012). Prognostic impact of lymph node harvest and lymph node ratio in patients with colon cancer. *Diseases of the Colon & Rectum*, 55(3), 307−315.

Sjövall, A., Granath, F., Cedermark, B., Glimelius, B., & Holm, T. (2007). Loco-regional recurrence from colon cancer: a population-based study. *Annals of Surgical Oncology*, 14(2), 432−440.

Søndenaa, K., Quirke, P., Hohenberger, W., Sugihara, K., Kobayashi, H., Kessler, H., Brown, G., Tudyka, V., D'Hoore, A., Kennedy, R. H., West, N. P., Kim, S. H., Heald, R., Storli, K. E., Nesbakken, A., & Moran, B. (2014). The rationale behind complete mesocolic excision (CME) and a central vascular ligation for colon cancer in open and laparoscopic surgery: Proceedings of a consensus conference. *International Journal of Colorectal Disease*, 29(4), 419−428. Available from https://doi.org/10.1007/s00384-013-1818-2.

Spasojevic, M., Stimec, B. V., Gronvold, L. B., Nesgaard, J.-M., Edwin, B., & Ignjatovic, D. (2011). The anatomical and surgical consequences of right colectomy for cancer. *Diseases of the Colon & Rectum*, 54(12), 1503−1509.

Spinoglio, G., Bianchi, P. P., Marano, A., Priora, F., Lenti, L. M., Ravazzoni, F., Petz, W., Borin, S., Ribero, D., & Formisano, G. (2018). Robotic versus laparoscopic right colectomy with complete mesocolic excision for the treatment of colon cancer: Perioperative outcomes and 5-year survival in a consecutive series of 202 patients. *Annals of Surgical Oncology*, 25(12), 3580−3586.

Storli, K. E., Søndenaa, K., Furnes, B., & Eide, G. E. (2013). Outcome after introduction of complete mesocolic excision for colon cancer is similar for open and laparoscopic surgical treatments. *Digestive Surgery*, 30(4−6), 317−327.

Storli, K., Søndenaa, K., Furnes, B., Nesvik, I., Gudlaugsson, E., Bukholm, I., & Eide, G. (2014). Short term results of complete (D3) vs. standard (D2) mesenteric excision in colon cancer shows improved outcome of complete mesenteric excision in patients with TNM stages I-II. *Techniques in Coloproctology*, *18*(6), 557–564.

Tagliacozzo, S., & Tocchi, A. (1997). Extended mesenteric excision in right hemicolectomy for carcinoma of the colon. *International Journal of Colorectal Disease*, *12*(5), 272–275.

Tan, K. Y., Kawamura, Y. J., Mizokami, K., Sasaki, J., Tsujinaka, S., Maeda, T., Nobuki, M., & Konishi, F. (2010). Distribution of the first metastatic lymph node in colon cancer and its clinical significance. *Colorectal Disease: The Official Journal of the Association of Coloproctology of Great Britain and Ireland*, *12*(1), 44–47. Available from https://doi.org/10.1111/j.1463-1318.2009.01924.x.

Toyota, S., Ohta, H., & Anazawa, S. (1995). Rationale for extent of lymph node dissection for right colon cancer. *Diseases of the Colon and Rectum*, *38*(7), 705–711. Available from https://doi.org/10.1007/bf02048026.

Trastulli, S., Coratti, A., Guarino, S., Piagnerelli, R., Annecchiarico, M., Coratti, F., Di Marino, M., Ricci, F., Desiderio, J., & Cirocchi, R. (2015). Robotic right colectomy with intracorporeal anastomosis compared with laparoscopic right colectomy with extracorporeal and intracorporeal anastomosis: A retrospective multicentre study. *Surgical Endoscopy*, *29*(6), 1512–1521.

Ueno, H., Shirouzu, K., Eishi, Y., Yamada, K., Kusumi, T., Kushima, R., Ikegami, M., Murata, A., Okuno, K., & Sato, T. (2013). Characterization of perineural invasion as a component of colorectal cancer staging. *The American Journal of Surgical Pathology*, *37*(10), 1542–1549.

Vandamme, J.-P., & Bonte, J. (1990). *Vascular anatomy in abdominal surgery*. Thieme Medical Publishers.

Wang, C., Gao, Z., Shen, K., Shen, Z., Jiang, K., Liang, B., Yin, M., Yang, X., Wang, S., & Ye, Y. (2017). Safety, quality and effect of complete mesocolic excision vs non-complete mesocolic excision in patients with colon cancer: a systemic review and meta-analysis. *Colorectal Disease*, *19*(11), 962–972.

Weber, K., Merkel, S., Perrakis, A., & Hohenberger, W. (2013). Is there a disadvantage to radical lymph node dissection in colon cancer? *International Journal of Colorectal Disease*, *28*(2), 217–226.

West, N. P., Hohenberger, W., Weber, K., Perrakis, A., Finan, P. J., & Quirke, P. (2010). Complete mesocolic excision with central vascular ligation produces an oncologically superior specimen compared with standard surgery for carcinoma of the colon. *Journal of Clinical Oncology: Official Journal of the American Society of Clinical Oncology*, *28*(2), 272–278. Available from https://doi.org/10.1200/jco.2009.24.1448.

West, N. P., Kobayashi, H., Takahashi, K., Perrakis, A., Weber, K., Hohenberger, W., Sugihara, K., & Quirke, P. (2012). Understanding optimal colonic cancer surgery: comparison of Japanese D3 resection and European complete mesocolic excision with central vascular ligation. *Journal of Clinical Oncology: Official Journal of the American Society of Clinical Oncology*, *30*(15), 1763–1769. Available from https://doi.org/10.1200/jco.2011.38.3992.

West, N., Kennedy, R., Magro, T., Luglio, G., Sala, S., Jenkins, J., & Quirke, P. (2014). Morphometric analysis and lymph node yield in laparoscopic complete mesocolic excision performed by supervised trainees. *British Journal of Surgery*, *101*(11), 1460–1467.

West, N. P., Morris, E. J., Rotimi, O., Cairns, A., Finan, P. J., & Quirke, P. (2008). Pathology grading of colon cancer surgical resection and its association with survival: A retrospective observational study. *The Lancet Oncology*, *9*(9), 857–865.

West, N. P., Sutton, K. M., Ingeholm, P., Hagemann-Madsen, R. H., Hohenberger, W., & Quirke, P. (2010). Improving the quality of colon cancer surgery through a surgical education program. *Diseases of the Colon & Rectum*, *53*(12), 1594–1603.

Wibe, A., Møller, B., Norstein, J., Carlsen, E., Wiig, J. N., Heald, R. J., Langmark, F., Myrvold, H. E., & Søreide, O. (2002). A national strategic change in treatment policy for rectal cancer—implementation of total mesorectal excision as routine treatment in Norway. A national audit. *Diseases of the Colon & Rectum*, *45*(7), 857–866.

Widmar, M., Keskin, M., Strombom, P., Beltran, P., Chow, O. S., Smith, J. J., Nash, G. M., Shia, J., Russell, D., & Garcia-Aguilar, J. (2017). Lymph node yield in right colectomy for cancer: A comparison of open, laparoscopic and robotic approaches. *Colorectal Disease*, *19*(10), 888–894.

World Cancer Research Fund. (2020). *Worldwide Cancer data: Global cancer statistics for the most common cancers* [Online] (Vol. 2020). https://www.wcrf.org/dietandcancer/cancer-trends/worldwide-cancer-data

Xie, Y., Jin, C., Zhang, S., Wang, X., & Jiang, Y. (2015). CT features and common causes of arc of Riolan expansion: an analysis with 64-detector-row computed tomographic angiography. *International Journal of Clinical and Experimental Medicine*, *8*(3), 3193.

Yamamoto, S., Inomata, M., Katayama, H., Mizusawa, J., Etoh, T., Konishi, F., Sugihara, K., Watanabe, M., Moriya, Y., & Kitano, S. (2014). Short-term surgical outcomes from a randomized controlled trial to evaluate laparoscopic and open D3 dissection for stage II/III colon cancer: Japan Clinical Oncology Group Study JCOG 0404. *Annals of Surgery*, *260*(1), 23–30.

Yozgatli, T. K., Aytac, E., Ozben, V., Bayram, O., Gurbuz, B., Baca, B., Balik, E., Hamzaoglu, I., Karahasanoglu, T., & Bugra, D. (2019). Robotic complete mesocolic excision versus conventional laparoscopic hemicolectomy for right-sided colon cancer. *Journal of Laparoendoscopic & Advanced Surgical Techniques*, *29*(5), 671–676.

12

Japanese D3 dissection in cancer of the colon: technique and results

Yuichiro Tsukada and Masaaki Ito

Department of Colorectal Surgery, National Cancer Center Hospital East, Kashiwa, Japan

12.1 Introduction

Colorectal cancer (CRC) remains a major cause of morbidity and mortality worldwide, despite significant improvements in its management. In Japan, the incidence of CRC is rapidly increasing, and CRC is the second most common cause of cancer mortality in this country (Kotake et al., 2003; Muto et al., 2001).

Curative treatment for colon cancer includes surgical resection of the primary tumor with en bloc removal of regional lymph nodes. In recent years, excellent outcomes of complete mesocolic excision (CME) with central vascular ligation (CVL) have been reported. Thus CME with CVL has been recognized worldwide as an appropriate surgical procedure for colon cancer (Hohenberger et al., 2009). CME involves dividing the embryological layers sharply and precisely, mobilizing the mesocolon along Toldt's fascia while preserving its integrity, and removing all the containing lymph nodes; this procedure is similar to the total mesorectal excision (TME) procedure for rectal cancers (Heald, 1988). CVL involves ligating the feeding vessels at their origin (Hohenberger et al., 2009).

Japanese D3 lymph node dissection has also been recognized as an appropriate surgical procedure for colon cancer. The difference between CME with CVL and Japanese D3 is a topic of much interest among colorectal surgeons (Kobayashi et al., 2014; West et al., 2012).

In this chapter, the history, definition, technique, and results of Japanese D3 dissection for colon cancer, and the differences between Japanese D3 and European CME with CVL are described.

12.2 History of lymphadenectomy for colon cancer in Japan

12.2.1 Japanese classification and Japanese guidelines

The Japanese Society for Cancer of the Colon and Rectum (JSCCR) was established in 1973 to improve the quality of diagnosis and treatment of CRC and promote associated research. For these purposes, the JSCCR has published both the "Japanese Classification of Colorectal, Appendiceal, and Anal Carcinoma" (Japanese classification) and the "JSCCR Guidelines for the Treatment of Colorectal Cancer" (Japanese guidelines). The 1st edition of the Japanese classification was published in 1977, and the latest edition was published in 2018 (9th Japanese edition.; 3rd English edition; Japanese Society for Cancer of the Colon and Rectum, 2018, 2019). The 1st edition of the Japanese guidelines was published in 2005, and the latest edition was published in 2019 (Hashiguchi et al., 2020). Generally, the treatment strategy for CRC in Japan is decided according to these guidelines (Ishiguro et al., 2014).

12.3 Basic principles of lymph node dissection in Japan

In Japan, the basic principles of lymph node dissection have been "removing lymph nodes with lymph ducts in one package, enveloped with mesocolon." Therefore separation of the mesocolon from the retroperitoneum has been performed along with the embryological layer, even though this layer has not been clearly described in Japanese guidelines. Cancer cells are thought to spread in two directions (central and pericolic) through lymph systems (lymph nodes and lymph ducts) that accompany arteries; therefore lymph node dissection in these two directions is thought to be essential (Fig. 12.1). According to this cancer cell spreading pattern, regional lymph nodes of colon cancer have been classified into three groups: pericolic, intermediate, and main lymph nodes (Fig. 12.1, Table 12.1).

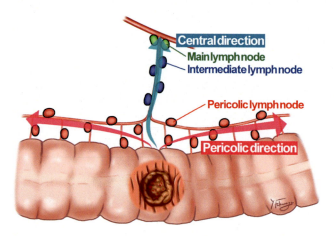

FIGURE 12.1 The two directions in which cancer cells may spread and the basic classification of regional lymph nodes. Cancer cells are thought to spread in two directions (central and pericolic) through lymph systems (lymph nodes and lymph ducts) which accompany arteries. In Japan, regional lymph nodes in colon cancer have been classified into three groups: pericolic, intermediate, and main lymph nodes.

TABLE 12.1 Basic classification of the regional lymph nodes in colon cancer, in Japan.

Name	Definition
Pericolic lymph nodes	Lymph nodes located along straight arteries (vasa recta) and marginal arteries (Pericolic direction)
Intermediate lymph nodes	Lymph nodes located along the trunk of the feeding artery (Central direction)
Main lymph nodes	Lymph nodes located around the origin of the feeding artery (Central direction)

TABLE 12.2 The older Japanese classification of lymph nodes (1994). (Japanese Classification of Colorectal, Appendiceal, and Anal Carcinoma 4th edition, Japanese Society for Cancer of the Colon and Rectum, 1994).

Group	Regional or extra-regional	Definition
Group 1	Regional	Pericolic lymph nodes located within 5 cm from the primary tumor
Group 2	Regional	Pericolic lymph nodes located 5–10 cm from the primary tumor/ Intermediate lymph nodes
Group 3	Regional	Main lymph nodes
Group 4	Extra-regional	Lymph nodes other than those in Groups 1–3

12.4 Changes in the recommended area of lymph node dissection in Japan

Lymph node dissection has been performed according to both the Japanese classification and the Japanese guidelines in Japan, and the area of standard lymphadenectomy has changed according to data from several studies.

In the 1st edition of the Japanese classification, lymph nodes were classified into four groups (Groups 1–4) according to the distance of the lymph nodes from the primary tumor (Japanese Society for Cancer of the Colon and Rectum, 1994) (Table 12.2). Group 1–3 lymph nodes were defined as "regional lymph nodes" and Group 4 lymph nodes were defined as "extra-regional lymph nodes." At that time, pericolic lymph nodes located within 5 cm of the primary tumor were defined as Group 1, those located 5–10 cm from the primary tumor were defined as Group 2, and those located over 10 cm from the primary tumor were defined as Group 4 lymph nodes. Intermediate lymph nodes were defined as Group 2, main lymph nodes were defined as Group 3, and lymph nodes other than Groups 1–3 were defined as Group 4 lymph nodes.

In the older version of the Japanese classification, lymph node dissection in an area wider than the area, including the involved lymph nodes was defined as "curative surgery"; therefore the resecting line for the bowel was more than 10 cm from the primary tumor if the involvement of Group 2 lymph nodes was suspected. However, in the 1st edition of Japanese guidelines published in 2005, it was mentioned that metastasis in the lymph nodes located more than 10 cm from the primary tumor was rare (Hida et al., 1997). Subsequently, the classification of regional lymph nodes was changed from Group 1 to 3 lymph nodes to pericolic, intermediate, and main lymph nodes, and regional pericolic lymph nodes were defined as those located within 10 cm of the primary tumor in the 7th edition of the Japanese classification published in 2006. Due to these changes, the standard resection line of the bowel was set at 10 cm from the primary tumor (termed the 10 cm rule). Before the Japanese guidelines were published, the range of central lymph node dissection was decided according to each patient's tumor stage. However, Kotake et al. (2014) reported a better prognosis in the D3 dissection group (dissection of Group 1–3 lymph nodes) than in the D2 dissection group (dissection of Group 1–2 lymph nodes) in a propensity score-matching study of 6850 T3–4 colon cancer patients who were registered in the JSCCR database. Therefore D3 dissection for T3–4 or lymph node-positive colon cancer patients was recommended in the 1st edition of the Japanese guidelines published in 2005.

12.5 Current classifications of lymph node metastasis (N) and lymph node dissection (D) in Japan

12.5.1 Basic principles of regional lymph node classification

Regional lymph nodes are classified into three groups: pericolic, intermediate, and main lymph nodes, as shown in Fig. 12.1 and Table 12.1. The Japanese classification takes into account the difference between right-sided colon cancers and left-sided colon cancers in terms of the association between the main lymph nodes and the feeding arteries (Japanese Society for Cancer of the Colon and Rectum, 2018, 2019) (Fig. 12.2A and B). In right-sided colon cancers, the main lymph nodes are located around the origin of each

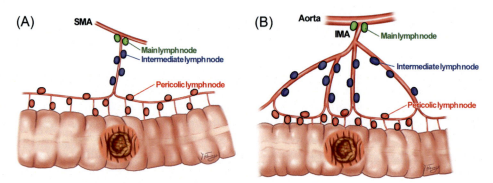

FIGURE 12.2 Regional lymph nodes in right-sided colon cancers and left-sided colon cancers. Locations of the pericolic lymph nodes, intermediate lymph nodes, and the main lymph nodes in right-sided colon cancers are shown. (A) In right-sided colon cancers, the main lymph nodes are located around the origin of each feeding artery, which branches directly from the superior mesenteric artery. (B) In all left-sided colon cancers, the main lymph nodes are located around the root of the inferior mesenteric artery. *SMA*, Superior mesenteric artery; *IMA*, inferior mesenteric artery.

feeding artery: the ileocolic artery, right colic artery (RCA), and middle colic artery (MCA) which branches directly from the superior mesenteric artery (SMA). On the other hand, the main lymph nodes are located around the root of the inferior mesenteric artery (IMA) in all left-sided colon cancers because the feeding arteries [the left colic artery (LCA) and the sigmoid artery (SA)] are branches of the IMA.

The Japanese classification also classifies the range of pericolic lymph nodes into four types on the basis of the positional relationship between the tumor and the feeding arteries (Fig. 12.3A−D).

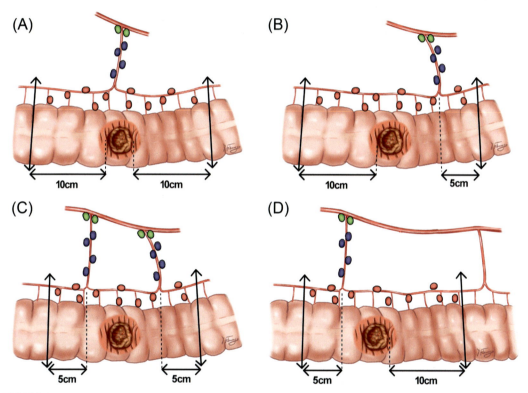

FIGURE 12.3 Four types of pericolic lymph nodes according to the positional relationship between the tumor and the feeding arteries. The pericolic lymph nodes are classified into four types on the basis of the positional relationship between the tumor and the feeding arteries, as follows: (A) When the feeding artery is directly below the tumor, the dissection area extends 10 cm in both the oral and the anal directions from the tumor margins. (B) When there is one feeding artery within 10 cm of the tumor margin, the dissection area extends up to 5 cm beyond the arterial inflow site and the opposite side extends up to 10 cm from the tumor margin. (C) When there are two dominant arteries within 10 cm from the tumor margin, the dissection area extends up to 5 cm beyond the arterial inflow site in both the oral and the anal sides. (D) When there are no feeding arteries within 10 cm of the tumor margin, the dissection area extends up to 5 cm beyond the artery closest to the tumor margin, and the opposite side extends up to 10 cm from the tumor margin.

12.6 Lymph node groups and station numbers

In Japan, each lymph node is named, numbered, and classified according to the accompanying artery (Japanese Society for Cancer of the Colon and Rectum, 2018, 2019) (Table 12.3, Fig. 12.4).

12.7 Classification of lymph node metastases (N)

The extent of lymph node metastasis is classified as N0–N3 according to the Japanese classification (N classification). Previously, N1 was defined as lymph node involvement of Group 1 lymph nodes, N2 as the involvement of Group 2 lymph nodes, and N3 as the involvement of Group 3 lymph nodes; however, the Japanese N classification has been changed to be similar to the TNM classification to promote the internationality of Japanese research. Despite this change, *the classification of N3 remains* as was defined earlier because of its oncological importance (Japanese Society for Cancer of the Colon and Rectum, 2018, 2019) Box 12.1.

TABLE 12.3 Lymph node groups and nodal station numbers at the colon.

	Superior mesenteric artery	**Inferior mesenteric artery**
Pericolic lymph nodes	Lymph nodes along the marginal arteries and near the bowel wall •Pericolic lymph nodes (201, 211, 221)	Lymph nodes along the marginal arteries, near the bowel wall, and along the terminal sigmoid artery •Pericolic lymph nodes (231, 241: 241-1, 241-2, 241-t)
Intermediate lymph nodes	Lymph nodes along with the ileocolic, right colic, and middle colic arteries. •Ileocolic nodes (202) •Right colic nodes (212) •Right middle colic nodes (222-rt) •Left middle colic nodes (222-lt)	Lymph nodes along with the left colic and sigmoid arteries and those along the inferior mesenteric artery between the origin of the left colic artery and the terminal sigmoid artery •Left colic nodes (232) •Sigmoid colic nodes (242: 242-1, 242-2) •Inferior mesenteric trunk nodes (252)
Main lymph nodes	Lymph nodes at the origin of the ileocolic, right colic, and middle colic arteries •Ileocolic root nodes (203) •Right colic root nodes (213) •Middle colic root nodes (223)	Lymph nodes along the inferior mesenteric artery from the origin of the inferior mesenteric artery to that of the left colic artery •Inferior mesenteric root nodes (253)
Lymph nodes proximal to the main lymph nodes	Lymph nodes at the origin of the superior mesenteric artery and those along the aorta •Superior mesenteric arterial root nodes (214) •Para-aortic nodes (216)	Lymph nodes along the aorta •Para-aortic nodes (216)
Other lymph nodes	•Sub-pyloric nodes (206) •Gastroepiploic nodes (204) •Splenic hilar nodes (210)	

The sigmoid artery commonly comprises the first, second, and terminal arteries, with pericolic lymph nodes designated as 241-1, 241-2, and 241-t respectively; the intermediate lymph nodes are designated as 242-1 and 242-2.

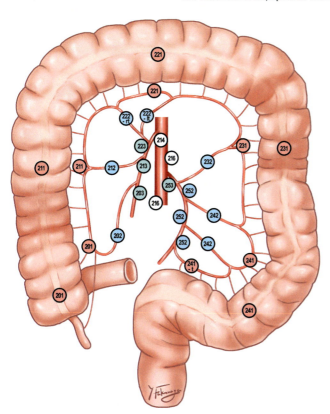

FIGURE 12.4 Lymph node groups and station number at colon. The locations and numbers of regional lymph nodes are shown. Pericolic lymph nodes are indicated in *red*; intermediate lymph nodes are indicated in *blue*; main lymph nodes are indicated in *green*; and the other lymph nodes are indicated in *white*.

BOX 12.1

Japanese classification of lymph node metastasis (N) (2018).

NX: Lymph node metastasis cannot be assessed

N0: No evidence of lymph node metastasis

N1: Metastasis in 1–3 pericolic/perirectal or intermediate lymph nodes
 N1a: Metastasis in 1 lymph node
 N1b: Metastasis in 2–3 lymph node

N2: Metastasis in 4 or more pericolic/perirectal or intermediate lymph nodes
 N2a: Metastasis in 4–6 lymph nodes
 N2b: Metastasis in 7 or more lymph nodes

N3: Metastasis in the main lymph node (s); in lower rectal cancer, metastasis in the main and/or lateral lymph node(s)

12.8 Classification of lymph node dissection (D)

The extent of lymph node dissection (D) is classified into D0−D3 categories. D3 dissection involves complete dissection of the pericolic, intermediate, and main lymph nodes (Japanese Society for Cancer of the Colon and Rectum, 2018, 2019) Box 12.2.

12.9 Technique of Japanese D3 dissection for colon cancer

The area and the technique of Japanese D3 dissection for each tumor location are described in this section.

12.10 Cecum cancer

The ICA is the main feeder of cecum cancer in almost all cases. Ileocecal resection is usually selected for cecum cancers (Fig. 12.5). The ICA is found in all cases, and about half of the ICAs run ventrally to the superior mesenteric vein (SMV) (Hamabe et al., 2018). After separating the right-sided colon and the mesocolon from the retroperitoneum along Toldt's fascia, ligating the ICA at its root is necessary for D3 dissection. The ileocecal vein is also ligated at its root. Exposing the surface of the SMV is essential for safe detection of the ICA root, and for the complete removal of the main lymph nodes located around the origin of the ICA. The hepatic flexure is fully mobilized, and the ascending colon and ileum are resected at 10 cm from the primary tumor. The specimen usually becomes "fan-shaped" (Fig. 12.6A and B). After removal of the specimen, the ileum and the ascending colon are anastomosed.

12.11 Ascending colon cancer

The ICA, RCA, or both of these arteries can be the main feeders of ascending colon cancer. The RCA is found in approximately 30% of the cases, and more than 90% of the RCAs run

---

BOX 12.2

Japanese classification of lymph node dissection (D) (2018).

DX: Extent of lymph node dissection cannot be assessed

D0: Incomplete pericolic/perirectal lymph node dissection

D1: Complete pericolic/perirectal lymph node dissection

D2: Complete pericolic/perirectal and intermediate lymph node dissection

D3: Pericolic/perirectal, intermediate, and main lymph node dissection

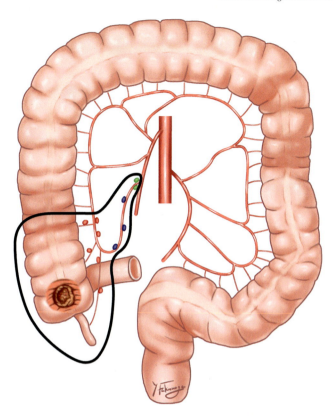

FIGURE 12.5 Regional lymph nodes and resection line in cecum cancer. The regional lymph nodes and bowel resection lines in cecum cancer are shown. Pericolic lymph nodes are indicated in *red*; intermediate lymph nodes are indicated in *blue*; and main lymph nodes are indicated in *green*. Black line shows the resection line of the ileocecal resection.

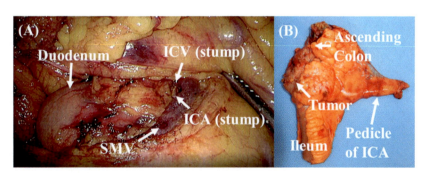

FIGURE 12.6 (A) D3 dissection for cecum cancer and (B) "fan-shaped" specimen. (A) Surgical photograph of a D3 dissection for cecum cancer is shown. (B) The ileocecal resection specimen is usually "fan-shaped."

ventrally to the SMV (Hamabe et al., 2018). If the ICA is the main feeder, ligation of the ICA at its root is necessary for D3 dissection. If the RCA is the main feeder, ligation of the RCA at its root is necessary for D3 dissection; however, the ICA does not need to be ligated at its root. If both the ICA and the RCA are the main feeders, they need to be ligated at their roots (Fig. 12.7). Exposing the surface of the SMV is essential for not only identifying the roots of the ICA and

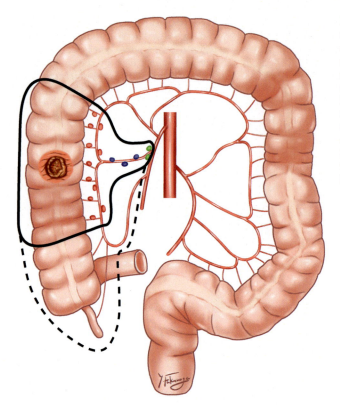

FIGURE 12.7 Regional lymph nodes and resection lines in ascending colon cancer. The regional lymph nodes and bowel resection lines in ascending colon cancer are shown. Pericolic lymph nodes are indicated in: *red*; intermediate lymph nodes are indicated in: *blue*; and main lymph nodes are indicated in: *green*. If the RCA is the main feeder, ascending colectomy can be chosen (solid line), however, right hemicolectomy is usually chosen to make an anastomosis safely (dotted line). *RCA*, right colic artery.

the RCA but also for en-bloc removal of the main lymph nodes on the SMV. After separating the right-sided colon and the mesocolon from the retroperitoneum along Toldt's fascia and mobilizing the hepatic flexure, the transverse colon is resected at 10 cm from the primary tumor. At the oral side, a 10 cm bowel resection is enough for removing regional pericolic lymph nodes; however, the ileum (not cecum) is often resected to make an anastomosis safely (Fig. 12.7). Therefore the length of the oral bowel usually becomes more than 10 cm. After removal of the specimen, the ileum and the transverse colon are anastomosed.

12.12 Transverse colon cancer

The main feeder of transverse colon cancer is the MCA, and it is found in all cases (Hamabe et al., 2018). Eighty percent of cases have one MCA and 20% of cases have two MCAs (right branch of the MCA (rt-MCA) and left branch of the MCA (lt-MCA)) which branch off separately from the SMA (Hamabe et al., 2018). Ligation of these MCAs at their roots is necessary for D3 dissection. To identify these roots safely, exposing the SMV surface is essential. Additionally, identifying the roots of drainage veins of the transverse and the ascending colon, such as the middle colic vein, accessory right colic vein, and the right colic vein (RCV) is also important for a safe D3 dissection. These right-sided veins show some variations among individuals, and surgeons need to be aware of such variations (Hamabe et al., 2018; Yamaguchi et al., 2002).

If the primary tumor is located on the right side of the transverse colon, a right hemicolectomy is usually performed (Figs. 12.8 and 12.9). If there is one MCA, the root of the

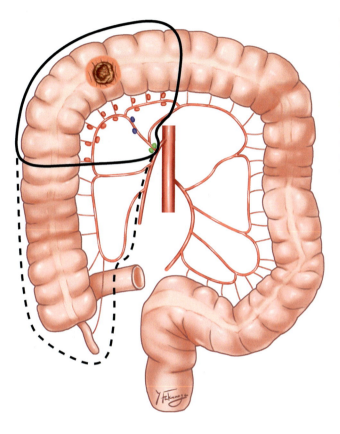

FIGURE 12.8 Regional lymph nodes and resection lines in right-sided transverse colon cancer. The regional lymph nodes and bowel resection lines in right-sided transverse colon cancer are shown. Pericolic lymph nodes are indicated in *red*; intermediate lymph nodes are indicated in *blue*; and main lymph nodes are indicated in *green*. The line of transverse colectomy is indicated with a solid line, and that of right hemicolectomy is indicated with a dotted line.

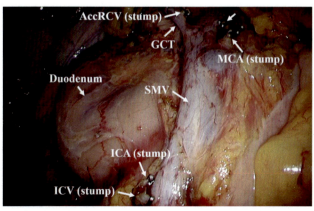

FIGURE 12.9 D3 dissection for right-sided transverse colon cancer (right hemicolectomy). A surgical photograph of a D3 dissection (right hemicolectomy) for right-sided transverse colon cancer is shown. The SMV is exposed and the MCA, ICA, and drainage veins are clipped at their roots. *SMV*, Superior mesenteric vein; *MCA*, middle colic artery; *ICA*, Ileocolic Artery.

3. Treatment

MCA is ligated for a D3 dissection, but ligating the root of the right branch of the MCA is acceptable for early-stage cancer to preserve the blood perfusion of the anal colon. If there are two MCAs, the main MCA (usually the main feeder is the rt-MCA for right-sided transverse colon cancer) is ligated at its root for a D3 dissection. If the RCA is the co-main feeder, its root is ligated for a D3 dissection as well; otherwise, ligation of its root is not necessary. The ICA is rarely the main feeder of transverse colon cancer; therefore ligation of its roots is not necessary. In spite of these principles of D3 dissection for transverse colon cancer, the RCA and the ICA are often ligated near their roots for two reasons: exposing the surface of the SMV is essential to identify the root of the MCA and both the ICA and the RCA are often ligated at their roots in the process of exposing the surface of the SMV, and right hemicolectomy is usually chosen in order to make an anastomosis safely (Fig. 12.8). Lymph nodes around the head of the pancreas or those along the gastroepiploic arcade at the greater curvature of the stomach are regarded as "extra-regional" lymph nodes (see Table 12.3). Therefore these lymph nodes are not routinely dissected as long as metastases are suspected, unlike in CME with CVL. After mobilization of the right-sided colon and the hepatic flexure (and sometimes the splenic flexure), the transverse colon is resected at 10 cm from the primary tumor. On the oral side, a 10 cm bowel resection is sufficient for removing regional pericolic lymph nodes; however, the ileum

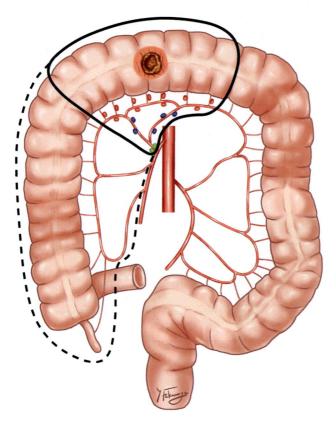

FIGURE 12.10 Regional lymph nodes and resection lines in middle transverse colon cancer. The regional lymph nodes and bowel resection lines in middle transverse colon cancer are shown. Pericolic lymph nodes are indicated in *red*; intermediate lymph nodes are indicated in *blue*; and main lymph nodes are indicated in *green*. The line of transverse colectomy is indicated using a solid line, and that of right hemicolectomy is indicated using a dotted line.

(not the cecum) is usually resected to ensure safe anastomosis (Fig. 12.8). As a result, the length of the oral bowel often becomes longer than 10 cm. After removal of the specimen, the ileum and the transverse colon are anastomosed.

If the primary tumor is located at the center of the transverse colon, both the rt-MCA and the lt-MCA are usually main feeders. Transverse colectomy is sometimes selected to preserve the right-sided colon and the terminal ileum. After ligating the root of the MCA(s) and mobilizing both the hepatic and splenic flexures, the transverse colon at 10 cm from both the oral and the anal sides of the primary tumor is resected in a "fan-shaped" manner, and a colon anastomosis is made (Fig. 12.10). Right hemicolectomy can also be selected for such cases.

In left-sided transverse colon cancers and splenic flexural cancers, the MCA (or lt-MCA), accessory middle colic artery (AMCA), and the LCA can be the main feeders. The AMCA feeds the splenic flexure and is observed in 14.3%–36.4% of cases (Hamabe et al., 2018; Miyake et al., 2018). The root of the main feeder needs to be ligated for D3 dissection; however, the root of the main feeder is usually preserved if the main feeder is the LCA because the main lymph nodes are located around the root of the IMA (Fig. 12.2). The IMA should be preserved for the blood perfusion of the anal bowel. In such cases, the lymph nodes located around the root of the IMA are usually removed and the origin of the LCA is ligated. After separating the left-sided colon from the retroperitoneum along

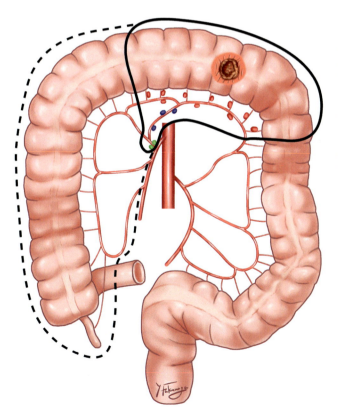

FIGURE 12.11 Regional lymph nodes and resection lines in left-sided transverse colon cancer. The regional lymph nodes and bowel resection lines in left-sided transverse colon cancer are shown. Pericolic lymph nodes are indicated in *red*; intermediate lymph nodes are indicated in *blue*; and main lymph nodes are indicated in *green*. The line of transverse colectomy is indicated using a solid line and that of extended right hemicolectomy is indicated using a dotted line.

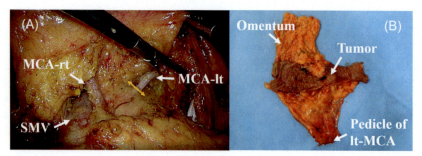

FIGURE 12.12 D3 dissection for left-sided transverse colon cancer. (A) A surgical photograph of a D3 dissection for left-sided transverse colon cancer is shown. There are two MCAs, and the root of the lt-MCA was ligated (yellow two-way arrow) and transverse colectomy was performed in this case. The SMV was exposed in order to identify the root of the MCAs and the drainage veins. (B) The specimen appeared to be "fan-shaped" after transverse colectomy. *MCAs*, Middle colic artery; *SMV*, superior mesenteric vein.

Toldt's fascia and mobilizing the splenic flexure, the transverse colon and the descending colon at 10 cm from the primary tumor are resected and anastomosed (Figs. 12.11 and 12.12). If a colon anastomosis is thought to be unsafe, the terminal ileum is resected, and an ileum-colon anastomosis is made (termed as "extended right hemicolectomy") (Fig. 12.11).

Splenic flexural cancer has several characteristics to be considered during lymphadenectomy, such as the presence of a wide variety of feeding arteries, and the long-distance between the root of the artery and the vein. Watanabe et al. reported that it may not be necessary to ligate both the lt-MCA and the LCA for splenic flexural cancer; they also noted the importance of lymph node dissection at the root of the inferior mesenteric vein (IMV) (Watanabe et al., 2017). The appropriate dissection area for splenic flexural cancer is being researched by a multicenter cohort study in Japan (UMIN000037195).

12.13 Descending colon cancer

The main feeder of descending colon cancer is usually the LCA, and the main lymph nodes are located around the root of the IMA; however, the root of the IMA is often persevered to maintain blood perfusion in the distal colon; the lymph nodes located around the root of the IMA are removed and the origin of the LCA is ligated (Fig. 12.13). Surgeons should be familiar with the three branching patterns of the LCA, as described below: the LCA branches directly from the IMA in 46% of the patients, the common trunk of the LCA, and the SA branches from the IMA in 24% of the patients, and the superior rectal artery (SRA), the SA, and the LCA branch directly from the IMA in 30% of the patients (Wang et al., 2018).

After ligating the LCA and separating the left-sided colon from the retroperitoneum along Toldt's fascia with splenic flexure mobilization, the transverse colon and the sigmoid colon are resected and anastomosed at 10 cm from the primary tumor.

12.15 Preservation of the LCA for left-sided colon cancer 207

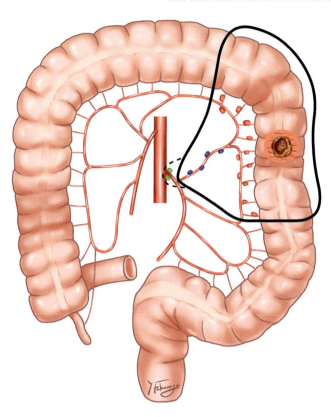

FIGURE 12.13 Regional lymph nodes and resection lines in descending colon cancer. The regional lymph nodes and bowel resection lines in descending colon cancer are shown. Pericolic lymph nodes are indicated in *red*; intermediate lymph nodes are indicated in *blue*; and main lymph nodes are indicated in *green*. Usually the main lymph nodes located around the root of the IMA are removed (dotted line), and the origin of the LCA is ligated (solid line). *IMA*, Inferior mesenteric artery; *LCA*, left colic artery.

12.14 Sigmoid colon cancer

The main feeder of sigmoid colon cancer is the SA, and the main lymph nodes are located around the root of the IMA; therefore the root of the IMA is ligated for D3 dissection. For early cancer, the root of the IMA is sometimes preserved for the purpose of preserving the lumbar splanchnic nerves (Mari et al., 2019) or preserving the blood supply to the oral colon (Komen et al., 2011). In these cases, the lymph nodes located around the root of the IMA are removed, and the origins of the SA and the SRA are ligated. The descending colon and the rectum are resected and anastomosed at 10 cm from the primary tumor (Fig. 12.14).

12.15 Preservation of the LCA for left-sided colon cancer

A basic concept in Japanese D3 dissection is the ligation of the root of the main feeder for the complete removal of the main lymph nodes. However, in the case of left-sided colon cancer, the root of the main feeder (that of the IMA) is sometimes preserved to avoid ischemia of the distal colon, particularly in cases of descending colon cancers (Fig. 12.13). In such cases, the main lymph nodes are removed, and the origin of the LCA or the SA is

3. Treatment

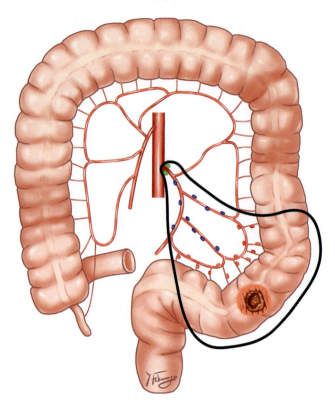

FIGURE 12.14 Regional lymph nodes and resection line in sigmoid colon cancer. The regional lymph nodes and bowel resection lines in sigmoid colon cancer are shown. Pericolic lymph nodes are indicated in *red*; intermediate lymph nodes are indicated in *blue*; and main lymph nodes are indicated in *green*.

ligated; however, whether such lymphadenectomy can be termed as "D3 dissection" is controversial in Japan.

Several studies have compared the results of preserving the LCA (termed as the "low-tie of IMA") and ligating the root of the IMA (termed as the "high-tie of IMA"). Komen et al. (2011) reported better blood perfusion of the colon, and Bonnet et al. (2012) reported less tension at the anastomotic site in "low-tie" groups compared to "high-tie" groups. The benefit of "low-tie" for decreasing anastomotic leakage is controversial; however, prognosis is reported to be similar to that achieved using "high-tie" (Zeng & Su, 2018). In Japan, Akagi et al. (2020) compared D3 lymph node dissection with LCA preservation (low-tie) and D3 lymph node dissection without LCA preservation (high-tie) for stage 2–3 sigmoid colon or recto-sigmoid cancers. The authors reported less postoperative complication rates and better recurrence-free survival in the "low-tie" group than in the "high-tie" group. Therefore they concluded that D3 with LCA preservation (low-tie) can be an alternative treatment for Stage 2–3 left-sided colon cancers.

12.16 Outcomes of Japanese D3 dissection for colon cancer

12.16.1 D3 lymphadenectomy versus D2 lymphadenectomy

There are several reports which compare D3 and D2 dissections.

Kotake et al. (2014) reported the results of a retrospective multicenter cohort study from a population of 10,098 prospectively registered pT3–4 colon cancer patients who underwent R0 resection between 1985 and 1994 in Japan. A total of 3425 propensity score-matched pairs were extracted from the entire cohort, and overall survival was compared between the D3 lymphadenectomy group and the D2 lymphadenectomy group. A significant difference in OS was reported between the D3 and the D2 groups in both the entire cohort (estimated hazard ratio [HR]: 0.827; 95% confidence interval [CI]: 0.757–0.904, $P = .00003$), and in the matched cohort (HR: 0.814; 95% CI: 0.734–0.904; $P = .0001$). Therefore the authors recommended D3 dissection for patients with T3 and T4 colon cancer.

Kotake et al. (2015) also compared the results of D3 dissection and D2 dissection for pT2 colon cancer patients in the same way as above, using 1433 patients who underwent major resections for pT2 colon cancer between 1995 and 2004 in multiple Japanese institutions. Four hundred and sixty-three matched pairs were extracted from the entire cohort, and OS was compared between the D3 and D2 groups. Results showed that there was no significant difference between the D3 and D2 groups in OS, either among the propensity score-matched cohort (HR: 0.85; 95% CI: 0.536–1.346; $P = .484$), or in the cohort as a whole (HR: 0.720; 95% CI: 0.492–1.052; $P = .089$). Therefore the authors concluded that D2 dissection is sufficient for T2 colon cancer patients.

In terms of elderly patients, Numata et al. (2019) than the prognosis and postoperative complication rate of 378 pStage 2–3 elderly patients aged 75 years or older who underwent primary resection with either D2 or D3 lymph node dissection in three Japanese hospitals using the propensity score-matching method. They reported significantly better recurrence-free survival in the D3 group than in the D2 group ($P = .01$), and a similar incidence of postoperative complications in the two groups. However, OS and cancer-specific survival (CSS) were similar in both groups, suggesting the necessity of further investigation to clarify the impact of D3 dissection on OS and CSS in elderly patients.

12.17 Current recommendations of the Japanese guidelines, 2019

According to the results of previous studies, the current Japanese guidelines (the 2019 edition) recommend the following strategies of lymphadenectomy for colon cancers (Hashiguchi et al., 2020):

- D3 dissection is recommended if lymph node metastasis is recognized or suspected based on the preoperative/intraoperative findings (Kotake et al., 2014).
- Lymphadenectomy according to T stage as follows is recommended if *no* lymph node metastases are observed based on the preoperative/intraoperative diagnostic findings:
- For pTis cancer, lymphadenectomy is not necessary because lymph node metastasis is not present in this type of cancer. However, D1 dissection can be considered for *c*Tis cancer if colon resection is performed.

TABLE 12.4 Frequency of pathological lymph node metastasis of colon cancer (cecum-sigmoid colon).

Depth of tumor invasion	Number of patients	N0 (%)	N1 (%)	N2 (%)	N3 (%)	N4 (%)
SM (T1)	1957	91.4	6.8	1.8	0	0
MP (T2)	1747	79.3	16.3	3.5	0.6	0.3
SS (T3)	7333	56.6	28.1	11.7	2.4	1.2
SE (T4a)	3363	37.4	34.0	19.3	5.6	3.7
SI (T4b)	960	44.6	28.6	14.7	5.5	6.6

Note 1: N classification is according to the older version of the Japanese classification of lymph nodes (Table 12.2); N1–4 denotes metastases in Group 1–4 lymph nodes respectively. Note 2: *SM (T1)*, Tumor is confined to the submucosa; *MP (T2)*, tumor invasion too, but not beyond, the muscularis propria; *SS (T3)*, tumor grows into the subserosa; *SE (T4a)*, tumor invades or perforates the serosa; *SI (T4b)*, tumor directly invades other organs or structures.

- For cT1 cancer, D2 dissection is necessary because the lymph node metastasis rate is approximately 10%; the metastasis rate of the intermediate lymph node (= N2) is approximately 2% in pT1 cancer (Table 12.4).
- For cT2 cancer, at least D2 dissection is necessary, although there is insufficient evidence describing the extent of lymph node dissection for this type of cancer (Kotake et al., 2015). D3 dissection can be performed because approximately 1% of pT2 cancers showed metastases at the main lymph node (= N3) (Table 12.4) (Hashiguchi et al., 2020), and the accuracy of the preoperative T stage diagnosis is not adequate.

12.18 Comparison of Japanese D3 dissection with European CME with CVL

The procedures for Japanese D3 and European CME with CVL are similar. CME involves resection of the mesocolon along the embryological layers to avoid injury to the mesocolon; this is similar to the TME procedure for rectal cancer (Heald, 1988; Hohenberger et al., 2009). In Japan, although it is not explicitly recommended by the Japanese guidelines, a similar procedure is performed according to the concept of "remove lymph nodes with lymph ducts in one package, enveloped with mesocolon." CVL involves the ligation of feeding vessels at their roots, and this procedure is being performed as D3 dissection in Japan.

The differences between CME with CVL and Japanese D3 involve the length of the resected bowel and the area of lymphadenectomy. In Japan, the regional pericolic lymph nodes are defined as those within 10 cm of the primary tumor (10 cm rule). This rule was framed in line with a study which reported a lower frequency of metastasis in the pericolic lymph nodes, which were more than 10 cm away from the primary tumor (Hida et al., 1997). Therefore the distant pericolic lymph nodes are considered as "extra-regional" and are not routinely dissected, as long as metastases are suspected. Similarly, lymph nodes located around the head of the pancreas or along the gastroepiploic arcade at the greater curvature of the stomach are regarded as "extra-regional" (Table 12.3); therefore these lymph nodes are not routinely dissected as long as metastases are suspected. On the other hand, in CME procedure, lymph nodes are removed to the extent possible; therefore the length of the resected bowel is longer than that after Japanese D3, and the lymph nodes located around the head of the pancreas or along the gastroepiploic arcade are dissected; such a procedure is based on data which indicated that the possibility of metastases was 4%–5% in these lymph nodes (Hohenberger et al., 2009).

In Japan, ileocecal resection is usually selected for cecum cancers, and transverse colectomy is sometimes selected for transverse colon cancers, according to the "10 cm rule." On the other hand, right hemicolectomy is the standard for such right-sided colon cancers in European countries according to the concept of "removing as many lymph nodes as possible." For tumors of the left-sided transverse colon and the splenic flexure, transverse colectomy has been the standard procedure in Japan, and the resected specimens have been "fan-shaped" (Fig. 12.12B). Lymph nodes along the inferior edge of the pancreatic tail are also targets of lymphadenectomy in the CME procedure because the lymphatic drainage root is multidirectional with respect to the splenic flexure (Hohenberger et al., 2009). As described above, the appropriate dissection area for splenic flexural cancer is being researched by a multicenter cohort study in Japan. This study aims to verify the importance of the inferior edge of the pancreatic tail and the root of the IMV as the potential area of lymph node metastasis (UMIN000037195). For sigmoid colon cancer, the procedures for lymphadenectomy in Japan and in European countries are similar: central ligation of the IMA with transection of the distal descending colon and the upper rectal third; such a procedure is indicated because lymph node metastases from the sigmoid colon travel unidirectionally and follow the supplying sigmoid arteries (Hohenberger et al., 2009).

West et al. (2012) investigated the differences between the Japanese D3 and European CME with CVL procedures by comparing 165 cases from two Japanese centers and 143 cases from one European center (cases of cancers of all stages). The mesocolic plane resection rates from both groups were high, and the distance from the high vascular tie to the bowel wall was equivalent (100 vs 99 mm). Specimens obtained from resections using the Japanese D3 procedure were significantly shorter (162 vs 324 mm), and the Japanese D3 procedure resulted in a smaller amount of mesentery (8309 vs 17,957 mm^2) and nodal yield (median, 18 vs 32); however, the number of tumor-involved nodes was similar (median, 0 vs 0).

Kobayashi et al. (2014) researched the differences among three types of conventional surgical procedures for *Stage 3* colon cancer by comparing 19 cases from one center in England, 26 cases from one center in Germany, and 60 cases from two centers in Japan (Table 12.5). The conventional surgery was hemicolectomy with low vascular tie in England, CME with CVL in Germany, and D3 dissection in Japan. The length of the resected bowel and the area of the resected mesentery of the English and German specimens were significantly greater than those of the Japanese specimens. The length of the vascular tie to the bowel wall was similar between the German and Japanese specimens, which was longer than that of the English specimens. The number of retrieved lymph nodes in the German specimens was the highest among the three groups; however, the number of tumor-involved nodes was similar. The rates of mesocolic plane surgery were higher in the German and Japanese specimens (88.5% and 71.7%) than in the English specimens (47.4%).

These comparative studies have revealed the differences in surgical results between procedures performed in England, Germany, and Japan; however, the influence of these differences on long-term oncologic outcomes is unknown.

TABLE 12.5 Comparison between complete mesocolic excision (CME) and central vascular ligation (CVL) and Japanese D3 (Kobayashi et al., 2014).

	Right-sided and transverse colon		Left-sided colon	
	CME with CVL	Japanese D3	CME with CVL	Japanese D3
Length of the resected bowel (mm)[a]	305 (213–348)	184 (137–230)	355 (322–399)	146 (132–176)
Area of the mesentery (mm^2)[a]	17,641 (13,382–24,620)	8705 (6900–12,017)	17,762 (14,938–24,560)	7807 (6089–10,742)
Tumor to vascular tie (mm)[a]	115 (102–128)	103 (78–115)	128 (120–146)	120 (100–138)
Retrieved lymph nodes[a]	38 (26–50)	24 (17–30)	24 (18–31)	20 (16–26)
Involved lymph nodes[a]	3 (1–4)	2 (1–7)	2 (1–4.5)	2 (1.5–4)

[a]*Median (interquartile range)*.
CME, Complete mesocolic excision; *CVL*, central vascular ligation.
Kobayashi, H., West, N.P., Takahashi, K., Perrakis, A., Weber, K., Hohenberger, W., Quirke, P., & Sugihara, K. (2014). *Quality of surgery for stage III colon cancer: Comparison between England, Germany, and Japan. Annals of Surgical Oncology, 21 Suppl 3, S398–S404. https://doi.org/10.1007/s00384-011-1188-6.*

12.19 Future perspective

Currently, several clinical studies are ongoing, and aim to clarify the appropriate areas of lymphadenectomy and surgical procedures for colon cancers. In Japan, a multicenter prospective observational study termed the T-REX study (UMIN000030331) is ongoing and aims to clarify the frequency and distribution of lymph node metastases to decide the appropriate length of bowel resection (Hashiguchi et al., 2011). Other trials such as the COLD trial (D3 dissection vs D2 dissection for colon cancer, NCT03009227) (Karachun et al., 2019), the LCME trial (Laparoscopic CME vs Laparoscopic D3 dissection for colon cancer, NCT01628250), the RELARC trial (D2 dissection vs CME for right-sided colon cancer, NCT02619942), and the SLRC trial (Laparoscopic CME vs Laparoscopic D3 dissection for right-sided colon cancer, NCT02942238) are also ongoing to clarify the appropriate surgical procedure for colon cancers worldwide. Throughout these and subsequent studies, the influence of surgical procedures on long-term oncologic outcomes of colon cancers will be clarified.

12.20 Summary

We described the history, definition, technique, and results of Japanese D3 dissection for colon cancer, and discussed the difference between Japanese D3 and European CME with CVL. Both Japanese D3 and European CME with CVL place high importance on the complete resection of the mesocolon and vessel ligation at the root. These procedures have contributed to the improvement of surgical and oncological results in colon cancer. More appropriate procedures or areas of lymphadenectomy according to the tumor stages or locations will be clarified by further research in the future.

Acknowledgment

We would like to thank Yutaka Fukunaga (National Cancer Center Hospital East, Japan) for drawing figures, and Editage (www.editage.com) for English language editing.

Disclosure

Conflict of Interest: All authors declare that they have no conflicts of interest.

References

Akagi, T., Inomata, M., Hara, T., Mizusawa, J., Katayama, H., Shida, D., Ohue, M., Ito, M., Kinugasa, Y., Saida, Y., Masaki, T., Yamamoto, S., Hanai, T., Yamaguchi, S., Watanabe, M., Sugihara, K., Fukuda, H., Kanemitsu, Y., & Kitano, S. (2020). Clinical impact of D3 lymph node dissection with left colic artery (LCA) preservation compared to D3 without LCA preservation: Exploratory subgroup analysis of data from JCOG0404. *Annals of Gastroenterological Surgery*, *4*(2), 163–169. Available from https://doi.org/10.1002/ags3.12318.

Bonnet, S., Berger, A., Hentati, N., Abid, B., Chevallier, J.-M., Wind, P., Delmas, V., & Douard, R. (2012). High tie versus low tie vascular ligation of the inferior mesenteric artery in colorectal cancer surgery: impact on the gain in colon length and implications on the feasibility of anastomoses. *Diseases of the Colon and Rectum*, *55*(5), 515–521. Available from https://doi.org/10.1097/DCR.0b013e318246f1a2.

Hamabe, A., Park, S., Morita, S., Tanida, T., Tomimaru, Y., Imamura, H., & Dono, K. (2018). Analysis of the vascular interrelationships among the first jejunal vein, the superior mesenteric artery, and the middle colic artery. *Annals of Surgical Oncology*, *25*(6), 1661–1667. Available from https://doi.org/10.1245/s10434-018-6456-z.

Hashiguchi, Y., Hase, K., Ueno, H., Mochizuki, H., Shinto, E., & Yamamoto, J. (2011). Optimal margins and lymphadenectomy in colonic cancer surgery. *The British Journal of Surgery*, *98*(8), 1171–1178. Available from https://doi.org/10.1002/bjs.7518.

Hashiguchi, Y., Muro, K., Saito, Y., Ito, Y., Ajioka, Y., Hamaguchi, T., Hasegawa, K., Hotta, K., Ishida, H., Ishiguro, M., Ishihara, S., Kanemitsu, Y., Kinugasa, Y., Murofushi, K., Nakajima, T. E., Oka, S., Tanaka, T., Taniguchi, H., ... Tsuji, A. K. (2020). Japanese Society for Cancer of the Colon and Rectum (JSCCR) guidelines 2019 for the treatment of colorectal cancer. *International Journal of Clinical Oncology*, *25*(1), 1–42. Available from https://doi.org/10.1007/s10147-019-01485-z.

Heald, R. J. (1988). The "Holy Plane" of rectal surgery. *Journal of the Royal Society of Medicine*, *81*(9), 503–508.

Hida, J., Yasutomi, M., Maruyama, T., Fujimoto, K., Uchida, T., & Okuno, K. (1997). The extent of lymph node dissection for colon carcinoma: The potential impact on laparoscopic surgery. *Cancer*, *80*(2), 188–192.

Hohenberger, W., Weber, K., Matzel, K., Papadopoulos, T., & Merkel, S. (2009). Standardized surgery for colonic cancer: Complete mesocolic excision and central ligation—technical notes and outcome. *Colorectal Disease: The Official Journal of the Association of Coloproctology of Great Britain and Ireland*, *11*(4), 354–364, discussion 364-5. Available from https://doi.org/10.1111/j.1463-1318.2008.01735.x.

Ishiguro, M., Higashi, T., Watanabe, T., & Sugihara, K. (2014). Changes in colorectal cancer care in japan before and after guideline publication: A nationwide survey about D3 lymph node dissection and adjuvant chemotherapy. *Journal of the American College of Surgeons*, *218*(5), 969–977, e1. Available from https://doi.org/10.1016/j.jamcollsurg.2013.12.046.

Japanese Society for Cancer of the Colon and Rectum. (1994). Japanese Classification of Colorectal, Appendiceal, and Anal Carcinoma.

Japanese Society for Cancer of the Colon and Rectum. (2018). Japanese Classification of Colorectal, Appendiceal, and Anal Carcinoma (ninth ed.). Kanehara & CO., Ltd.

Japanese Society for Cancer of the Colon and Rectum. (2019). Japanese Classification of Colorectal, Appendiceal, and Anal Carcinoma (third English ed.). Kanehara & CO., Ltd.

Karachun, A., Petrov, A., Panaiotti, L., Voschinin, Y., & Ovchinnikova, T. (2019). D3 lymph node dissection in colonic cancer (COLD trial). *BJS Open*, *3*(3), 288–298. Available from https://doi.org/10.1002/bjs5.50142.

Kobayashi, H., West, N. P., Takahashi, K., Perrakis, A., Weber, K., Hohenberger, W., Quirke, P., & Sugihara, K. (2014). Quality of surgery for stage III colon cancer: Comparison between England, Germany, and Japan. *Annals of Surgical Oncology*, 21(Suppl 3), S398–S404. Available from https://doi.org/10.1245/s10434-014-3578-9.

Komen, N., Slieker, J., de Kort, P., de Wilt, J. H. W., van der Harst, E., Coene, P.-P., Gosselink, M. P., Gosselink, M., Tetteroo, G., de Graaf, E., van Beek, T., den Toom, R., van Bockel, W., Verhoef, C., & Lange, J. F. (2011). High tie versus low tie in rectal surgery: Comparison of anastomotic perfusion. *International Journal of Colorectal Disease*, 26(8), 1075–1078. Available from https://doi.org/10.1007/s00384-011-1188-6.

Kotake, K., Honjo, S., Sugihara, K., Kato, T., Kodaira, S., Takahashi, T., Yasutomi, M., Muto, T., & Koyama, Y. (2003). Changes in colorectal cancer during a 20-year period: An extended report from the multi-institutional registry of large bowel cancer, Japan. *Diseases of the Colon and Rectum*, 46(10 Suppl), S32–S43.

Kotake, K., Kobayashi, H., Asano, M., Ozawa, H., & Sugihara, K. (2015). Influence of extent of lymph node dissection on survival for patients with pT2 colon cancer. *International Journal of Colorectal Disease*, 30(6), 813–820. Available from https://doi.org/10.1007/s00384-015-2194-x.

Kotake, K., Mizuguchi, T., Moritani, K., Wada, O., Ozawa, H., Oki, I., & Sugihara, K. (2014). Impact of D3 lymph node dissection on survival for patients with T3 and T4 colon cancer. *International Journal of Colorectal Disease*, 29(7), 847–852. Available from https://doi.org/10.1007/s00384-014-1885-z.

Mari, G. M., Crippa, J., Cocozza, E., Berselli, M., Livraghi, L., Carzaniga, P., Valenti, F., Roscio, F., Ferrari, G., Mazzola, M., Magistro, C., Origi, M., Forgione, A., Zuliani, W., Scandroglio, I., Pugliese, R., Costanzi, A. T. M., & Maggioni, D. (2019). Low ligation of inferior mesenteric artery in laparoscopic anterior resection for rectal cancer reduces genitourinary dysfunction: Results from a randomized controlled trial (HIGHLOW Trial). *Annals of Surgery*, 269(6), 1018–1024. Available from https://doi.org/10.1097/SLA.0000000000002947.

Miyake, H., Murono, K., Kawai, K., Hata, K., Tanaka, T., Nishikawa, T., Otani, K., Sasaki, K., Kaneko, M., Emoto, S., & Nozawa, H. (2018). Evaluation of the vascular anatomy of the left-sided colon focused on the accessory middle colic artery: A single-centre study of 734 patients. *Colorectal Disease: The Official Journal of the Association of Coloproctology of Great Britain and Ireland*, 20(11), 1041–1046. Available from https://doi.org/10.1111/codi.14287.

Muto, T., Kotake, K., & Koyama, Y. (2001). Colorectal cancer statistics in Japan: Data from JSCCR registration, 1974–1993. *International Journal of Clinical Oncology*, 6(4), 171–176.

Numata, M., Sawazaki, S., Aoyama, T., Tamagawa, H., Sato, T., Saeki, H., Saigusa, Y., Taguri, M., Mushiake, H., Oshima, T., Yukawa, N., Shiozawa, M., Rino, Y., & Masuda, M. (2019). D3 lymph node dissection reduces recurrence after primary resection for elderly patients with colon cancer. *International Journal of Colorectal Disease*, 34(4), 621–628. Available from https://doi.org/10.1007/s00384-018-03233-7.

Wang, K.-X., Cheng, Z.-Q., Liu, Z., Wang, X.-Y., & Bi, D.-S. (2018). Vascular anatomy of inferior mesenteric artery in laparoscopic radical resection with the preservation of left colic artery for rectal cancer. *World Journal of Gastroenterology*, 24(32), 3671–3676. Available from https://doi.org/10.3748/wjg.v24.i32.3671.

Watanabe, J., Ota, M., Suwa, Y., Ishibe, A., Masui, H., & Nagahori, K. (2017). Evaluation of lymph flow patterns in splenic flexural colon cancers using laparoscopic real-time indocyanine green fluorescence imaging. *International Journal of Colorectal Disease*, 32(2), 201–207. Available from https://doi.org/10.1007/s00384-016-2669-4.

West, N. P., Kobayashi, H., Takahashi, K., Perrakis, A., Weber, K., Hohenberger, W., Sugihara, K., & Quirke, P. (2012). Understanding optimal colonic cancer surgery: Comparison of Japanese D3 resection and European complete mesocolic excision with central vascular ligation. *Journal of Clinical Oncology: Official Journal of the American Society of Clinical Oncology*, 30(15), 1763–1769. Available from https://doi.org/10.1200/JCO.2011.38.3992.

Yamaguchi, S., Kuroyanagi, H., Milsom, J. W., Sim, R., & Shimada, H. (2002). Venous anatomy of the right colon: Precise structure of the major veins and gastrocolic trunk in 58 cadavers. *Diseases of the Colon and Rectum*, 45(10), 1337–1340.

Zeng, J., & Su, G. (2018). High ligation of the inferior mesenteric artery during sigmoid colon and rectal cancer surgery increases the risk of anastomotic leakage: A meta-analysis. *World Journal of Surgical Oncology*, 16(1), 157. Available from https://doi.org/10.1186/s12957-018-1458-7.

Management of para-aortic nodal disease in colon cancer

Alexander De Clercq and Gabrielle H. van Ramshorst

Department of Gastrointestinal Surgery, Ghent University Hospital, Ghent, Belgium

13.1 Introduction

Cancers of the colon and the rectum are the third most common type of cancer worldwide, accounting for 630,000 deaths annually (Albandar et al., 2016). Colon cancer is more prevalent than rectal cancer: the ratio of colon to rectal cancer cases is 2:1 or more in industrialized countries, while the rates are mostly similar in nonindustrialized countries (Labianca et al., 2010). Every year, around 250,000 new colon cases are diagnosed in Europe, which accounts for around 9% of all the malignancies (Labianca et al., 2010).

It is presumed that lymphatic drainage of the colon and rectum progresses orderly from the submucosal lymphoid follicles through the bowel wall to epicolic, paracolic, intermediate, and para-aortic lymph nodes (Merrie et al., 2001). Roughly 10%−20% of patients who are newly diagnosed with colorectal cancer present with coexistent distant metastasis (Nakai et al., 2017). Anatomic dissections have indicated that locoregional metastases from rectal cancer follow a sequential pattern (Merrie et al., 2001). As a rule, the regional lymph glands are the first to be affected by metastases in rectal cancer. Irregular or interrupted spread is very rare, and lateral or downward lymphatic spread is found in later stages (Gabriel et al., 1935).

The para-aortic lymph nodes are also known as the left lumbar lymph nodes. They consist of three primary groups of nodes: the preaortic lymph nodes, the lateral aortic lymph nodes, and the retro-aortic lymph nodes. The preaortic lymph nodes are located anterior to the abdominal aorta, positioned in clusters surrounding the branches of the abdominal aorta. To the left of the abdominal aorta, the lateral aortic lymph nodes are found anterior to the medial border of the left psoas major muscle, the sympathetic trunk, and the left diaphragmatic crura. Lastly, the retro-aortic lymph nodes are considered the smallest group of para-aortic lymph nodes. Although there is not a specified set of areas that drain to these nodes, every so often they receive tributaries from the paraspinal posterior abdominal wall. Otherwise, they function as an intermediary group that drains to the lateral aortic lymph nodes. Sometimes, they can also obtain lymph drained from the common iliac lymph nodes (Crumbie & Zehra, 2020).

Patients who have advanced colorectal cancer are at a higher risk of developing lateral pelvic lymph node metastasis along with para-aortic lymph node metastasis (to some extent) as it follows the inferior or superior mesenteric artery, respectively (Albandar et al., 2016). Still, involvement of para-aortic lymph nodes is rare in colorectal cancer, with a reported incidence of less than 2% (Wong et al., 2016). Right now, the definition of para-aortic lymph node metastasis states the presence of an undoubtedly enlarged lymph node next to the abdominal aorta in the absence of distal metastasis at any other site (Albandar et al., 2016). Para-aortic lymph node metastasis is classified according to location opposed to the renal vein. The Japanese Society of Clinical Oncology refers to 16A as above the renal vessels and to 16B as below the renal vessels. The American Joint Committee on Cancer staging system though recognizes para-aortic lymph node metastasis as an M1 disease (Albandar et al., 2016).

For most studies, patients with a history of colorectal cancer who present with lymph nodes with a diameter of 7 mm or larger with or without fluorodeoxyglucose (FDG)-avidity on positron emission tomography scan were considered indicative of metastatic involvement (Ushigome et al., 2020). Cases series with unclear mixed populations of retroperitoneal or mesenteric lymph node recurrences were excluded for this chapter.

13.2 Imaging and implications for prognosis

Lu et al. (2015) compared 5-year overall survival in patients with and without visible para-aortic lymph nodes (>2 mm short axis) on computed tomography or magnetic resonance imaging. In the absence of distant metastasis, visibility of para-aortic lymph nodes in 409 patients was associated with significantly shorter overall survival of 67% versus 76% in comparison with 2979 patients without visible para-aortic lymph nodes (Lu et al., 2015). However, this difference did not remain in multivariate analysis. Independent prognostic factors for overall survival were age, male gender, rectal cancer, lymphovascular invasion, perineural invasion, high-grade histopathology, pT3/4, and preoperative carcinoembryonic antigen (CEA) level >10 ng/mL. After patients with visible lymph nodes were divided and compared according to size (smaller or equal to 10 mm, lymph node size >10 mm), elevated serum CEA, pN+, and lymphovascular invasion were identified as independent prognostic variables for overall survival (Lu et al., 2015).

During operative resection, fluorescence-guided imaging has been reported to be helpful in the identification and resection of para-aortic lymph nodes in colon cancer (Kim et al., 2021; Liberale et al., 2015).

13.3 Differences between right-sided and left-sided para-aortic node involvement

Para-aortic lymph nodes lie between the renal veins and the bifurcation of the aorta into the common iliac arteries, including the aortocaval area and paravertebral plane. On the right side, affected lymph nodes can concern nodes that would normally be removed during D3 resections.

Involvement of the apical lymph node in primary resection has been described as a risk factor for para-aortic lymph node metastasis in multiple studies (Kang et al., 2011; Tsai et al., 2019; Wang et al., 2020). Kang et al. evaluated a database of prospectively collected data on 625 patients with rectosigmoid cancer, with a median follow-up of 52 months. They identified a disease-free survival rate of 31.9% in patients with inferior mesenteric artery lymph node metastasis versus 69.4% in patients with negative inferior mesenteric artery lymph nodes. The presence of inferior mesenteric artery lymph node metastasis was an independent predictive variable with a hazard ratio (HR) of 11.8 (95% CI: 2.7–52.2) for para-aortic lymph node recurrences (Kang et al., 2011). Wang et al. (2020) also identified a significantly increased risk of para-aortic lymph node recurrences in a retrospective cohort of 498 patients undergoing D3 resections for stage III left-sided colorectal cancer, 15.7% versus 2.0% after propensity-score matching.

This effect, however, was not found in 254 patients with right-sided stage III disease in a separate publication, of them 28 had apical node involvement (Wang et al., 2020). After propensity score matching, no differences remained between both groups in recurrence-free survival or overall survival. Regression analysis showed that depth of invasion and lymphatic vessel infiltration had independent negative effects on overall survival in right-sided colon cancer (Wang et al., 2020).

The effects of inferior mesenteric lymph node versus para-aortic lymph node metastatic involvement were compared in a study by Lee et al. (2017). This retrospective cohort consisted of 27 patients with rectosigmoid cancer with para-aortic node involvement and 47 patients with inferior mesenteric lymph node involvement. Recurrence rates —high at 70% and 64%, respectively—nor the site of recurrence showed statistically significant differences. No differences were found in overall or 5-year disease-free survival rates—37% versus 39% and 28% versus 30%, respectively. Independent risk factors for disease-free survival were perineural invasion, positive inferior mesenteric or para-aortic lymph nodes, and the presence of four or more metastatic lymph nodes (Lee et al., 2017).

In rectal cancer, lung and para-aortic lymph node metastases are more common in low-to-mid rectal tumors than upper rectal tumors, presenting more frequently with liver metastasis (Yeo et al., 2013). A retrospective comparative study from China showed that combined total pelvic lymph node adenectomy and para-aortic lymphadenectomy contributed to prolonged disease-free survival in stage III rectal cancer patients with noninfiltrating type histology and 1 or 2 positive pelvic lymph node sites (Liu et al., 2012).

13.4 Metachronous isolated lymph node metastasis

Han et al., (2020) described a retrospective cohort of 1326 patients with recurrent colorectal cancer, of them 301 patients were diagnosed with isolated lymph node metastases. Para-aortic (48.8%), pelvic (29.9%), and lung hilum (10%) were the most common sites. Almost 80% were diagnosed within 3 years after primary resection. Multimodal management, including surgery and chemotherapy, with or without radiotherapy proved more effective than single modality treatment (including chemotherapy, radiotherapy, or operation alone). Other independent predictive variables for survival were diagnosis time to lymph node metastasis, histological type (high-grade lesions), and T-stage (Han et al., 2020).

Kim et al. (2021) described a retrospective series of patients with isolated para-aortic lymph node recurrences treated between 2004 and 2014. A total of 35 patients had isolated recurrences, of them 16 underwent resection and 19 received chemotherapy. Three out of 16 operated patients proved false-positive on definitive histopathology. Median survival after recurrence was significantly longer in these 16 operated patients compared to nonoperated patients (71 vs 39 months, $P = .017$). All patients received chemotherapy at some point during their treatment (Kim et al., 2020). Eight patients developed secondary recurrence in the pelvis, lungs, or distant lymph nodes.

Wong et al. (2016) performed a systematic review in 2016 on this topic and included several studies on surgical resection of metachronous para-aortic lymph node metastases. Most of the studies were published before the introduction of modern multimodal treatment, including immunotherapy. Shibata et al. described a series of 25 patients who were scheduled for resection of retroperitoneal local recurrences of colorectal cancer. Median survival in resectable patients (20/25) was 31 months. Prognosis was worse in patients with incomplete resections and tumor size of 5 cm or greater. Survival improved with disease-free intervals of 24 months or more (median survival, 30 vs 48 months, $P = .02$). Overall, 2- and 5-year survival was 60% and 15%, respectively (Shibata et al., 2002).

The study by Min et al., (2008) showed that surgical resection of isolated metachronous para-aortic lymph node metastases followed by adjuvant chemotherapy improved survival significantly in comparison to chemotherapy or chemoradiation alone, with a median survival time of 34 months versus 12 months for unresected patients. Most patients eventually developed liver, lung, bone, or brain metastases.

13.5 Synchronous metastases

Patients who present with synchronous disease have shown worse survival than patients with metachronous disease in some studies (Arimoto et al., 2015). Ushigome et al. (2020) described a cohort of 20 patients with synchronous metastatic para-aortic lymph node metastases below the renal vein who underwent radical surgical resection. Sixty percent of patients underwent adjuvant chemotherapy. Five-year overall survival was 39%. Recurrence-free survival was 25%, and 76% recurred within 1 year after primary resection. In multivariate analysis, T-stage, time to diagnosis of recurrence, and resection of recurrence proved prognostic for overall survival (Ushigome et al., 2020).

Sahara et al. (2019) investigated the influence of para-aortic lymphadenectomy in a retrospective cohort study of 322 patients with left-sided colorectal cancer. Sixty-two patients had involved para-aortic lymph nodes. Independent risk factors for overall survival were elevated serum CEA (>10 ng/L; HR: 2.1; 95% CI: 1.11–4.27), the number of para-aortic lymph nodes of four or more (HR: 3.34; 95% CI: 1.53–7.31), incomplete resection (HR: 3.61; 95% CI: 1.85–7.06) and undifferentiated type histology (poor/mucinous/signet cell; HR: 4.51; 95% CI: 2.22–9.19). Patients who underwent complete resection of fewer than four involved para-aortic nodes and well/moderately differentiated adenocarcinoma had a 5-year overall survival of 54% (Sahara et al., 2019). Ogura et al. (2017) similarly described 16 patients who underwent complete synchronous para-aortic lymph node resection for left-sided colorectal cancer. Patients with bead-like extra-regional lymph node metastasis were only considered for palliative surgery without para-aortic lymphadenectomy. At 5 years, the cancer-specific survival rate was 70% and the relapse-free survival rate was 61%. The number of metastatic extra-regional lymph nodes was low at 1 (range 0–4) and proved a significant risk factor for relapse-free survival in multivariate analyses. This highly selected patient group also demonstrated a high rate of perioperative chemotherapy use (Ogura et al., 2017). It is noteworthy that a subset of this group who developed nodal or liver recurrence was resected again and was disease-free at the last follow-up.

Sakamoto et al. (2020) found recurrence rates of 79% after a median follow-up of 30 months in 29 patients who underwent simultaneous resections of the primary tumor and para-aortic lymph nodes. The 3-year overall survival rate was 50.5%, with a 3-year recurrence-free survival rate of 17.2%. Patients with pM1a metastases performed better in terms of 3-year overall survival compared to the combined group of patients with pM1b and pM1c metastases (63% vs 24%). Hazard ratios for the pM1a group for overall survival and recurrence-free survival were 5.15 (95% CI: 1.5–17.5) and 2.5 (95% CI: 1.1–5.9), respectively (Sakamoto et al., 2020).

Similar observations were made by Yamada et al. (2019), who found that 5-year recurrence-free survival was better in patients with M1a metastases compared to patients with M1b and M1c metastases (27% vs 0%). Patients with a maximum of two involved para-aortic lymph nodes had significantly higher 5-year recurrence-free survival compared to patients with three or more involved lymph nodes (42% vs 0%). A total of 81% patients (29/36) developed recurrences after a median follow-up of 2.1 years; 26% were local recurrences (Yamada et al., 2019). This patient cohort dated from 1984 to 2011; 2 patients received neoadjuvant chemotherapy, 25 patients received adjuvant chemotherapy, and 11 patients received no perioperative chemotherapy.

The aforementioned review by Wong et al. (2016) included four studies investigating synchronous removal of metastatic para-aortic lymph node metastases. The 5-year overall survival ranged between 23% and 66% in included studies.

Bae et al. (2018) performed a retrospective analysis of 49 patients who underwent resection of isolated synchronous para-aortic lymph node metastases in the period 1988–2009. In multivariate analyses, the number of involved para-aortic lymph nodes proved an independent risk factor for overall survival (HR: 3.29; 95% CI: 1.31–8.28). Preoperative CEA level and a number of involved para-aortic lymph nodes were borderline independent risk factors for disease-free survival (HR: 1.95; 95%CI: 0.94–4.06; $P = .073$ and HR: 2.48; 95%CI: 0.99–6.21, $P = .052$, respectively). Five-year overall survival was 36.5% for patients with seven or fewer para-aortic lymph node metastases versus 14.3% in patients with more than seven lymph node metastases, with postoperative chemotherapy received by 47/49 patients. Notably, the median survival time was 37 months (range 6–169), illustrating that prolonged survival may occur after para-aortic lymphadenectomy (Bae et al., 2018).

Min et al. performed a retrospective study from a prospectively collected database involving rectal cancer patients who underwent simultaneous resection of para-aortic lymph node metastases and the primary tumor, with or without lateral pelvic wall lymphadenectomy ($n = 86$ and $n = 50$, respectively) (Mi et al., 2008). Patients with combined lateral pelvic wall and para-aortic lymph node metastases showed the worst cancer-specific and disease-free survival (15.4% and 11.1%, respectively), followed by para-aortic lymph node metastases (22.7% and 17.6%, respectively) and lateral pelvic wall metastases (39.7% and 26.3%, respectively) (Min et al., 2008).

Choi et al. (2010) reported on 24 patients who were subjected to para-aortic lymph node resection for isolated metastasis (synchronous or metachronous). Outcomes were compared with those of 53 control group patients who were managed nonsurgically for para-aortic lymph node metastases. The overall survival rate at 5 years was significantly better in the resection group at 53% versus 12% in the control group, with a better prognosis for patients with one or two affected lymph nodes. Median survival was 64 and 33 months, respectively. Patients included in this study were treated between January 1993 and March 2006, that is, before the era of targeted therapy. Although not statistically significant, the majority of patients in the resection group was treated with adjuvant chemotherapy (23/24) in contrast to 40/53 patients in the control group. After para-aortic lymphadenectomy, 67% of patients developed recurrences after a mean follow-up of 37 months, with seven patients metastasizing to para-aortic lymph nodes (Choi et al., 2010).

Song et al. (2016) investigated a total laparoscopic approach for simultaneous resection of primary colorectal cancer and para-aortic lymph node metastases below the renal veins in 40 patients without the distant disease. Three-year disease-free survival in patients with metastatic para-aortic lymph nodes was 55.6% in patients with three or fewer involved lymph nodes versus 0% in patients with four or more affected lymph nodes (Song et al., 2016).

Yamamoto et al. (2019) also described the outcomes of a laparoscopic approach for combined resection of left-sided primary colorectal cancer and suspected para-aortic lymph node metastases. Conversion was necessary for 1/11 patients due to rib invasion, morbidity was limited (ileus, chylous ascites, neurogenic bladder, all $n = 1$). Para-aortic lymph node metastases were confirmed by histopathology in five cases, all of whom were treated with adjuvant chemotherapy. One out of five patients developed recurrent para-aortic lymph node metastasis and lateral pelvic nodes after 1 year and died after developing distant solid organ metastases after 25 months. Two patients did not develop recurrences after 2 and 17 months, respectively, while two were alive with recurrences after 33 and 44 months (Yamamoto et al., 2019).

13.6 Morbidity of surgery

Sakamoto et al. (2020) described their experience with surgical resection of the primary tumor and synchronous para-aortic metastases in 29 patients. The population's median age was 60 years, with approximately half of the population male. The median number of resected para-aortic lymph node metastasis was 12 (1–81), with the metastatic disease found in 4 (1–71). The vast majority of patients were operated through an open approach (28/29), with a median operating time of 248 min (range 110–645). Median estimated blood loss was 628 mL (range 20–4900), with grade I–III complications occurring in 31% of patients. Surgical site infections were most common, no high-grade complications occurred. Comparable complication rates were reported by Choi et al. (2010) at 28%. Yamada et al. (2019) reported 39% grade I–III complications in a series of 36 patients undergoing synchronous resection of the primary tumor and para-aortic lymph node metastases.

Nakai et al. (2017) performed para-aortic lymphadenectomy in 30 patients with left-sided colorectal cancer and involved lymph nodes. Twenty-seven percent of patients developed grade II—IV morbidity include anastomotic leakage, intraabdominal abscess formation, ileus, urinary tract infections/retention, surgical site infections, and chylous ascites (Nakai et al., 2017).

In a series described by Arimoto et al., seven patients underwent resection of para-aortic lymph nodes only, resulting in a median operating time of 324 min (range 280—514) and a median blood loss of 365 mL (range 42—528 mL). Median postoperative stay was 11 days (range 7—17 days). In patients undergoing simultaneous resection of the primary tumor and para-aortic metastases, operation time, blood loss, and hospitalization were negatively affected [588 min (range 506—1053), 1300 mL (range 850—3300), and 34 days (range 21—50)]. Four patients developed lymph node recurrences after a median follow-up of 33 months (29%) (Arimoto et al., 2015).

Song et al. reported the only series of patients who underwent simultaneous laparoscopic resection of the primary tumor and synchronous para-aortic lymph node metastases. Operations involved right and left hemicolectomy and (low) anterior resections, with a mean operating time of 192 ± 69 min. Mean blood loss was 66 mL (range 20—210 mL), with 15% of patients developing postoperative complications. There was no mortality in this series, and the mean hospital stay was 9.8 days (5—35) (Song et al., 2016).

13.7 The role of chemotherapy

Many of the aforementioned series were treated before 2010. Although most of the chemotherapy regimens were comparable nowadays, targeted therapy was not part of regular treatment. The series by Han et al. that involved patients with metachronous metastases prescribed capecitabine, FOLFOX, and FOLFIRI. Bae et al. achieved a high rate of postoperative chemotherapy (FOLFOX, 47/49 patients), administered in 3- to 4-week schedules over 6 months time.

Although adjuvant chemotherapy was described by most authors, neoadjuvant chemotherapy was less frequently prescribed. Ushigome et al. (2020) reported that two patients were treated with neoadjuvant chemotherapy (oxaliplatin), and adjuvant chemotherapy was used in 12 patients ($n = 9$ for oxaliplatin or camptothecin-11). Song et al. did not use neoadjuvant chemotherapy in any of the patients with synchronous metastases, but the vast majority of patients with stage III and stage IV disease received adjuvant chemotherapy.

There is a tendency to change treatment algorithms for both locally advanced colon cancer and rectal cancer, moving to neoadjuvant chemotherapy treatment (Bahadoer et al., 2021; Cercek et al., 2018; Karoui et al., 2020). In most studies, this results in a higher overall chemotherapy treatment rate, as patients with postoperative complications are often not able to start adjuvant chemotherapy on time, or at all. Although higher pathological response rates are frequently found in these studies, the effects on overall survival are still unclear. A small proportion of these patients will suffer from acute and/or chronic toxicity, which will impact the patients' quality of life. The neoadjuvant strategy seems to gain popularity for other patient groups too, for instance in the case of peritoneal metastases or pelvic recurrences of colorectal cancer. A similar change of approach can be expected to occur for patients who present with para-aortic lymph node metastases, presenting with either metachronous or with synchronous disease.

The combination of chemoradiation (capecitabine/oxaliplatin) with bevacizumab has shown effectiveness in one patient with para-aortic lymph node recurrence (Miyazawa et al., 2012). With regard to immunotherapy, Tzovaras et al. (2011) described a case series of three patients who had undergone intense chemotherapy for metastatic colorectal cancer. All patients had K-RAS (Kirsten rat sarcoma virus) wild-type-expressing tumors, and had failed bevacizumab. After starting treatment with panitumumab monotherapy, treatment responses were found in all patients, including one patient with a 7-month partial response of metastatic aorto-iliac nodes (Tzovaras et al., 2011).

13.8 Conclusion and future perspectives

The literature supports the surgical resection of metastatic para-aortic lymph nodes in selected patients. Overall, the fewer nodes affected, the better the results. Currently, there is little evidence for an aggressive surgical approach for patients with para-aortic metastatic disease in combination with distant solid-organ metastases (e.g., liver, lung, brain, or bone metastases). Only a few case series have been published on the topic, mainly on the combination of para-aortic metastatic disease and colorectal liver metastases (Adam et al., 2008; Nanji et al., 2017; Okuno et al., 2018; Pulitanò et al., 2012; Uemura et al., 2016). The ORCHESTRA trial will provide data on the additional value of debulking compared to chemotherapy alone in patients with metastatic disease in multiple organ systems, including para-aortic lymph nodes (ClinicalTrials.gov Identifier: NCT01792934).

Most of the series describe negligible mortality and limited morbidity associated with surgical resection, although it must be realized that these patients were highly selected and operated in expert centers. Expertise in retroperitoneal surgery and the ability to manage vascular and hollow viscus injuries are prerequisites for performing this type of surgical resection. In general, patients with a poor performance status are considered unfit for surgical resection. In case of encasement of the superior mesenteric artery/celiac axis, or involvement of the pancreatic head, body, duodenum, or bile ducts necessitating a Whipple procedure, it would often be considered too morbid to consider resection (Sasaki et al., 2020).

A multimodal approach with neoadjuvant chemotherapy can prevent performing surgery in patients with unfavorable tumor biology, who progress in spite of adequate treatment. As the majority of patients recur at some point during the treatment process, the benefits of postoperative chemotherapy need to be weighed against its side effects and the patient's performance status. As an alternative, improved radiotherapy techniques including intraoperative electron radiotherapy may help to radiate affected lymph nodes effectively while reducing treatment-related side effects. Especially in patients with considerable morbidity, radiotherapy can prove beneficial over chemotherapy alone (Isozaki et al., 2017). A trimodality therapy has been proposed by Johnson et al. (2018), involving neoadjuvant chemotherapy, external beam radiotherapy with chemotherapy, followed by lymphadenectomy, and intraoperative radiotherapy. The treatment results of targeted next-generation sequencing and newly developed immune checkpoint inhibitors are yet to be awaited, as patients included in the reported series were often treated before the introduction of these techniques and drugs.

Acknowledgment

We are grateful to Miss Amanda Verstraete for the provided administrative support.

References

Adam, R., de Haas, R. J., Wicherts, D. A., Aloia, T. A., Delvart, V., Azoulay, D., Bismuth, H., & Castaing, D. (2008). Is hepatic resection justified after chemotherapy in patients with colorectal liver metastases and lymph node involvement? *Journal of Clinical Oncology: Official Journal of the American Society of Clinical Oncology*, 26(22), 3672–3680. Available from https://doi.org/10.1200/JCO.2007.15.7297.

Albandar, M. H., Cho, M. S., Bae, S. U., & Kim, N. K. (2016). Surgical management of extra-regional lymph node metastasis in colorectal cancer. *Expert Review of Anticancer Therapy*, 16(5), 503–513. Available from https://doi.org/10.1586/14737140.2016.1162718.

Arimoto, A., Uehara, K., Kato, T., Nakamura, H., Kamiya, T., & Nagino, M. (2015). Clinical significance of para-aortic lymph node dissection for advanced or metastatic colorectal cancer in the current era of modern chemotherapy. *Digestive Surgery*, 32(6), 439–444. Available from https://doi.org/10.1159/000439547.

Bae, S. U., Hur, H., Min, B. S., Baik, S. H., Lee, K. Y., & Kim, N. K. (2018). Which patients with isolated para-aortic truly benefit from extended lymph node dissection for colon cancer? *Cancer Research and Treatment*, 50(3), 712–719. Available from https://doi.org/10.4143/crt.2017.100.

Bahadoer, R. R., Dijkstra, E. A., van Etten, B., Marijnen, C. A. M., Putter, H., Kranenbarg, E. M.-K., Roodvoets, A. G. H., Nagtegaal, I. D., Beets-Tan, R. G. H., Blomqvist, L. K., Fokstuen, T., Ten Tije, A. J., Capdevila, J., Hendriks, M. P., Edhemovic, I., Cervantes, A., Nilsson, P. J., Glimelius, B., van de Velde, C. J. H., & Hospers, G. A. P. (2021). Short-course radiotherapy followed by chemotherapy before total mesorectal excision (TME) versus preoperative chemoradiotherapy, TME, and optional adjuvant chemotherapy in locally advanced rectal cancer (RAPIDO): A randomised, open-label, phase 3 trial. *The Lancet Oncology*, 22(1), 29–42. Available from https://doi.org/10.1016/S1470-2045(20)30555-6.

Cercek, A., Roxburgh, C. S. D., Strombom, P., Smith, J. J., Temple, L. K. F., Nash, G. M., Guillem, J. G., Paty, P. B., Yaeger, R., Stadler, Z. K., Seier, K., Gonen, M., Segal, N. H., Reidy, D. L., Varghese, A., Shia, J., Vakiani, E., Wu, A. J., Crane, C. H., & Weiser, M. R. (2018). Adoption of total neoadjuvant therapy for locally advanced rectal cancer. *JAMA Oncology*, 4(6), e180071. Available from https://doi.org/10.1001/jamaoncol.2018.0071.

Choi, P. W., Kim, H. C., Kim, A. Y., Jung, S. H., Yu, C. S., & Kim, J. C. (2010). Extensive lymphadenectomy in colorectal cancer with isolated para-aortic lymph node metastasis below the level of renal vessels. *Journal of Surgical Oncology*, 101(1), 66–71. Available from https://doi.org/10.1002/jso.21421.

Crumbie, L., & Zehra, U. (2020). Lymphatics of the retroperitoneal space. <https://www.kenhub.com/en/library/anatomy/lymphatics-of-the-retroperitoneal-space>; (Original work published 2020).

Gabriel, W. B., Dukes, C., & Bussey, H. J. R. (1935). Lymphatic spread in cancer of the rectum. *British Journal of Surgery*, 23(90), 395–413. Available from https://doi.org/10.1002/bjs.1800239017.

Han, J., Lee, K. Y., Kim, N. K., & Min, B. S. (2020). Metachronous metastasis confined to isolated lymph node after curative treatment of colorectal cancer. *International Journal of Colorectal Disease*, 35(11), 2089–2097. Available from https://doi.org/10.1007/s00384-020-03695-8.

Isozaki, Y., Yamada, S., Kawashiro, S., Yasuda, S., Okada, N., Ebner, D., Tsuji, H., Kamada, T., & Matsubara, H. (2017). Carbon-ion radiotherapy for isolated para-aortic lymph node recurrence from colorectal cancer. *Journal of Surgical Oncology*, 116(7), 932–938. Available from https://doi.org/10.1002/jso.24757.

Johnson, B., Jin, Z., Haddock, M. G., Hallemeier, C. L., Martenson, J. A., Smoot, R. L., Larson, D. W., Dozois, E. J., Nagorney, D. M., & Grothey, A. (2018). A curative-intent trimodality approach for isolated abdominal nodal metastases in metastatic colorectal cancer: Update of a single-institutional experience. *The Oncologist*, 23(6), 679–685. Available from https://doi.org/10.1634/theoncologist.2017-0456.

Kang, J., Hur, H., Min, B. S., Kim, N. K., & Lee, K. Y. (2011). Prognostic impact of inferior mesenteric artery lymph node metastasis in colorectal cancer. *Annals of Surgical Oncology*, 18(3), 704–710. Available from https://doi.org/10.1245/s10434-010-1291-x.

Karoui, M., Rullier, A., Piessen, G., Legoux, J. L., Barbier, E., De Chaisemartin, C., Lecaille, C., Bouche, O., Ammarguellat, H., Brunetti, F., Prudhomme, M., Regimbeau, J. M., Glehen, O., Lievre, A., Portier, G., Hartwig, J., Goujon, G., Romain, B., Lepage, C., & Taieb, J. (2020). Perioperative FOLFOX 4 versus FOLFOX 4 plus cetuximab versus immediate surgery for high-risk stage ii and iii colon cancers: A phase ii multicenter randomized controlled trial (PRODIGE 22). *Annals of Surgery, 271*(4), 637−645. Available from https://doi.org/10.1097/SLA.0000000000003454.

Kim, H. J., Song, S. H., Choi, G.-S., Park, J. S., & Park, S. Y. (2021). Laparoscopic multivisceral resection with fluorescence-guided para-aortic lymph node dissection for advanced t4b colon cancer. *Diseases of the Colon and Rectum, 64*(2), e23−e24. Available from https://doi.org/10.1097/DCR.0000000000001902.

Kim, Y. I., Park, I. J., Park, J.-H., Kim, T. W., Ro, J.-S., Lim, S.-B., Yu, C. S., & Kim, J. C. (2020). Management of isolated para-aortic lymph node recurrence after surgery for colorectal cancer. *Annals of Surgical Treatment and Research, 98*(3), 130−138. Available from https://doi.org/10.4174/astr.2020.98.3.130.

Labianca, R., Beretta, G. D., Kildani, B., Milesi, L., Merlin, F., Mosconi, S., Pessi, M. A., Prochilo, T., Quadri, A., Gatta, G., de Braud, F., & Wils, J. (2010). Colon cancer. *Critical Reviews in Oncology/Hematology, 74*(2), 106−133. Available from https://doi.org/10.1016/j.critrevonc.2010.01.010.

Lee, S. H., Lee, J. L., Kim, C. W., Lee, H. I., Yu, C. S., & Kim, J. C. (2017). Oncologic significance of para-aortic lymph node and inferior mesenteric lymph node metastasis in sigmoid and rectal adenocarcinoma. *European Journal of Surgical Oncology: The Journal of the European Society of Surgical Oncology and the British Association of Surgical Oncology, 43*(11), 2076−2083. Available from https://doi.org/10.1016/j.ejso.2017.08.014.

Liberale, G., Vankerckhove, S., Galdon, M. G., Donckier, V., Larsimont, D., & Bourgeois, P. (2015). Fluorescence imaging after intraoperative intravenous injection of indocyanine green for detection of lymph node metastases in colorectal cancer. *European Journal of Surgical Oncology: The Journal of the European Society of Surgical Oncology and the British Association of Surgical Oncology, 41*(9), 1256−1260. Available from https://doi.org/10.1016/j.ejso.2015.05.011.

Liu, Y.-L., Wang, Y.-H., Yang, Y.-M., Li, M.-Q., Jiang, S.-X., & Wang, X.-S. (2012). The role of para-aortic lymphadenectomy in surgical management of patients with stage N+ rectal cancer below the peritoneal reflection. *Cell Biochemistry and Biophysics, 62*(1), 41−46. Available from https://doi.org/10.1007/s12013-011-9256-7.

Lu, H.-J., Lin, J.-K., Chen, W.-S., Jiang, J.-K., Yang, S.-H., Lan, Y.-T., Lin, C.-C., Liu, C.-A., & Teng, H.-W. (2015). The prognostic role of para-aortic lymph nodes in patients with colorectal cancer: Is it regional or distant disease? *PLoS One, 10*(6), e0130345. Available from https://doi.org/10.1371/journal.pone.0130345.

Merrie, A. E., Phillips, L. V., Yun, K., & McCall, J. L. (2001). Skip metastases in colon cancer: Assessment by lymph node mapping using molecular detection. *Surgery, 129*(6), 684−691.

Min, B. S., Kim, N. K., Sohn, S. K., Cho, C. H., Lee, K. Y., & Baik, S. H. (2008). Isolated paraaortic lymph-node recurrence after the curative resection of colorectal carcinoma. *Journal of Surgical Oncology, 97*(2), 136−140.

Miyazawa, T., Ebe, K., Koide, N., & Fujita, N. (2012). Complete response of isolated para-aortic lymph node recurrence from rectosigmoid cancer treated by chemoradiation therapy with capecitabine/oxaliplatin plus bevacizumab: A case report. *Case Reports in Oncology, 5*(2), 216−221. Available from https://doi.org/10.1159/000338840.

Nakai, N., Yamaguchi, T., Kinugasa, Y., Shiomi, A., Kagawa, H., Yamakawa, Y., Numata, M., & Furutani, A. (2017). Long-term outcomes after resection of para-aortic lymph node metastasis from left-sided colon and rectal cancer. *International Journal of Colorectal Disease, 32*(7), 999−1007. Available from https://doi.org/10.1007/s00384-017-2806-8.

Nanji, S., Tsang, M. E., Wei, X., & Booth, C. M. (2017). Regional lymph node involvement in patients undergoing liver resection for colorectal cancer metastases. *European Journal of Surgical Oncology: The Journal of the European Society of Surgical Oncology and the British Association of Surgical Oncology, 43*(2), 322−329. Available from https://doi.org/10.1016/j.ejso.2016.10.033.

Ogura, A., Akiyoshi, T., Takatsu, Y., Nagata, J., Nagasaki, T., Konishi, T., Fujimoto, Y., Nagayama, S., Fukunaga, Y., & Ueno, M. (2017). The significance of extended lymphadenectomy for colorectal cancer with isolated synchronous extraregional lymph node metastasis. *Asian Journal of Surgery, 40*(4), 254−261. Available from https://doi.org/10.1016/j.asjsur.2015.10.003.

Okuno, M., Goumard, C., Mizuno, T., Kopetz, S., Omichi, K., Tzeng, C.-W. D., Chun, Y. S., Lee, J. E., Vauthey, J.-N., & Conrad, C. (2018). Prognostic impact of perihepatic lymph node metastases in patients with resectable colorectal liver metastases. *The British Journal of Surgery, 105*(9), 1200−1209. Available from https://doi.org/10.1002/bjs.10822.

Pulitanò, C., Bodingbauer, M., Aldrighetti, L., Choti, M. A., Castillo, F., Schulick, R. D., Gruenberger, T., & Pawlik, T. M. (2012). Colorectal liver metastasis in the setting of lymph node metastasis: Defining the benefit of surgical resection. *Annals of Surgical Oncology, 19*(2), 435−442. Available from https://doi.org/10.1245/s10434-011-1902-1.

References

Sahara, K., Watanabe, J., Ishibe, A., Suwa, Y., Suwa, H., Ota, M., Kunisaki, C., & Endo, I. (2019). Long-term outcome and prognostic factors for patients with para-aortic lymph node dissection in left-sided colorectal cancer. *International Journal of Colorectal Disease, 34*(6), 1121–1129. Available from https://doi.org/10.1007/s00384-019-03294-2.

Sakamoto, J., Ozawa, H., Nakanishi, H., & Fujita, S. (2020). Oncologic outcomes after resection of para-aortic lymph node metastasis in left-sided colon and rectal cancer. *PLoS One, 15*(11), e0241815. Available from https://doi.org/10.1371/journal.pone.0241815.

Sasaki, K., Nozawa, H., Kawai, K., Hata, K., Tanaka, T., Nishikawa, T., Shuno, Y., Kaneko, M., Murono, K., Emoto, S., Sonoda, H., & Ishihara, S. (2020). Management of isolated para-aortic lymph node recurrence of colorectal cancer. *Surgery Today, 50*(9), 947–954. Available from https://doi.org/10.1007/s00595-019-01872-z.

Shibata, D., Paty, P. B., Guillem, J. G., Wong, W. D., & Cohen, A. M. (2002). Surgical management of isolated retroperitoneal recurrences of colorectal carcinoma. *Diseases of the Colon and Rectum, 45*(6), 795–801.

Song, S. H., Park, S. Y., Park, J. S., Kim, H. J., Yang, C.-S., & Choi, G.-S. (2016). Laparoscopic para-aortic lymph node dissection for patients with primary colorectal cancer and clinically suspected para-aortic lymph nodes. *Annals of Surgical Treatment and Research, 90*(1), 29–35. Available from https://doi.org/10.4174/astr.2016.90.1.29.

Tsai, H.-L., Chen, Y.-T., Yeh, Y.-S., Huang, C.-W., Ma, C.-J., & Wang, J.-Y. (2019). Apical lymph nodes in the distant metastases and prognosis of patients with stage iii colorectal cancer with adequate lymph node retrieval following folfox adjuvant chemotherapy. *Pathology Oncology Research: POR, 25*(3), 905–913. Available from https://doi.org/10.1007/s12253-017-0381-5.

Tzovaras, A. A., Karagiannis, A., Margari, C., Barla, G., & Ardavanis, A. (2011). Effective panitumumab treatment in patients with heavily pre-treated metastatic colorectal cancer: A case series. *Anticancer Research, 31*(3), 1033–1037.

Uemura, M., Kim, H. M., Ikeda, M., Nishimura, J., Hata, T., Takemasa, I., Mizushima, T., Yamamoto, H., Doki, Y., & Mori, M. (2016). Long-term outcome of adrenalectomy for metastasis resulting from colorectal cancer with other metastatic sites: A report of 3 cases. *Oncology Letters, 12*(3), 1649–1654.

Ushigome, H., Yasui, M., Ohue, M., Haraguchi, N., Nishimura, J., Sugimura, K., Yamamoto, K., Wada, H., Takahashi, H., Omori, T., Miyata, H., & Takiguchi, S. (2020). The treatment strategy of R0 resection in colorectal cancer with synchronous para-aortic lymph node metastasis. *World Journal of Surgical Oncology, 18*(1), 229. Available from https://doi.org/10.1186/s12957-020-02007-2.

Wang, L., Hirano, Y., Heng, G., Ishii, T., Kondo, H., Hara, K., Obara, N., Asari, M., & Yamaguchi, S. (2020). Prognostic utility of apical lymph node metastasis in patients with left-sided colorectal cancer. *In Vivo (Athens, Greece), 34*(5), 2981–2989. Available from https://doi.org/10.21873/invivo.12129.

Wong, J. S. M., Tan, G. H. C., & Teo, M. C. C. (2016). Management of para-aortic lymph node metastasis in colorectal patients: A systemic review. *Surgical Oncology, 25*(4), 411–418. Available from https://doi.org/10.1016/j.suronc.2016.09.008.

Yamada, K., Tsukamoto, S., Ochiai, H., Shida, D., & Kanemitsu, Y. (2019). Improving selection for resection of synchronous para-aortic lymph node metastases in colorectal cancer. *Digestive Surgery, 36*(5), 369–375. Available from https://doi.org/10.1159/000491100.

Yamamoto, S., Kanai, T., Yo, K., Hongo, K., Takano, K., Tsutsui, M., Nakanishi, R., Yoshikawa, Y., & Nakagawa, M. (2019). Laparoscopic para-aortic lymphadenectomy for colorectal cancer with clinically suspected lymph node metastasis. *Asian Journal of Endoscopic Surgery, 12*(4), 417–422. Available from https://doi.org/10.1111/ases.12666.

Yeo, S.-G., Kim, M.-J., Kim, D. Y., Chang, H. J., Kim, M. J., Baek, J. Y., Kim, S. Y., Kim, T. H., Park, J. W., & Oh, J. H. (2013). Patterns of failure in patients with locally advanced rectal cancer receiving pre-operative or post-operative chemoradiotherapy. *Radiation Oncology (London, England), 8*, 114. Available from https://doi.org/10.1186/1748-717X-8-114.

CHAPTER 14

Lateral lymph node dissection in rectal cancer

Tania C. Sluckin[1], Sanne-Marije J.A. Hazen[1], Takashi Akiyoshi[2] and Miranda Kusters[1]

[1]Department of Surgery, Cancer Center Amsterdam, Amsterdam University Medical Centers, Vrije Universiteit Amsterdam, Amsterdam, the Netherlands
[2]Department of Gastroenterological Surgery, Cancer Institute Hospital of the Japanese Foundation for Cancer Research, Tokyo, Japan

14.1 Introduction

Rectal cancer is a complex disease, primarily due to its location. Tapering of the mesorectum means that the mesorectal fascia is more easily involved, resulting in incomplete resection margins (R1), compared to higher tumors. In addition, many crucial nerves, reproductive organs, blood vessels, and other organs are in close proximity, increasing the chances of significant morbidity. Distal rectal cancers, situated below the peritoneal reflection, also have a greater tendency to spread to lymph nodes positioned laterally in the pelvis (lateral lymph nodes) (Steup et al., 2002). These lateral lymph nodes are located along the (internal) iliac and obturator vessels in the pelvis and are not resected during standard total mesorectal excision (TME) surgery (Christou et al., 2019).

Current research is, increasingly, presenting evidence for the importance of lateral lymph nodes in the accurate diagnosis and subsequent treatment of rectal cancer and their important role in lateral local recurrences. An important issue is when and which lateral lymph nodes should be considered a risk for developing a local recurrence.

In the era before the TME procedure, most local recurrences of rectal cancer occurred centrally in the pelvis. Although the introduction of neoadjuvant (chemo)radiotherapy and the TME procedure has successfully lowered overall local recurrence rates after 5 years to 5%–10% (Kusters et al., 2010; Martling et al., 2001), recent research provides evidence for high rates of lateral local recurrence for patients with primarily enlarged lateral lymph nodes, ranging from 19.5% to 52% (Kusters et al., 2017; Ogura, Konishi, Beets, et al., 2019;

Ogura, Konishi, Cunningham, et al., 2019). Currently, approximately 50% of the local recurrences are located in the lateral pelvic compartments, most likely caused by lateral nodal disease (Iversen et al., 2018).

This evidence creates an important question: if lateral lymph nodes carry such a considerable risk for long-term local recurrence outcomes, how can these lymph nodes be correctly identified, adequately irradiated, and, if necessary, effectively surgically removed? This chapter addresses this by first describing the historical differences in the treatment of rectal cancer between Eastern and Western physicians. Thereafter the current understanding and awareness of lateral nodal disease is discussed along with advice for future perspectives.

14.2 East versus West

Throughout the development of Western medicine, a significant part of the oncological treatment for rectal cancer involves (chemo)radiotherapy. Western surgeons primarily rely on (chemo)radiotherapy as neoadjuvant treatment to sterilize the lateral compartments and often interpret a lateral local recurrence as a sign of metastatic disease (Kusters et al., 2017; Ogura, Konishi, Beets, et al., 2019; Ogura, Konishi, Cunningham, et al., 2019). This is in contrast to Eastern medicine, predominantly Japan, which relies more solely on surgical techniques; the TME, which can be accompanied by a lateral lymph node dissection (LLND). Eastern surgeons have traditionally adopted the (prophylactic) LLND into their general practice, highlighting the early divergence of Eastern and Western medical practices for patients with low, locally advanced, rectal cancer (Nakamura & Watanabe, 2013; Obara et al., 2012; Tamura et al., 2017). This divergence is apparent when discussing the interpretation of what an enlarged lateral lymph node represents. For Eastern surgeons, the presence of enlarged lateral lymph nodes represents an advanced, yet resectable, disease, for which extensive surgery is the answer. Western surgeons often, however, interpret enlarged lateral lymph nodes as a sign of metastatic disease (Kusters et al., 2017; Ogura, Konishi, Beets, et al., 2019; Ogura, Konishi, Cunningham, et al., 2019) and even the latest Tumour-Node-Metastasis classification by the American Joint Committee on Cancer (Weiser, 2018) describes lateral lymph nodes as an indication of distant metastatic disease (Christou et al., 2019). This significant difference in interpretation between Eastern and Western surgeons is of vital importance and should be addressed through comparative international research.

In the era before "good quality" magnetic resonance imaging (MRI), when the size of lateral lymph nodes was not relevant during diagnostics, one study compared Japanese patients undergoing a rectal resection and LLND to Dutch patients from the Dutch TME trial who underwent a TME procedure with or without preoperative radiotherapy (5×5 Gy). Local recurrence rates after 5 years were 6.9%, 5.8%, and 12.1% for the Japanese group, Dutch TME + RT group, and Dutch TME group, respectively (Kusters et al., 2009). These numbers are relatively low due to the fact that patients without enlarged lymph nodes also underwent a prophylactic LLND, but the similar rates between the Japanese group and TME + RT suggest that irradiation and prophylactic LLND result in similar outcomes. Furthermore, the recurrence rates in the lateral pelvis were 2.2%, 0.8%, and 2.7%, respectively, indicating the important role of radiotherapy to sterilize the lateral pelvic compartments (Kusters et al., 2009). As similar local recurrence rates were found between patients undergoing prophylactic lateral node dissections and those receiving radiotherapy and TME surgery, Western surgeons were validated in their belief that neoadjuvant treatment and traditional surgery were acceptable, without performing LLND with its concurrent risks of bleeding and nerve damage in obese western patients.

On the other hand, an increasing number of Eastern studies have emerged demonstrating alarming local recurrence rates of approximately 40% after 5 years for lateral lymph nodes larger than 10 mm treated with TME only, confirming Eastern surgeons' view that prophylactic LLND is a vital aspect of appropriate treatment (Kim et al., 2008, 2014, 2015). However, emerging evidence suggests a shift in the traditional Eastern and Western paradigms. For example, several recent Western studies suggest that only neoadjuvant treatment and TME surgery may be insufficient for particular cases. Patients with lateral lymph nodes larger than 10 mm had a 33.3% chance for lateral local recurrence after 4 years versus those with lateral lymph nodes smaller than 10 mm, resulting in a lateral local recurrence of 10.1% (Kusters et al., 2017; Ogura, Konishi, Beets, et al., 2019; Ogura, Konishi, Cunningham, et al., 2019). This was despite sterilization of the lateral compartments with neoadjuvant radiotherapy. Similarly, patients in a Western cohort with lateral lymph nodes larger than 10 mm resulted in a 37% lateral local recurrence rate after 5 years compared to those with lateral lymph nodes smaller than 10 mm [7.7%; (Schaap et al., 2018)]. Considering this influx of research with study populations that are applicable to Western patients, Western physicians are beginning to acknowledge the importance of lateral nodal disease and its representation of local disease. None of the results from the aforementioned studies found an increase in distant metastases; an aspect which Western physicians have related to lateral nodal disease for many years. These results, among others, encourage a shift by Western clinicians to the interpretation of lateral, local, nodal disease (Kusters et al., 2017; Ogura, Konishi, Beets, et al., 2019; Ogura, Konishi, Cunningham, et al., 2019).

There is also a shift in some Japanese institutions to apply neoadjuvant chemoradiotherapy, and promising results are arising from various Eastern research groups investigating the application of selective LLND procedures after neoadjuvant treatment. A Japanese study researched 127 patients with clinical Stage 2—3 low rectal cancer who underwent both neoadjuvant treatment and a curative resection (Table 14.1). Of thes, 38 patients with suspected lateral pelvic lymph node metastases (equal or greater than 7 mm) also underwent a LLND. These results are very encouraging: in 66% of the patients undergoing a LLND, metastases were confirmed and the local recurrence rate after 3 years was 3.4% (TME alone) versus 0% (LLND) with a relapse-free survival of 74.6% (TME alone) versus 83.8% (LLND) (Akiyoshi et al., 2014). Furthermore, the same research group later compared lateral lymph nodes sizes on pre- and posttreatment MR images and found correlations; patients with persistently enlarged lateral lymph nodes (>8 and >5 mm) had a significantly higher percentage of lymph node metastases (75% vs 20%, $P < .0001$) (Akiyoshi et al., 2015). Another study also incorporated posttreatment MRI scans into their decision protocols and after evaluating 64 patients they concluded that a posttreatment lymph node size of equal or greater than 5 mm was strongly associated with a pathological activity. Following a selective LLND for these patients, none developed a lateral recurrence after a median of 39 months of follow-up (Songphol et al., 2019). In this manner, research is providing excellent local recurrence rates for situations in which neoadjuvant

TABLE 14.1 Local recurrence rates with LLNs evaluated on the primary MRI.

Article	Design	No. of patients (n)	Patient population	Neoadjuvant therapy: (C)RT	LLND indication	Primary endpoint	Results
Akiyoshi et al. (2014)	Retrospective	127	Stage 2/3 low rectal cancer	Yes	LLND performed when suspected metastasis (>7 mm) before CRT	LR at 3 years	Pathological LLN in 66% (25) LLND. LR 3-year 3.4% TME and 0% LLND.
Akiyoshi et al. (2015)	Retrospective	77	Advanced low rectal cancer and underwent LLND for suspicious LLN	Yes	LLND performed when suspected metastasis (>7 mm) before CRT	Predictors of LLN metastasis as indication for LLND	Metastasis higher for LLN ≧8 mm versus <8 mm pre-MRI (75% vs 20%). Metastasis higher for LLN >5 mm versus ≦5 mm post-MRI (75% vs 20%).
Kim et al. (2015)	Retrospective	900	Primary rectal cancer <10 cm anal verge, stage 2/3	Yes	—	Patterns of LR for curative resection and CRT at 5 years	LR 7.2%, LLR: 64.6%. LLN <5/5–10/10 mm 5 year: RFS: 98%, 91%, and 40%. LRFS: 95%, 87%, 40%. OS: 86%, 83%, 57%.
Fujita et al. (2017)	RCT: TME versus TME + LLND	701	Clinical stage 2/3 rectal cancer	No	Standard treatment in Japan	Relapse-free survival	RFS 5-year 73.4% (TME + LLND) versus 73.3% (TME). LRFS 5-year 87.7% (TME + LLND) versus 82.4% (TME).
Ogura, Konishi, Cunningham, et al. (2019)	Retrospective	1216	cT3/T4 rectal cancer, <8 cm anal verge	Yes	Japan: prophylactic or based on >7 mm primary imaging, USA and Karolinska: >5 mm or malignant features	LLR at 5 years	>7 mm: HR for LLR: 2.060. LR: 108 (10%) [LLR 59 (54%)]. >7 mm primary MRI + (C)RT + TME: 19.5% 5-year

EMVI, Extramural venous invasion; *(C)RT*, (chemo)radiotherapy; *DFS*, disease-free survival; *DM*, distant metastasis; *LLN*, lateral lymph node; *LLND*, lateral lymph node dissection; *LLR*, lateral local recurrence; *LRFS*, local recurrence-free survival; *LR*, local recurrence; *MRI*, magnetic resonance imaging; *OS*, overall survival; *RCT*, randomized controlled trial; *RFS*, relapse-free survival; *TME*, total mesorectal excision.

treatment is administered and lateral lymph node restaging is properly evaluated, allowing the LLND procedure only to be advised for a select group of 'high-risk' patients.

These results indicate a movement in both Western and Eastern treatment paradigms for rectal cancer, in which the usefulness of neoadjuvant treatment for small lymph nodes appears to be noticed and the application of selective LLND procedures is advised for particular patients. Together, the gap between the understanding, evaluation, and treatment of lateral nodal disease appears to be diminishing, and the bridge toward universal definitions and guidelines may be nearing.

14.3 Defining lateral nodal disease

For the appropriate application of current understanding, certain aspects of lateral nodal disease are important to discuss; both lateral lymph node size and anatomical location are proving to be essential aspects of the current knowledge regarding lateral nodal disease.

Currently, many standard and international guidelines predominantly focus on the "malignant" features of lymph nodes, such as shape, border irregularity, or internal heterogeneity (Beets-Tan et al., 2018; Brown et al., 2003; Kim et al., 2004). These factors are justified when considering mesorectal lymph nodes, however, various studies have shown that these characteristics do not apply to lateral lymph nodes, where the size of a lateral lymph node is the most important criterion in assessing lateral nodal disease and its potential malignancy (Kusters et al., 2017; Ogura, Konishi, Beets, et al., 2019; Ogura, Konishi, Cunningham, et al., 2019). One possibility is to consider the original size as measured on primary MRI; the Lateral Node Consortium study found that patients with lateral lymph nodes equal or greater than 7 mm on the primary MRI had an increased chance of local recurrence, 19.5% after 5 years (see Table 14.1; Ogura, Konishi, Cunningham, et al., 2019). This is a twofold increase in the local recurrence risk that relates purely to the primary lateral lymph node size. Many publications from Eastern clinicians also rely on lymph node sizes measured on the primary MRI as these patients often do not receive neoadjuvant treatment (Table 14.1). For example, the indication for LLND may be based on a lymph node size of greater than 7 mm on the primary MRI (Kim et al., 2008, 2014, 2015) or separated into categories such as 5 , 5–10 , or larger than 10 mm when considering patterns of recurrence (Fujita et al., 2017; Nakamura & Watanabe, 2013; Obara et al., 2012).

Another possibility is to evaluate the behavior of lateral lymph nodes on the restaging MRI after neoadjuvant treatment, allowing for the evaluation of the different reactions of lateral lymph nodes to neoadjuvant treatment (Table 14.2). Again, the Lateral Nodal Consortium found that the lateral local recurrence rate increased dramatically when lateral lymph nodes did not significantly decrease in size after neoadjuvant treatment. In fact, for internal iliac lateral lymph nodes which remained larger than 4 mm on the restaging MRI, lateral local recurrence rates surged to 52% (Ogura, Konishi, Cunningham, et al., 2019). This is in contrast to the lateral lymph nodes which adequately downsized. The internal iliac lymph nodes that downsized to less than 4 mm on the restaging MRI resulted in 0% lateral local recurrences after 5 years. Similarly, obturator lymph nodes which downsized to less than 6 mm also resulted in 0% lateral local recurrences (Ogura, Konishi, Beets,

TABLE 14.2 Local recurrence rates with LLNs evaluated on the restaging MRI.

	Design	No. of patients (n)	Patient population	Neoadjuvant therapy: (C)RT	LLND indication	Primary endpoint	Results
Kim et al. (2008)	Retrospective	366	Primary rectal cancer <8 cm anal verge, cT3/T4 and MRI evaluation	Yes and 303/366 adjuvant chemotherapy	—	LR at 5 years	29 (7.9%) LR at 5 years - 20.7% central pelvis - 82.7% lateral pelvis Lateral pelvic recurrences: 1.4% <5 mm LLN (SA) post-CRT; 2.9% 5–9.9 mm LLN (SA) post-CRT; 50% ≥10 mm LLN (SA) post-CRT.
Kim et al. (2014)	Retrospective	443	Stage 2/3 rectal cancer <15 cm anal verge	Yes	—	LR, LLR and LRFS at 5 years	LR in 53 patients (11.9%). LLR 28/54 (52.8%). LLN >10 mm and >2 LLN risk factor for LLR
Schaap et al. (2018)	Retrospective	192	T3–4 rectal cancer <8 cm anal verge	Yes (15.6% 5 × 5, 75.5% CRT), No (8.9%).	—	LLR at 5 years	LLR: LLN >10 mm (SA) = 37% (7.7% LLN <10 mm). DM: LLN >10 mm 23% versus 27% <10 mm. EMVI+: DM rate 43% versus 26.3% EMVI-
Malakorn et al. (2019)	Retrospective	64	Rectal cancer and suspected LLN metastasis and restaging MRI	Yes	All	LLN positivity	LLN <5 mm post-CRT (20%): 0% positivity. LLN ≥5 mm post-CRT (79%): 64% positivity. OS 5-year: 79% (LLN-) versus 61% (LLN+). DSS 5-year: 84% (LLN-) versus 66% (LLN+). 39 months later: <5 mm LLN = no deaths
Ogura, Konishi, Cunningham, et al. (2019)	Retrospective	741	cT3/T4 rectal cancer, <8 cm anal verge and restaging MRI	Yes	—	LR at 5 years	>7 mm primary MRI: 17.9% 5-year. >4 mm iliac LLN restaging + TME: 52.3% LR. >4 mm iliac LLN restaging + TME + LLND: 8.7%. >6 mm obt LLN restaging + TME: 17.8%. >6 mm obt LLN restaging + TME + LLND: 0%.

CD, Clavien and Dindo; *(C)RT*, (chemo)radiotherapy; *DSS*, disease-specific survival; *DM*, distant metastasis; *EMVI*, extramural venous invasion; *LLND*, lateral lymph node dissection; *LLN*, lateral lymph node; *LLR*, lateral local recurrence; *LR*, local recurrence; *OS*, overall survival; *RCT*, randomized controlled trial; *TME*, total mesorectal excision.

et al., 2019). Various other studies have stratified their patient populations for varying lateral lymph node sizes on the restaging MRI, such as larger or smaller than 10 mm (Kusters et al., 2017; Ogura, Konishi, Beets, et al., 2019; Ogura, Konishi, Cunningham, et al., 2019), or the individual reaction of lateral lymph nodes to neoadjuvant treatment (Ogura, Konishi, Beets, et al., 2019; Songphol et al., 2019). The studies shown in Table 14.2 are a selection of examples, many with dramatic local recurrence rates when considering lateral lymph nodes that do not adequate respond to neoadjuvant treatment.

The location of enlarged lateral lymph nodes is also important. The lateral pelvic compartments are often described as the internal iliac, external iliac and obturator compartments, related respectively to the vessels which they encompass. The obturator and internal iliac compartments are divided by the lateral border of the main trunk of internal iliac vessels (Figs. 14.1 14.2) and these compartments are important when evaluating lateral nodal disease. Various studies have examined the roles of lymph nodes related to the external and communal iliac vessels; however, enlargement of these lymph nodes appears to result in an increase in distant metastases, but not be involved in increased local recurrences and is therefore not an indication of LLND (Maeda et al., 2003; Ogura, Konishi, Beets, et al., 2019; Steup et al., 2002).

The location of lateral lymph nodes relates directly to their behavior. According to results from the Lateral Node Consortium, internal iliac lymph nodes behaved more aggressively and shrunk less often; they had a lower cut-off size value for their risk of local recurrence and a higher local recurrence rate compared to obturator lymph nodes. This is in contrast to lymph nodes in the external iliac compartment; these lymph nodes significantly influenced the distant recurrence risk, with a twofold increase, but did not result in any local recurrences (Ogura, Konishi, Beets, et al., 2019). This would suggest that the location of a lateral lymph node could be used as the starting point for predicting the chance of local recurrence rates.

Although lymph nodes in the internal iliac and obturator compartments both influence local recurrence rates, their behavior differs. Specifically, lateral lymph nodes in the internal iliac compartment only had a 22% chance of becoming less than 4 mm, while this chance was 36% for obturator lateral lymph nodes. Moreover, while internal

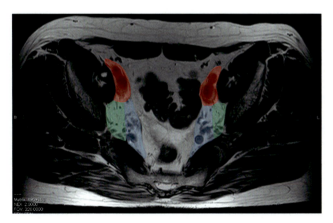

FIGURE 14.1 A surgical atlas for the lateral compartments. *Green*: obturator compartment; *blue*: internal iliac compartment; *red*: external iliac compartment. Source: *From Ogura, A., Konishi, T., Beets, G. L., Cunningham, C., Garcia-Aguilar, J., Iversen, H., Toda, S., Lee, I. K., Lee, H. X., Uehara, K., Lee, P., Putter, H., van der Velde, C. J. H., Rutten, H. J. T., Tuynman, J. B., & Kusters, M. (2019). Lateral nodal features on restaging magnetic resonance imaging associated with lateral local recurrence in low rectal cancer after neoadjuvant chemoradiotherapy or radiotherapy. JAMA Surgery, 154(9). https://doi.org/10.1001/jamasurg.2019.2172.*

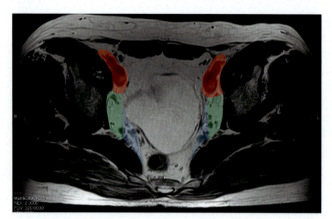

FIGURE 14.2 A surgical atlas for the lateral compartments. *Green*: obturator compartment; *blue*: internal iliac compartment; *red*: external iliac compartment. Source: *From Ogura, A., Konishi, T., Beets, G. L., Cunningham, C., Garcia-Aguilar, J., Iversen, H., Toda, S., Lee, I. K., Lee, H. X., Uehara, K., Lee, P., Putter, H., van der Velde, C. J. H., Rutten, H. J. T., Tuynman, J. B., & Kusters, M. (2019). Lateral nodal features on restaging magnetic resonance imaging associated with lateral local recurrence in low rectal cancer after neoadjuvant chemoradiotherapy or radiotherapy. JAMA Surgery, 154(9). https://doi.org/10.1001/jamasurg.2019.2172.*

iliac lateral lymph nodes which remained equal or greater than 4 mm after neoadjuvant treatment resulted in a 5-year lateral local recurrence rate of 52.3%, obturator lateral lymph nodes that remained equal or greater than 6 mm resulted in a lateral local recurrence rate of 17.8% (Ogura, Konishi, Beets, et al., 2019). This highlights that lateral lymph nodes in separate compartments behave and respond independently from each other. These results further indicate the important role of the restaging of lateral lymph nodes after neoadjuvant treatment and the importance of determining their exact location.

Enlarged lymph nodes located in the internal iliac or obturator compartments, which do not sufficiently respond to neoadjuvant treatment, can benefit from a LLND. This procedure can significantly reduce local recurrence rates, decreasing the rate of local recurrence for enlarged internal iliac lymph nodes from 52% to 8.7% or obturator lymph nodes from 17.8% to 0% (Ogura, Konishi, Cunningham, et al., 2019). Similar results are also found in Eastern studies researching the selective LLND procedure with local recurrence rates of 0% when appropriately selecting "high-risk" patients (Akiyoshi et al., 2014). Together, these results reiterate the necessity of evaluating the size and location of lateral lymph nodes and their behavior, specifically whether they reduce in size or not, after neoadjuvant treatment. In the specific cases in which lateral lymph nodes do not sufficiently respond to neoadjuvant treatment, a LLND procedure may be vital.

14.4 The future

The studies mentioned in this chapter highlight the need for appropriate definitions and universal guidelines. The behavior of lateral lymph nodes is substantial, and their location, size, and reaction to neoadjuvant treatment are essential for physicians when deciding on treatment strategies. Eastern and Western physicians are, in increasing tempo, moving toward the selective application of LLNDs in specific cases, with the realization that adequate irradiation can be sufficient in many cases, meaning that an extensive

LLND, with the associated risks (described later in this chapter), could be omitted for certain patients.

Answers may be sought in new international prospective research, which combines Western and Eastern approaches. The Lateral Nodal Recurrence in Rectal Cancer (LaNoReC) study is a prospective registration study attempting to address the need for a multidisciplinary approach to lateral nodal disease. Patients receive standardized neoadjuvant therapy, allowing for the appropriate sterilization of lateral compartments. All patients undergo restaging imaging after neoadjuvant therapy and the subsequent treatment decisions depend on the lateral lymph node size and their reaction to neoadjuvant therapy. Patients with insufficient downsizing of internal iliac or obturator lymph nodes are advised to undergo a TME and LLND procedure. In this manner, the LaNoReC study hopes to provide evidence that through proper irradiation, maximal response can be achieved and, if necessary, the additional surgical removal of lateral lymph nodes, which do not appropriate respond to irradiation, can decrease the local recurrence rates to less than 6%.

An approach in this manner, which incorporates the necessity for standardized neoadjuvant treatment to sterilize enlarged lateral lymph nodes and requires the adequate evaluation of lymph node size, location, and reaction to neoadjuvant treatment, may provide the necessary answers for the treatment of lateral nodal disease. Furthermore, the LLND procedure can be offered to "high-risk" patients but also be omitted for patients where the LLND would provide more harm than care.

14.5 The lateral lymph node dissection

A LLND can be performed unilaterally or bilaterally and involves the complete removal of lateral lymphatic tissue with preservation of the hypogastric nerve plexuses. The procedure includes the visualization and medial retraction of the ureter (Fig. 14.3) to ensure that it is removed from the operating field and is separated together with the plexus along

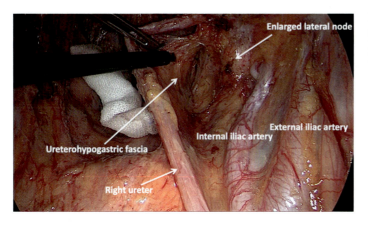

FIGURE 14.3 Visualization and medial retraction of the ureter. A unilateral, right-sided LLND. *LLND*, Lateral lymph node dissection.

the ureterohypogastric fascia to form the medial border of the dissection. This medialization is necessary to prevent injury to the hypogastric nerve plexus.

The initial step is to remove lymphatic tissue from the obturator compartment. Lymphatic tissue needs to be dissected from the surface of the external iliac artery and vein (Fig. 14.4), which is continued deeper and laterally until the internal obturator muscle can be seen, forming the lateral margin of the dissection. This is continued distally until the levator ani muscle is visualized and the space of the previous TME resection is opened. Anteriorly, the obturator nerve can be identified reaching toward the obturator foramen, with its proximal end originating between the bifurcation of the internal and external iliac veins (Fig. 14.5). The obturator nerve is routinely preserved and only sacrificed in situations in which the nerve is grossly involved by metastatic lymph nodes. Once the obturator nerve has been separated from the lymphatic tissue, both the proximal and distal end of the obturator vein and artery can be ligated. The deep circumflex iliac vein represents the most caudal part of the anterior dissection plane. The entire obturator compartment is cleared out by dissecting along the side branches of the internal iliac vessels anteriorly toward the bladder, along the vesicohypogastric fascia (Fig. 14.6).

The next step is to remove lymphatic tissue from the internal iliac compartment. If necessary, all remaining branches of the internal iliac artery (the umbilical artery, uterine artery in females, superior and inferior vesicle arteries) can be dissected and ligated, though in certain cases surgeons may decide to preserve certain branches (Fig. 14.7: a female patient with aggressive lateral nodal disease extending to the vessels, requiring resection of these branches). This allows for the removal of lymphatic tissue between the internal iliac vein and the ureterohypogastric fascia. By dissecting the superior and inferior vesicle arteries, the most caudal part of lymphatic tissue can be reached (Fig. 14.8), between the levator ani muscle and pelvic plexus.

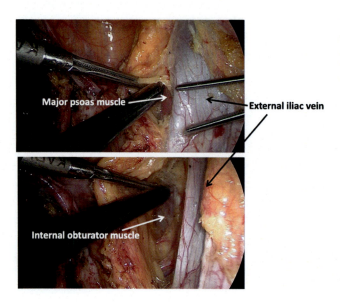

FIGURE 14.4 Dissection of lymphatic tissue from the surface of the external iliac artery and vein. A unilateral, right-sided LLND. *LLND*, Lateral lymph node dissection.

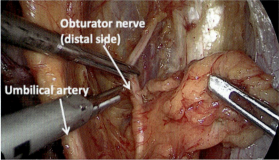

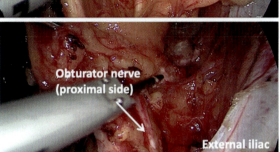

FIGURE 14.5 Identification of the obturator nerve. A unilateral, right-sided LLND. *LLND*, Lateral lymph node dissection.

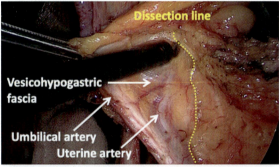

FIGURE 14.6 Complete clearance of the obturator compartment along the branches of the internal iliac vessels. A unilateral, right-sided LLND. *LLND*, Lateral lymph node dissection.

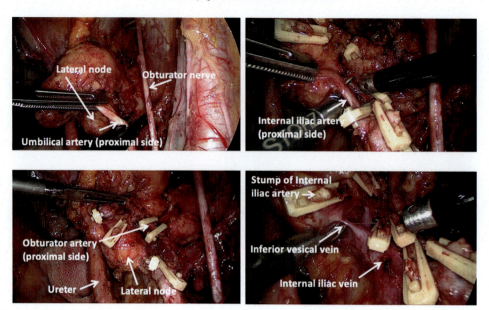

FIGURE 14.7 Dissection of the proximal branches of the internal iliac artery. A female patient with aggressive lateral nodal disease extending to the vessels requires resection of these branches.

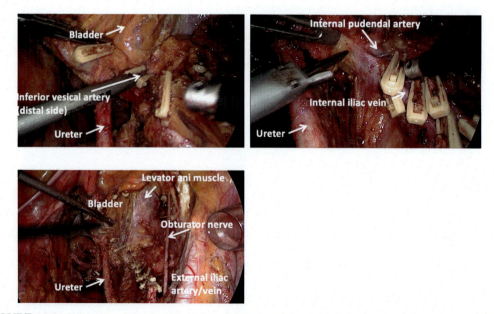

FIGURE 14.8 Dissection of the distal branches of the internal iliac artery. A female patient with aggressive lateral nodal disease extending to the vessels requires resection of these branches.

Once completed, lymphatic tissue can be carefully removed from the pelvis. If necessary, this procedure is then repeated for the other side.

14.6 Risks

There are risks involved in performing a LLND and these remain some of the strongest arguments Western surgeons use as why not to adopt the LLND method as employed by Eastern surgeons.

The only randomized controlled trial regarding the role of the LLND in Japan was published in 2017 and randomized patients without enlarged lymph nodes between a TME with or without a LLND. While this trial is otherwise not applicable to Western patients, as small lymph nodes are treated with neoadjuvant therapy in Western clinics, the study revealed interesting details concerning the risks associated with LLND. Both groups suffered from erectile dysfunction (for patients who had no erectile dysfunction before the surgery: 71% after LLND vs 59% after TME-only) and urinary dysfunction (5% LLND vs 3% TME-only). Furthermore, a substantial amount of grade 3–4 Clavien and Dindo complications were present in both groups (22% LLND vs 16% TME-only) (Fujita et al., 2017; Nakamura & Watanabe, 2013; Obara et al., 2012). This not only shows the continuously higher risk for patients undergoing a LLND, but also illustrates the risks associated with a TME operation, the standard surgical procedure for rectal cancer (Fujita et al., 2012; Ito et al., 2018; Saito et al., 2016).

The LLND procedure has led to numerous studies reviewing the risks involved (Kim & Oh, 2018), such as a 3.7 times greater risk of urinary dysfunction (Georgiou et al., 2009) and an 11%–27% rate of sexual dysfunction (Maeda et al., 2003; Ogura, Konishi, Beets, et al., 2019). Furthermore, the degree to which urinary and sexual dysfunction occur has been significantly linked to the extent of autonomic nerve resection and the extent of the LLND. For example, when unilateral hypogastric nerves or pelvic plexuses were ligated, rates of normal urinary and sexual function decreased by 50% (Akasu et al., 2009), meaning that the identification and preservation of autonomic nerves are essential when performing a LLND to prevent urinary and/or sexual dysfunction (Table 14.3).

14.7 Procedural variation

A significant problem in the West, most likely due to the lack of experience in performing a LLND, is heterogeneity in surgical techniques, resulting from the lack of standardized, international guidelines. Firstly, this poses a significant problem when researching the LLND. Many studies retrospectively reconstruct the LLND procedure from operation reports, meaning that it is almost impossible to guarantee if dissections were performed within formal anatomical landmarks and boundaries. This could mean that results from these retrospective studies discover such high local recurrence rates due to, in fact, the incomplete removal of lymphatic tissue during dissections. Secondly, while the evidence is scarce, there are indications that diverse hospitals and surgeons may perform individual node-sampling techniques instead of complete and formal LLNDs. In the Lateral Node

TABLE 14.3 Functional outcomes after LLND versus TME surgery.

	Design	No. of patients (n)	Patient population	Neoadjuvant therapy: (C)RT	Primary endpoint	Results
Maeda et al. (2003)	Retrospective	77	Mid/low rectal cancer: TME versus TME + LLND	No	Bladder and sexual dysfunction 1 year after surgery	Minor bladder dysfunction: 10/65 (LLND: 15%) versus 3/12 (TME: 25%). Sexual dysfunction: 10/37 (LLND: 27%) versus 1/5 (TME: 20%).
Akasu et al. (2009)	Retrospective	69 males	TME or TME + LLND with bilateral or unilateral pelvic plexus preservation (PPP)	No	Residual urine <50 mL, sexual function 1 year postoperatively	Residual urine <50 mL <14 days: TME: 96%, Bilateral PPP & LLND: 73%, Unilateral PPP & LLND: 23%, No PPP & LLND: 0%. Maintain sexual intercourse after 1 year: TME: 95%, Bilateral PPP & LLND: 56%, Unilateral PPP & LLND: 45%, No PPP & LLND: 0%.
Georgiou et al. (2009)	Meta-analysis	20 studies (5502 patients)	Comparative studies comparing LLND versus non-LLND for rectal cancer	14/20 studies: no	Overall survival 5 years and morbidity	OS 5-year (6 studies: 2260 patients): HR 1.09 LR 5-year (16 studies: 5383 patients): OR 0.83 Urinary dysfunction (3 studies: 139 patients): OR 3.70 (more prevalent in LLND group)
Saito et al. (2016)	RCT	472 males/701	Rectal cancer stage 2/3, no enlarged LLN	No	Sexual dysfunction 1 year after surgery	Dysfunction = TME: 68% versus TME + LLND: 79% Dysfunction in only those with no/mild dysfunction pre-op = TME: 59% versus TME + LLND: 71%
Fujita et al. (2017)	RCT	701	Rectal cancer stage 2/3, no enlarged LLN	No		Erectile dysfunction: 71% (LLND) versus 59% (TME) Urinary dysfunction: 5% (LLND) versus 3% (TME) Grade 3–4 CD: 22% (LLND) versus 16% (TME)
Ito et al. (2018)	RCT	551/701	Rectal cancer stage 2/3, no enlarged LLN	No	Urinary dysfunction postoperatively: <14 days	TME: 58% versus TME + LLND: 59%

CD, Clavien and Dindo; *HR*, hazard ratio; *LLND*, lateral lymph node dissection; *LR*, local recurrence; *RCT*, randomized controlled trial; *TME*, total mesorectal excision; *OS*, overall survival; *PPP*, pelvic plexus preservation; *OR*, odds ratio; *RCT*, randomised controlled trial.

Consortium study, 12 patients were identified to have undergone a node-sampling technique, in which individual lymph nodes are removed due to pre-or peri-operative suspicions of malignancy. For these 12 patients, 75% were indeed pathologically malignant, however, the 5-year local recurrence rate was 51.1% and all recurrences were still located in the lateral compartments (Ogura, Konishi, Cunningham, et al., 2019). This suggests that these techniques are insufficient in reducing the chances of lateral pelvic recurrence.

While this sample is extremely small, it provides a glimpse at the potential lack of effectiveness of lymph node sampling and reiterates the theory that LLND procedures with nerve-sparing techniques are required to be formally completed to effect recurrence rates.

14.8 Conclusion

The evidence cited in this chapter suggests that the evaluation of lateral lymph node downsizing on a restaging MRI after neoadjuvant treatment is of crucial importance. This evaluation of the response to neoadjuvant treatment can subsequently be considered as the single most significant factor when considering the indication for performing a LLND.

Issues that need to be addressed by future research include the following problems. Exact universal definitions, encompassing the different medical specialties involved, are needed when considering lateral compartments, lateral lymph nodes, and the LLND. The awareness of lateral nodal disease needs to be increased and there needs to be an ambition to reach a consensus between Western and Eastern treatment practices. Rectal cancer remains a multidisciplinary disease; radiologists need to be aware of lateral nodal disease, look for enlarged lateral lymph nodes, and state these in their reports and during multidisciplinary meetings. Radiation oncologists need to ensure the inclusion of lateral compartments in irradiation fields and surgeons need to correctly perform a LLND for patients with a "high-risk" of local recurrence. The multidisciplinary teamwork of these specialities is essential for the accurate treatment of this distinct patient population. Lastly, the selective LLND, with proven oncological benefits for patients that are selected purely based on the size of their lateral lymph node(s) after neoadjuvant treatment, needs to remain as a factor of the utmost importance for future research.

Continuing multidisciplinary research with the growing understanding of lateral lymph nodes should support the increased awareness, ability, and consistency in the treatment of lateral nodal disease, promoting the accurate treatment of patients with distal rectal cancer and lateral nodal disease.

References

Akasu, T., Sugihara, K., & Moriya, Y. (2009). Male urinary and sexual functions after mesorectal excision alone or in combination with extended lateral pelvic lymph node dissection for rectal cancer. *Annals of Surgical Oncology*, 16(10), 2779–2786.

Akiyoshi, T., Matsueda, K., Hiratsuka, M., Unno, T., Nagata, J., Nagasaki, T., Konishi, T., Fujimoto, Y., Nagayama, S., Fukunaga, Y., & Ueno, M. (2015). Indications for lateral pelvic lymph node dissection based on magnetic resonance imaging before and after preoperative chemoradiotherapy in patients with advanced low-rectal cancer. *Annals of Surgical Oncology*, 22, 614–620. Available from https://doi.org/10.1245/s10434-015-4565-5.

Akiyoshi, T., Ueno, M., Matsueda, K., Konishi, T., Fujimoto, Y., Nagayama, S., Fukunaga, Y., Unno, T., Kano, A., Kuroyanagi, H., Oya, M., Yamaguchi, T., Watanabe, T., & Muto, T. (2014). Selective lateral pelvic lymph node dissection in patients with advanced low rectal cancer treated with preoperative chemoradiotherapy based on pretreatment imaging. *Annals of Surgical Oncology, 21*(1), 189−196.

Beets-Tan, R. G. H., Lambregts, D. M. J., Maas, M., Barbaro, B., Curvo-Semedo, L., Fenlon, H. M., Gollub, M. J., Gourtsoyianni, S., Halligan, S., Hoeffel, C., Kim, S. H., Laghi, A., Maier, A., Rafaelsen, S. R., Stoker, J., Talyor, S. A., Torkzad, M. R., & Blomqvist, L. (2018). Magnetic resonance imaging for clinical management of rectal cancer: Updated recommendations from the 2016 european society of gastrointestinal and abdominal radiology (ESGAR) consensus meeting. *European Radiology, 28*, 1465−1475.

Brown, G., Richards, C. J., Bourne, M. W., Newcombe, R. G., Radcliffe, A. G., Dallimore, N. S., & Williams, G. T. (2003). Morphologic predictors of lymph node status in rectal cancer with use of high-spatial-resolution mr imaging with histopathologic comparison. *Radiology, 227*(2), 371−377.

Christou, N., Meyer, J., Toso, C., Ris, F., & Buchs, N. C. (2019). Lateral lymph node dissection for low rectal cancer: Is it necessary? *World Journal of Gastroenterology, 25*(31), 4294−4299.

Fujita, S., Akasu, T., Mizusawa, J., Saito, N., Kinugasa, Y., Kanemitsu, Y., Ohue, M., Fujii, S., Shiozawa, M., Yamaguchi, T., & Moriya, Y. (2012). Postoperative morbidity and mortality after mesorectal excision with and without lateral lymph node dissection for clinical stage II or stage III lower rectal cancer (JCOG0212): Results from a multicentre, randomised controlled, non-inferiority trial. *The Lancet Oncology, 13*(6), 616−621.

Fujita, S., Mizusawa, J., Kanemitsu, Y., Ito, M., Kinugasa, Y., Komori, K., Ohue, M., Ota, M., Akazai, Y., Shiozawa, M., Yamaguchi, T., Bandou, H., Katsumata, K., Murata, K., Akagi, Y., Takiguchi, N., Saida, Y., Nakamura, K., Haruhiko, F., & Moriya, Y. (2017). Mesorectal excision with or without lateral lymph node dissection for clinical stage ii/iii lower rectal cancer (JCOG0212) a multicenter, randomized controlled, noninferiority trial. *Annals of Surgery, 266*(2), 201−207.

Georgiou, P., Tan, E., Gouvas, N., Antoniou, A., Brown, G., Nicholls, R. J., & Tekkis, P. (2009). Extended lymphadenectomy versus conventional surgery for rectal cancer: A meta-analysis. *The Lancet Oncology, 10*(11), 1053−1062.

Ito, M., Kobayashi, A., Fujita, S., Mizusawa, J., Kanemitsu, Y., Kinugasa, Y., Komori, K., Ohoe, M., Ota, M., Akazai, Y., Shiozawa, M., Yamaguchi, T., Akasu, T., & Moriya, Y. (2018). Urinary dysfunction after rectal cancer surgery: Results from a randomized trial comparing mesorectal excision with and without lateral lymph node dissection for clinical stage II or III lower rectal cancer (Japan Clinical Oncology Group Study, JCOG0212). *European Journal of Surgical Oncology, 44*(4), 463−468.

Iversen, H., Martling, A., Johansson, H., Nilsson, P. J., & Holm, T. (2018). Pelvic local recurrence from colorectal cancer: Surgical challenge with changing preconditions. *Colorectal Disease, 20*(5), 399−406.

Kim, J. H., Beets, G. L., Kim, M. J., Kessels, A. G., & Beets-Tan, R. G. (2004). High-resolution MR imaging for nodal staging in rectal cancer: Are there any criteria in addition to the size? *European Journal of Radiology, 52*(1), 78−83.

Kim, M. J., & Oh, J. H. (2018). Lateral lymph node dissection with the focus on indications, functional outcomes, and minimally invasive surgery. *Annals of Coloproctology, 34*(5), 229−233. Available from https://doi.org/10.3393/ac.2018.10.26.

Kim, M. J., Kim, T. H., Kim, D. Y., Kim, S. Y., Baek, J. Y., Chang, H. J., Park, S. C., Park, J. W., & Oh, J. H. (2015). Can chemoradiation allow for omission of lateral pelvic node dissection for locally advanced rectal cancer? *Journal of Surgical Oncology, 111*(4), 459−464.

Kim, T. G., Park, W., Choi, D. H., Park, H. C., Kim, S. H., Cho, Y. B., Yun, S. H., Kim, H. C., Lee, W. Y., Lee, J., Park, J. O., Park, Y. S., Kim, H. Y., Kang, W. K., & Chun, H. (2014). Factors associated with lateral pelvic recurrence after curative resection following neoadjuvant chemoradiotherapy in rectal cancer patients. *International Journal of Colorectal Disease, 29*(2), 193−200.

Kim, T. H., Jeong, S. Y., Choi, D. H., Kim, D. Y., Jung, K. H., Moon, S. H., Chang, H. J., Lim, S., Choi, H. S., & Park, J. (2008). Lateral lymph node metastasis is a major cause of locoregional recurrence in rectal cancer treated with preoperative chemoradiotherapy and curative resection. *Annals of Surgical Oncology, 15*(3), 729−737.

Kusters, M., Beets, G. L., van de Velde, C. J. H., Beets-Tan, R. G. H., Marijnen, C. A. M., Rutten, H. J. T., Putter, H., & Moriya, Y. (2009). A comparison between the treatment of low rectal cancer in japan and the netherlands, focusing on the patterns of local recurrence. *Annals of Surgery, 249*(2), 229−235.

Kusters, M., Marijnen, C. A., van de Velde, C. J., Rutten, H. J., Lahaye, M. J., Kim, J. H., Beets-Tan, R. G. H., & Beets, G. L. (2010). Patterns of local recurrence in rectal cancer; A study of the dutch TME trial. *European Journal of Surgical Oncology, 36*(5), 470−476.

Kusters, M., Slater, A., Murihead, R., Hompes, R., Guy, R. J., Jones, O. M., George, B., Lindsey, I., Mortensen, N., & Cunningham, C. (2017). What to do with lateral nodal disease in low locally advanced rectal cancer? a call for further reflection and research. *Diseases of the Colon & Rectum, 60*(6), 577−585.

Maeda, K., Maruta, M., Utsumi, T., Sato, H., Toyama, K., & Matsuoka, H. (2003). Bladder and male sexual functions after autonomic nerve-sparing TME with or without lateral node dissection for rectal cancer. *Techniques in Coloproctology, 7*(1), 29−33.

Malakorn, S., Yang, Y., Bednarski, B. K., Kaur, H., You, Y. N., Holliday, E. B., Dasari, A., Skibber, J., Rodrigeuz-Bigas, M. A., & Chang, G. J. (2019). Who should get lateral pelvic lymph node dissection after neoadjuvant chemoradiation? *Diseases of the Colon & Rectum, 62*(10), 1158−1166.

Martling, A., Holm, T., Johansson, H., Rutqvist, L. E., & Cedermark, B. (2001). The Stockholm II trial on preoperative radiotherapy in rectal carcinoma: Long-term follow-up of a population-based study. *Cancer, 92*(4), 896−902.

Nakamura, T., & Watanabe, M. (2013). Lateral lymph node dissection for lower rectal cancer. *World Journal of Surgery, 37*(8), 1808−1813.

Obara, S., Koyama, F., Nakagawa, T., Nakamura, S., Ueda, T., Nishigori, N., & Nakajima, Y. (2012). Laparoscopic lateral pelvic lymph node dissection for lower rectal cancer: Initial clinical experiences with prophylactic dissection. *Gan to Kagaku Ryoho Cancer & Chemotherapy, 39*(12), 2173−2175.

Ogura, A., Konishi, T., Beets, G. L., Cunningham, C., Garcia-Aguilar, J., Iversen, H., Toda, S., Lee, I. K., Lee, H. X., Uehara, K., Lee, P., Putter, H., van der Velde, C. J. H., Rutten, H. J. T., Tuynman, J. B., & Kusters, M. (2019). Lateral nodal features on restaging magnetic resonance imaging associated with lateral local recurrence in low rectal cancer after neoadjuvant chemoradiotherapy or radiotherapy. *JAMA Surgery, 154*(9). Available from https://doi.org/10.1001/jamasurg.2019.2172.

Ogura, A., Konishi, T., Cunningham, C., Garcia-Aguilar, J., Iversen, H., Toda, S., Lee, I. K., Lee, H. X., Uehara, H., Lee, P., Putter, H., van de Velde, C. J. H., Beets, G. L., Rutten, H. J. T., & Kusters, M. (2019). Neoadjuvant (chemo)radiotherapy with total mesorectal excision only is not sufficient to prevent lateral local recurrence in enlarged nodes: Results of the multicenter lateral node study of patients with low ct3/4 rectal cancer. *Journal of Clinical Oncology, 37*(1), 33−43. Available from https://doi.org/10.1200/JCO.18.00032.

Saito, S., Fuijta, S., Mizusawa, J., Kanemitsu, Y., Saito, N., Kinugasa, Y., Akazai, Y., Ota, M., Ohue, M., Komori, K., Shiozawa, M., Yamaguchi, T., Akasu, T., & Moriya, Y. (2016). Male sexual dysfunction after rectal cancer surgery: Results of a randomized trial comparing mesorectal excision with and without lateral lymph node dissection for patients with lower rectal cancer: Japan clinical oncology group study JCOG0212. *European Journal of Surgical Oncology, 42*(12), 1851−1858.

Schaap, D., Ogura, A., Nederend, J., Maas, M., Cnossen, J. S., Creemers, G. J., van Lijnschoten, I., Nieuwenhuijzen, G. A. P., Rutten, H. J. T., & Kusters, M. (2018). Prognostic implications of MRI-detected lateral nodal disease and extramural vascular invasion in rectal cancer. *The. British Journal of Surgery, 105*(13), 1844−1852.

Songphol, M., Yun, Y., Bednarski., K, B., Harmeet, K., Nancy, Y. Y., B., H. E., Arvind, D., M., S. J., A., R.-B. M., & J., C. G. (2019). Who should get lateral pelvic lymph node dissection after neoadjuvant chemoradiation? *Diseases of the Colon & Rectum*, 1158−1166. Available from https://doi.org/10.1097/dcr.0000000000001465.

Steup, W. H., Moriya, Y., & van de Velde, C. J. H. (2002). Patterns of lymphatic spread in rectal cancer. A topographical analysis on lymph node metastases. *European Journal of Cancer, 38*(7), 911−918.

Tamura, H., Shimada, Y., Kameyama, H., Yagi, R., Tajima, Y., Okamura, T., Nakano, M., Nagahashi, M., Sakata, J., Kobayashi, T., Kosugi, S. I., Nogami, H., Maruyama, S., Takji, Y., & Wakai, T. (2017). Prophylactic lateral pelvic lymph node dissection in stage IV low rectal cancer. *World Journal of Clinical Oncology, 8*(5), 412−419. Available from https://doi.org/10.5306/wjco.v8.i5.412.

Weiser, M. R. (2018). AJCC 8th edition: Colorectal cancer. *Annals of Surgical Oncology, 25*(6), 1454−1455.

CHAPTER 15

Fluorescence-guided sentinel lymph node detection in colorectal cancer surgery

Ruben P.J. Meijer[1,2], Hidde A. Galema[3,4], Lorraine J. Lauwerends[3,4], Cornelis Verhoef[3], Jacobus Burggraaf[2], Stijn Keereweer[3,4], Merlijn Hutteman[1], Alexander L. Vahrmeijer[1] and Denise E. Hilling[1,3]

[1]Department of Surgery, Leiden University Medical Center, Leiden, The Netherlands
[2]Centre for Human Drug Research, Leiden, The Netherlands [3]Department of Surgical Oncology and Gastrointestinal Surgery, Erasmus MC Cancer Institute, Rotterdam, The Netherlands [4]Department of Otorhinolaryngology, Head and Neck Surgery, Erasmus MC Cancer Institute, Rotterdam, The Netherlands

15.1 Concept of sentinel lymph node mapping

Adequate lymph node staging is important in colorectal cancer (CRC) treatment, as lymph node metastases are an important determinant for patient prognosis and an indication for adjuvant systemic treatment. The sentinel lymph node (SLN) is defined as the first group of lymph node(s) draining a tumor. The identification, removal, and analysis of these SLNs (SLN mapping) can therefore be of added value for the staging of CRC patients, and subsequent treatment. SLN mapping was first described in 1960 for parotid cancer and is nowadays standard of care in breast cancer and melanoma patients (Gould et al., 1960; He et al., 2016; Valsecchi et al., 2011).

The SLN concept could also be of added value in CRC patients. Patients with stage I and II diseases (with no lymph node involvement) still develop distant metastases in up to 30% of cases (Figueredo et al., 2008). This could be the result of, among others, understaging of these patients due to missed lymph nodes with occult tumor cells and micrometastases during routine histopathological examination, or inadequate lymph node harvesting at the time of primary treatment. Routine histopathological examination currently exists of reviewing a single paraffin–embedded slide per lymph node, with the chance of missing tumor cells away from the slide's cutting edge. Extensive histopathological analysis of all lymph nodes using serial sectioning or reverse transcriptase polymerase chain reaction could result in more accurate lymph node staging. However, both methods are expensive and time consuming (Doekhie et al., 2010; Liefers et al., 1998; Yamamoto et al., 2016). As the SLN procedure identifies the lymph node(s) with the highest chance of containing metastases, more extensive histopathological analysis of only these lymph nodes is feasible. Furthermore, tumor-negative SLNs create an opportunity for local or endoscopic resection of CRC, especially in early-stage tumors (Cahill and Leroy, 2009).

Conventional methods for SLN mapping include the use of either blue dye, a radiocolloid tracer, or the combination of both (He et al., 2016; Valsecchi et al., 2011). The use of blue dye for SLN mapping in CRC is limited due to inadequate depth penetration and the utilization of radiocolloid tracers has some logistic hurdles (Bembenek et al., 2007). A nuclear medicine physician and an endoscopist are required for tracer injection. Moreover, the gamma probe used for localization does not enable real-time visualization. These shortcomings have increased the interest in novel techniques, such as near-infrared (NIR) fluorescence imaging.

15.2 Fluorescence-guided surgery

NIR fluorescence imaging provides the surgeon with real-time information of the surgical field and can aid in differentiation between malignant and healthy tissue during surgery (Vahrmeijer et al., 2013). NIR light (700–900 nm) is not visible to the human eye and has relatively deep (up to 10 mm) tissue penetration. Human tissue itself has low autofluorescence in the NIR light spectrum, resulting in a high signal-to-background ratio (Frangioni, 2008).

NIR fluorescence-guided surgery requires two components: an NIR camera system and a fluorescent agent (Fig. 15.1). These NIR camera systems are composed of an NIR excitation light source, collection optics (including optical filtration), and a camera that can detect emitted NIR light. These systems emit photons with a specific wavelength, which are absorbed by the fluorescent agent. Electrons within these agent's molecules transit to an excited state and fall back to their ground state (Fig. 15.2). This will release the stored energy in an emitted photon with a longer wavelength than the exciting light of the NIR camera system, the so-called Stokes shift. This emitted photon is subsequently captured by the camera system. The camera output is usually displayed on a monitor including a merged image of the fluorescence signal and the white light image. Both the camera systems and fluorescent agents have shown great improvements in the last decades and have resulted in the clinical use of NIR fluorescence for different purposes during surgery (e.g., bile duct detection, tissue perfusion) (Griffiths et al., 2016; van den Hoven et al., 2019; van Manen et al., 2018).

Several dedicated NIR fluorescence imaging systems are clinically used for open, laparoscopic, and robotic surgery. On the other hand, only two fluorescent agents, indocyanine green (ICG) and methylene blue, have been approved for clinical use by the US Food and Drug Administration (FDA) and European Medicines Agency (EMA). Both are

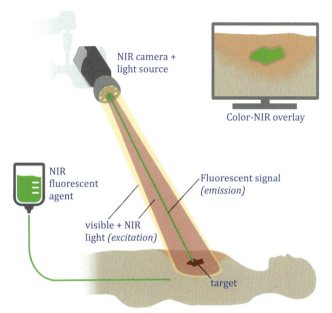

FIGURE 15.1 Fluorescence-guided surgery. NIR fluorescent agents are administered intravenously or locally. NIR fluorescence is visualized using a specialized imaging system for intraoperative imaging. The imaging system uses dedicated NIR excitation light to excite the fluorophore. Collection optics, emission filters, and an image sensor capable of detecting NIR fluorescence emission light. The NIR fluorescence signal is displayed on a monitor in the surgical theater. A simultaneous white light image, which can be merged with the NIR fluorescence image, is desirable.

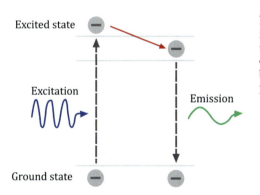

FIGURE 15.2 Schematic overview of the principle of fluorescence. A photon in the appropriate wavelength (excitation light) is absorbed by the fluorophore, elevating an electron to an excited state. When the electron transitions back to its ground state, a photon is emitted. This emitted photon is of a longer wavelength and lower energy.

nontargeted fluorescent agents, meaning these agents do not bind to a specific target. ICG, first described in 1957 for the determination of cardiac output, is the most used agent for fluorescence-guided SLN mapping in various cancer types and can also be used for the visualization of vital structures (e.g., bile ducts), liver tumors, and assessment of tissue perfusion (Fox et al., 1957; van Manen et al., 2018).

15.3 Fluorescence-guided sentinel lymph node detection in colorectal cancer

While peritumoral injection of ICG is the most used technique for NIR fluorescence-guided SLN mapping, alternative fluorescent dyes have also been assessed (Fig. 15.3).

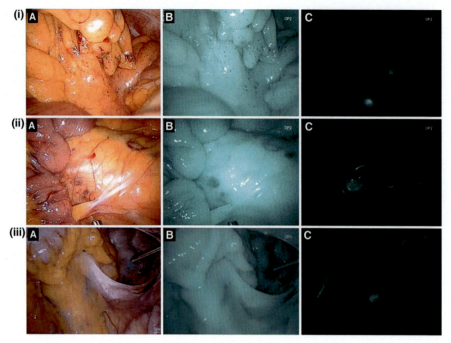

FIGURE 15.3 Intraoperative results of sentinel lymph node mapping using with indocyanine green. Three patients (i–iii) demonstrating NIR fluorescence-guided sentinel lymph node mapping with indocyanine green (A, a white light image; B, an NIR image without filter; and C, a filtered NIR image). (i) Two bright spots in the mesocolon were identified in the filtered view, consistent with sentinel lymph nodes. (ii) A bright spot is seen in the filtered view, consistent with a paraaortic sentinel lymph node. (iii) A bright spot is seen in the filtered view, alongside the right iliac artery, consistent with a sentinel lymph node. The injection site of the rectal tumor is clearly visible, as well as an efferent channel in the sigmoid mesentery which was found to lead to another sentinel lymph node. Source: *Reproduced by permission from Cahill, R. A., Anderson, M., Wang, L. M., Lindsey, I., Cunningham, C., Mortensen, N. J. (2012). Near-infrared (NIR) laparoscopy for intraoperative lymphatic road-mapping and sentinel node identification during definitive surgical resection of early-stage colorectal neoplasia. Surgical Endoscopy, 26(1), 197–204.*

In addition, different injection sites and variable timing of the injection have been investigated.

The injection site of ICG can be either subserosal or submucosal, with a slight preference for the latter (Andersen et al., 2017; Ankersmit et al., 2019; Hutteman et al., 2011). The submucosa houses an important part of the intestinal lymphatic system, which might improve the lymphatic uptake of the fluorescent agent from the tumor surrounding tissue (Miller et al., 2010). Submucosal injection is performed prior to or during surgery, via endoscopy. Subserosal injection, on the other hand, is performed intraoperatively by the surgeon. In minimally invasive surgery, this requires transcutaneous injection of the fluorescent agent. Correct positioning of the needle and maintaining this position during injection of the fluorescent agent is easier with the submucosal technique and therefore leads to less spillage of ICG (Ankersmit et al., 2019; Andersen et al., 2017).

Timing of fluorescent dye administration and assessment has been assessed directly pre- or intraoperatively (both referred to as in vivo), and postoperatively (ex vivo). In vivo administration has some practical drawbacks, particularly for the preferred submucosal injection. For in vivo submucosal injection, an endoscopy in the operating room is required directly before or during surgery, which poses a logistical challenge. Furthermore, bowel insufflation might alter the surgical field and therefore hamper the surgical procedure. The alternative, ex vivo imaging, is logistically simpler and enables the use of experimental agents. Ex vivo imaging also has some disadvantages. Lymphatic flow may be disrupted after resection, and altering the surgical plan (i.e., perform a more limited resection) based on histopathological analysis of the harvested lymph nodes is not possible. Moreover, ex vivo fluorescent agent injection and lymph node identification does not facilitate identifying SLNs in patients with aberrant lymph node drainage patterns (Tuech et al., 2004).

Table 15.1 summarizes studies that describe fluorescence-guided SLN mapping in CRC patients. A procedure is defined as successful if one or more lymph nodes were identified by fluorescence (the SLN). An upstaged patient is defined as a patient who was staged as N0 (all lymph nodes being tumor-negative) using conventional histopathology but showed tumor-positive SLNs after additional extensive histopathological assessment of the SLNs. These patients can consequently change from stage I or II CRC to stage III. The percentage of upstaged patients was calculated by dividing the number of upstaged patients by all lymph node-negative patients before extensive histopathological assessment of the SLN.

In all studies, intraoperative identification of the SLNs was performed using ICG and this was successful in most cases. Andersen et al. (2017) had a remarkably lower success rate of 65.5% in their multicenter trial, with all other studies being single center. This could be explained by a learning curve, which is suggested by Bembenek et al. (2007) to be more than 22 cases per center, a number none of their centers had reached. The sensitivity of SLN identification with ICG ranged from 0.33 to 1, and the negative predictive value was relatively low, with only three (33%) studies reporting an NPV above 0.9.

HSA800 (IRDye 800CW conjugated to human serum albumin) is another fluorescent agent used for ex vivo SLN mapping in CRC patients, which has not been approved by the FDA or EMA yet. Preclinical studies have shown an advantage of HSA800 over ICG regarding lymphatic entry, flow, fluorescence yield, and reproducibility. This is most likely a result of its bigger hydrodynamic diameter, resulting in improved retention in the SLN (Ohnishi et al., 2005). Clinical ex vivo studies with HSA800 have shown comparable results to in vivo assessment with ICG with a wide range in sensitivity (0.64–0.89) and negative predictive value (0.74–0.94).

15.4 Future perspectives

Fluorescence-guided SLN mapping has the potential to improve adequate staging in CRC patients. Despite its advantages and several published clinical studies, it is not used in common day practice. This might be the result of technical and logistic hurdles. Moreover, it is unknown if the upstaging of patients with micrometastatic lymph nodes and subsequent adjuvant treatment will lead to improved patient outcomes.

The number of early-stage CRC patients is expected to increase in the coming years, due to the introduction of nationwide screening programs (Cardoso et al., 2021;

TABLE 15.1 Results of clinical trials assessing fluorescence-guided SLN mapping for CRC.

	Agent	Patients	Diagnosis	Procedure	Injection site	No. of SLNs	Success rate (%)	Sensitivity	NPV	Upstaged patients* (%)
Intraoperative										
Andersen et al. (2017)	ICG: HA	29	CC	L	SS	1 (0–3)*	65.5	0.33	0.76	1 (12)
Ankersmit et al. (2019)	ICG: HA	29	CC	L	14 SS; 15 SM	2 (0–6)*	89.7	0.44	0.8	3 (13)
Cahill et al. (2012)	ICG	18	CRC	L	SM	3.6 (1–5)**	100	1	1	0 (0)
Carrara et al. (2020)	ICG	95	CRC	L	SS	1.5 (1–5)**	96.8	0.73	0.96	1 (1)
Currie et al. (2017)	ICG	30	CC	L	SM	3 (1–4)***	90	0.33	0.75	1 (5)
Hirche et al. (2012)	ICG	26	CC	O	SS	1.7 (0–5)**	96	0.82	0.87	3 (21)
Kusano et al. (2008)	ICG	26	CRC	O	SS	2.6 (±2.4)****	88.5	0.5	0.81	nr
Nagata et al. (2006)	ICG	48	CRC	L	SS	3.5 (±1.7)****	97.9	0.44	0.89	nr
Noura et al. (2010)	ICG	25	RC	O	SM	2.1 (±0.8)****	92	1	1	nr
Watanabe et al. (2017)	ICG	31	CC	L	SS	10.4 (±4.73)****	100	0.67	0.93	nr
Postoperative										
Hutteman et al. (2011)	HSA800	24	CRC	na	SM	3 (1–5)*	100	0.89	0.94	nr
Liberale et al. (2016)	ICG	20	CC	na	SS	1 (0–4)*	95	0.57	0.81	3 (23)
Schaafsma et al. (2013)	HSA800	22	CC	na	SM	3.5 (±1.9)****	95	0.8	0.94	nr
Weixler et al. (2017)	HSA800	50	CC	na	SS	4.4 (±2.2)****	98	0.64	0.74	5 (17)

Notes: The number of detected SLNs are presented as: *median with range; **mean with range; ***median with interquartile range; ****mean with standard deviation. The sensitivity is calculated by dividing the number of procedures with a tumor-positive SLN (true positives) by the sum of true positive and false negative procedures. The negative predictive value is determined by dividing the amount of true negative procedures by the sum of true negative and false negative procedures. Upstaged patients are defined as patients with no tumor involvement on conventional histopathology of all lymph nodes, but a tumor-positive SLN at advanced histopathology. The percentage of upstaged patients is calculated as: upstaged patients/upstaged patients + true negatives. CC, Colon cancer; CRC, colorectal cancer; HA, human albumin; ICG, indocyanine green; L, laparoscopic; NPV, negative predictive value; nr, not reported; O, open; RC, rectal cancer; SLN, sentinel lymph nodes; SM, submucosal; SS, subserosal.

Navarro et al., 2017). With this increasing number of early-stage CRC patients, the number of lymph node-negative patients is also expected to rise, since 90% of the T1 tumors are N0 (Fields et al., 2020; Ricciardi et al., 2006). Especially in these patients SLN mapping might be valuable. Because of the low incidence of lymph node metastases in these patients, a reliable SLN procedure showing a tumor-negative SLN enables the possibility for local excision without an extensive lymphadenectomy, thereby potentially lowering perioperative morbidity (Cahill and Leroy, 2009).

The relatively low negative predictive value of the SLNs (the probability that in case of a tumor-negative SLN, all other regional lymph nodes are tumor negative) is an important reason that this procedure is not yet implemented in daily practice. It withholds surgeons from performing a local excision and omitting an oncological resection based on a tumor-negative SLN. The low NPV is mainly a consequence of a high false negative rate (tumor-negative SLNs in the presence of a tumor-positive regional lymph node). One explanation for this high false negative rate is the occurrence of the so-called skip metastases, which are reported in 10%−22% of the cases (Bao et al., 2016; Saha et al., 2006). Tumor size could be another reason for this high false negative rate. T3−T4 tumors showed false negative results in 23% of the cases compared to 2% of the T1−T2 tumors (Burghgraef et al., 2021). It is suggested that these more invasive tumors (T3−T4) alter the lymphatic flow, resulting in skip metastases.

Based on the promising preliminary results, the interest in neoadjuvant treatment for colon cancer has increased in recent years (Karoui et al., 2020; Seymour and Morton, 2019). This novel treatment strategy could influence the success rate of SLN mapping, as research in other tumor types suggest altered lymphatic flow after neoadjuvant treatment (Kuehn et al., 2013). As a result, it could be preferable to perform SLN mapping prior to neoadjuvant therapy.

A meta-analysis by Ankersmit et al. (2019) showed a pooled upstaging (no tumor involvement on conventional histopathology, but a tumor-positive SLN at advanced histopathology) in 15% of the patients. This means that roughly one out of seven patients is wrongly classified as N0 without the use of fluorescence imaging and extensive histopathological assessment of the SLN. These patients would not have been upstaged to stage III and wrongfully been withheld adjuvant therapy, which theoretically leads to worse survival.

As emphasized, the use of fluorescence-guided SLN mapping with ICG increases the detection rate of SLNs in CRC patients and can result in upstaging in a substantial number of patients. Nevertheless, this concept still requires postoperative histopathological analysis. Direct intraoperative feedback regarding the malignancy status of any lymph node could be provided with the use of tumor-targeted fluorescence-guided surgery. Tumor-targeted agents consist of a fluorophore conjugated to a targeting component and therefore possess strong binding affinity for a specific cancer-associated molecular target or biomarker (Hernot et al., 2019). Unfortunately, tumor-targeted tracers tend to show a relatively high false positive rate (fluorescent lymph node without tumor localization) of 7%−33% for lymph node imaging, due to aspecific tracer localization (de Valk et al., 2020; Lu et al., 2020; Rosenthal et al., 2017). On the other hand, it is still debated whether a small tumor load (micrometastases and lymph nodes with isolated tumor cells) accumulate enough volume of the tracer to produce a sufficiently enhanced fluorescent signal. Nevertheless, tumor-targeted agents do not only allow for the identification of lymph node metastases but also other metastases, the primary tumor and tumor-positive resection margins (de Valk et al., 2021). Several tumor-targeted agents are currently studied in phase II and III trials (SGM-101 in Locally Advanced and Recurrent Rectal Cancer (SGM-LARRC [Internet]; Performance of SGM-101 for the Delineation of Primary and Recurrent Tumor and Metastases in Patients Undergoing Surgery for Colorectal Cancer [Internet]).

15.5 Conclusions

Fluorescence-guided SLN mapping in CRC can be a valuable addition to detect micrometastases and occult metastases in locoregional lymph nodes. It can result in upstaging in a significant part of the patients, whom otherwise would not have received adjuvant therapy. The low negative predictive value appears to be an important reason for the delayed introduction to current standard of care. Tumor-targeted fluorescent agents might overcome these shortcomings in the future.

Disclosure

The authors reported no disclosures.

References

Andersen, H. S., Bennedsen, A. L. B., Burgdorf, S. K., Eriksen, J. R., Eiholm, S., Toxværd, A., ... Gögenur, I. (2017). In vivo and ex vivo sentinel node mapping does not identify the same lymph nodes in colon cancer. *International Journal of Colorectal Disease, 32*(7), 983–990.

Ankersmit, M., Bonjer, H. J., Hannink, G., Schoonmade, L. J., van der Pas, M. H. G. M., & Meijerink, W. J. H. J. (2019). Near-infrared fluorescence imaging for sentinel lymph node identification in colon cancer: A prospective single-center study and systematic review with meta-analysis. *Techniques in Coloproctology, 23*(12), 1113–1126.

Bao, F., Zhao, L. Y., Balde, A. I., Liu, H., Yan, J., Li, T. T., ... Li, G. X. (2016). Prognostic impact of lymph node skip metastasis in stage III colorectal cancer. *Colorectal Disease: The Official Journal of the Association of Coloproctology of Great Britain and Ireland, 18*(9), O322–O329.

Bembenek, A. E., Rosenberg, R., Wagler, E., Gretschel, S., Sendler, A., Siewert, J. R., ... Schlag, P. M. (2007). Sentinel lymph node biopsy in colon cancer: A prospective multicenter trial. *Annals of Surgery, 245*(6), 858–863.

Burghgraef, T. A., Zweep, A. L., Sikkenk, D. J., van der Pas, M., Verheijen, P. M., & Consten, E. C. J. (2021). In vivo sentinel lymph node identification using fluorescent tracer imaging in colon cancer: A systematic review and meta-analysis. *Critical Reviews in Oncology/Hematology, 158*, 103149.

Cahill, R. A., Anderson, M., Wang, L. M., Lindsey, I., Cunningham, C., & Mortensen, N. J. (2012). Near-infrared (NIR) laparoscopy for intraoperative lymphatic road-mapping and sentinel node identification during definitive surgical resection of early-stage colorectal neoplasia. *Surgical Endoscopy, 26*(1), 197–204.

Cahill, R. A., Leroy, J., & Marescaux, J. (2009). Localized resection for colon cancer. *Surgical Oncology, 18*(4), 334–342.

Cardoso, R., Guo, F., Heisser, T., Hackl, M., Ihle, P., De Schutter, H., ... Brenner, H. (2021). Colorectal cancer incidence, mortality, and stage distribution in European countries in the colorectal cancer screening era: An international population-based study. *The Lancet Oncology, 22*(7), 1002–1013.

Carrara, A., Motter, M., Amabile, D., Pellecchia, L., Moscatelli, P., Pertile, R., ... Tirone, G. (2020). Predictive value of the sentinel lymph node procedure in the staging of non-metastatic colorectal cancer. *International Journal of Colorectal Disease, 35*(10), 1921–1928.

Currie, A. C., Brigic, A., Thomas-Gibson, S., Suzuki, N., Moorghen, M., Jenkins, J. T., ... Kennedy, R. H. (2017). A pilot study to assess near infrared laparoscopy with indocyanine green (ICG) for intraoperative sentinel lymph node mapping in early colon cancer. *European Journal of Surgical Oncology: The Journal of the European Society of Surgical Oncology and the British Association of Surgical Oncology*, 43(11), 2044–2051.

de Valk, K. S., Deken, M. M., Handgraaf, H. J. M., Bhairosingh, S. S., Bijlstra, O. D., van Esdonk, M. J., ... Vahrmeijer, A. L. (2020). First-in-human assessment of cRGD-ZW800-1, a zwitterionic, integrin-targeted, near-infrared fluorescent peptide in colon carcinoma. *Clinical Cancer Research: An Official Journal of the American Association for Cancer Research*, 26(15), 3990–3998.

de Valk, K. S., Deken, M. M., Schaap, D. P., Meijer, R. P., Boogerd, L. S., Hoogstins, C. E., ... Vahrmeijer, A. L. (2021). Dose-finding study of a CEA-targeting agent, SGM-101, for intraoperative fluorescence imaging of colorectal cancer. *Annals of Surgical Oncology*, 28(3), 1832–1844.

Doekhie, F. S., Mesker, W. E., Kuppen, P. J., van Leeuwen, G. A., Morreau, H., de Bock, G. H., ... Tollenaar, R. A. (2010). Detailed examination of lymph nodes improves prognostication in colorectal cancer. *International Journal of Cancer. Journal International du Cancer*, 126(11), 2644–2652.

Fields, A. C., Lu, P., Hu, F., Hirji, S., Irani, J., Bleday, R., ... Goldberg, J. E. (2020). Lymph node positivity in T1/T2 rectal cancer: A word of caution in an era of increased incidence and changing biology for rectal cancer. *Journal of Gastrointestinal Surgery: Official Journal of the Society for Surgery of the Alimentary Tract*, 25(4), 1029–1035.

Figueredo, A., Coombes, M. E., & Mukherjee, S. (2008). Adjuvant therapy for completely resected stage II colon cancer. *Cochrane Database of Systematic Reviews (Online)* (3), CD005390.

Fox, I. J., Brooker, L. G., Heseltine, D. W., Essex, H. E., & Wood, E. H. (1957). A tricarbocyanine dye for continuous recording of dilution curves in whole blood independent of variations in blood oxygen saturation. *Proceedings of the Staff Meetings. Mayo Clinic*, 32(18), 478–484.

Frangioni, J. V. (2008). New technologies for human cancer imaging. *Journal of Clinical Oncology: Official Journal of the American Society of Clinical Oncology*, 26(24), 4012–4021.

Gould, E. A., Winship, T., Philbin, P. H., & Kerr, H. H. (1960). Observations on a "sentinel node" in cancer of the parotid. *Cancer.*, 13, 77–78.

Griffiths, M., Chae, M. P., & Rozen, W. M. (2016). Indocyanine green-based fluorescent angiography in breast reconstruction. *Gland Surg*, 5(2), 133–149.

He, P. S., Li, F., Li, G. H., Guo, C., & Chen, T. J. (2016). The combination of blue dye and radioisotope vs radioisotope alone during sentinel lymph node biopsy for breast cancer: A systematic review. *BMC Cancer*, 16, 107.

Hernot, S., van Manen, L., Debie, P., Mieog, J. S. D., & Vahrmeijer, A. L. (2019). Latest developments in molecular tracers for fluorescence image-guided cancer surgery. *The Lancet Oncology*, 20(7), e354–e367.

Hirche, C., Mohr, Z., Kneif, S., Doniga, S., Murawa, D., Strik, M., & Hünerbein, M. (2012). Ultrastaging of colon cancer by sentinel node biopsy using fluorescence navigation with indocyanine green. *International Journal of Colorectal Disease*, 27(3), 319–324.

Hutteman, M., Choi, H. S., Mieog, J. S. D., Van Der Vorst, J. R., Ashitate, Y., Kuppen, P. J. K., ... Vahrmeijer, A. L. (2011). Clinical translation of ex vivo sentinel lymph node mapping for colorectal cancer using invisible near-infrared fluorescence light. *Annals of Surgical Oncology*, 18(4), 1006–1014.

Karoui, M., Rullier, A., Piessen, G., Legoux, J. L., Barbier, E., De Chaisemartin, C., ... for PRODIGE 22 investigators/collaborators. (2020). Perioperative FOLFOX 4 vs FOLFOX 4 plus cetuximab vs immediate surgery for high-risk stage II and III colon cancers: A phase II multicenter randomized controlled trial (PRODIGE 22). *Annals of Surgery*, 271(4), 637–645.

Kuehn, T., Bauerfeind, I., Fehm, T., Fleige, B., Hausschild, M., Helms, G., ... Untch, M. (2013). Sentinel-lymph-node biopsy in patients with breast cancer before and after neoadjuvant chemotherapy (SENTINA): A prospective, multicentre cohort study. *The Lancet Oncology*, 14(7), 609–618.

Kusano, M., Tajima, Y., Yamazaki, K., Kato, M., Watanabe, M., & Miwa, M. (2008). Sentinel node mapping guided by indocyanine green fluorescence imaging: A new method for sentinel node navigation surgery in gastrointestinal cancer. *Digestive Surgery*, 25(2), 103–108.

Liberale, G., Vankerckhove, S., Galdon, M. G., Larsimont, D., Ahmed, B., Bouazza, F., ... R&D Group for the Clinical Application of Fluorescence Imaging at the Jules Bordet Institute. (2016). Sentinel lymph node detection by blue dye vs indocyanine green fluorescence imaging in colon cancer. *Anticancer Research*, 36(9), 4853–4858.

Liefers, G. J., Cleton-Jansen, A. M., van de Velde, C. J., Hermans, J., van Krieken, J. H., Cornelisse, C. J., & Tollenaar, R. A. (1998). Micrometastases and survival in stage II colorectal cancer. *The New England Journal of Medicine, 339*(4), 223–228.

Lu, G., van den Berg, N. S., Martin, B. A., Nishio, N., Hart, Z. P., van Keulen, S., ... Poultsides, G. A. (2020). Tumour-specific fluorescence-guided surgery for pancreatic cancer using panitumumab-IRDye800CW: A phase 1 single-centre, open-label, single-arm, dose-escalation study. *Lancet Gastroenterologia y Hepatologia, 5*(8), 753–764.

Miller, M. J., McDole, J. R., & Newberry, R. D. (2010). Microanatomy of the intestinal lymphatic system. *Annals of the New York Academy of Sciences, 1207*(Suppl 1), E21–E28.

Nagata, K., Endo, S., Hidaka, E., Tanaka, J., Kudo, S. E., & Shiokawa, A. (2006). Laparoscopic sentinel node mapping for colorectal cancer using infrared ray laparoscopy. *Anticancer Research, 26*(3B), 2307–2311.

Navarro, M., Nicolas, A., Ferrandez, A., & Lanas, A. (2017). Colorectal cancer population screening programs worldwide in 2016: An update. *World Journal of Gastroenterology: WJG, 23*(20), 3632–3642.

Noura, S., Ohue, M., Seki, Y., Tanaka, K., Motoori, M., Kishi, K., ... Miyamoto, Y. (2010). Feasibility of a lateral region sentinel node biopsy of lower rectal cancer guided by indocyanine green using a near-infrared camera system. *Annals of Surgical Oncology, 17*(1), 144–151.

Ohnishi, S., Lomnes, S. J., Laurence, R. G., Gogbashian, A., Mariani, G., & Frangioni, J. V. (2005). Organic alternatives to quantum dots for intraoperative near-infrared fluorescent sentinel lymph node mapping. *Molecular Imaging: Official Journal of the Society for Molecular Imaging, 4*(3), 172–181.

Performance of SGM-101 for the Delineation of Primary and Recurrent Tumor and Metastases in Patients Undergoing Surgery for Colorectal Cancer [Internet]. Available from: https://clinicaltrials.gov/ct2/show/NCT03659448.

Ricciardi, R., Madoff, R. D., Rothenberger, D. A., & Baxter, N. N. (2006). Population-based analyses of lymph node metastases in colorectal cancer. *Clinical Gastroenterology and Hepatology: The Official Clinical Practice Journal of the American Gastroenterological Association, 4*(12), 1522–1527.

Rosenthal, E. L., Moore, L. S., Tipirneni, K., de Boer, E., Stevens, T. M., Hartman, Y. E., ... Zinn, K. R. (2017). Sensitivity and specificity of cetuximab-IRDye800CW to identify regional metastatic disease in head and neck cancer. *Clinical Cancer Research: An Official Journal of the American Association for Cancer Research, 23*(16), 4744–4752.

Saha, S., Sehgal, R., Patel, M., Doan, K., Dan, A., Bilchik, A., ... Yee, C. (2006). A multicenter trial of sentinel lymph node mapping in colorectal cancer: Prognostic implications for nodal staging and recurrence. *American Journal of Surgery, 191*(3), 305–310.

Schaafsma, B. E., Verbeek, F. P., van der Vorst, J. R., Hutteman, M., Kuppen, P. J., Frangioni, J. V., & Vahrmeijer, A. L. (2013). Ex vivo sentinel node mapping in colon cancer combining blue dye staining and fluorescence imaging. *The Journal of Surgical Research, 183*(1), 253–257.

Seymour, M. T., & Morton, D. (2019). FOxTROT: An international randomised controlled trial in 1052 patients (pts) evaluating neoadjuvant chemotherapy (NAC) for colon cancer. *Journal of Clinical Oncology, 37*, 3504.

SGM-101 in Locally Advanced and Recurrent Rectal Cancer (SGM-LARRC) [Internet]. Available from: https://clinicaltrials.gov/ct2/show/NCT04642924.

Tuech, J. J., Pessaux, P., Regenet, N., Bergamaschi, R., & Colson, A. (2004). Sentinel lymph node mapping in colon cancer. *Surgical Endoscopy, 18*(12), 1721–1729.

Vahrmeijer, A. L., Hutteman, M., van der Vorst, J. R., van de Velde, C. J., & Frangioni, J. V. (2013). Image-guided cancer surgery using near-infrared fluorescence. *Nature Reviews Clinical Oncology, 10*(9), 507–518.

Valsecchi, M. E., Silbermins, D., de Rosa, N., Wong, S. L., & Lyman, G. H. (2011). Lymphatic mapping and sentinel lymph node biopsy in patients with melanoma: A meta-analysis. *Journal of Clinical Oncology: Official Journal of the American Society of Clinical Oncology, 29*(11), 1479–1487.

van den Hoven, P., Ooms, S., van Manen, L., van der Bogt, K. E. A., van Schaik, J., Hamming, J. F., ... Mieog, J. S. D. (2019). A systematic review of the use of near-infrared fluorescence imaging in patients with peripheral artery disease. *Journal of vascular surgery, 70*(1), 286–297, e1.

van Manen, L., Handgraaf, H. J. M., Diana, M., Dijkstra, J., Ishizawa, T., Vahrmeijer, A. L., & Mieog, J. S. D. (2018). A practical guide for the use of indocyanine green and methylene blue in fluorescence-guided abdominal surgery. *Journal of Surgical Oncology, 118*(2), 283–300.

Watanabe, J., Ota, M., Suwa, Y., Ishibe, A., Masui, H., & Nagahori, K. (2017). Evaluation of lymph flow patterns in splenic flexural colon cancers using laparoscopic real-time indocyanine green fluorescence imaging. *International Journal of Colorectal Disease*, 32(2), 201−207.

Weixler, B., Rickenbacher, A., Raptis, D. A., Viehl, C. T., Guller, U., Rueff, J., . . . Zuber, M. (2017). Sentinel lymph node mapping with isosulfan blue or indocyanine green in colon cancer shows comparable results and identifies patients with decreased survival: A prospective single-center trial. *World Journal of Surgery*, 41(9), 2378−2386.

Yamamoto, H., Murata, K., Fukunaga, M., Ohnishi, T., Noura, S., Miyake, Y., . . . Masaki, M. (2016). Micrometastasis volume in lymph nodes determines disease recurrence rate of stage II colorectal cancer: A prospective multicenter trial. *Clinical Cancer Research: An Official Journal of the American Association for Cancer Research*, 22(13), 3201−3208.

CHAPTER 16

Systemic treatment of localized colorectal cancer

Dedecker Hans, Vandamme Timon, Teuwen Laure-Anne, Wuyts Laura, Prenen Hans, ten Tije, Albert Jan and Peeters Marc

Department of Oncology, University of Antwerp, Antwerp, Belgium

16.1 Introduction

In lymph-node-positive colorectal cancer, surgical resection alone might not be curative, and additional systemic chemotherapy might be indicated. Reported 5-year survival rates after surgical resection alone are 99% for stage I and 68%–83% for stage II, whereas in patients with stage III, lymph-node-positive disease a 5-year survival of 45%–65% was seen (Brierley et al., 2016). To determine who might benefit from additional systemic adjuvant therapy, an adequate assessment of the risk of relapse is essential. Next to the disease stage, this risk of relapse can be determined by clinicopathological features of the tumor, molecular markers, and DNA mismatch repair (dMMR)/microsatellite instability (MSI) status. The presence of circulating tumor cells and circulating tumor DNA (ctDNA) postresection can be a potential prognostic biomarker and is currently the subject of ongoing trials (Tie et al., 2019). Different chemotherapy regimens, mainly 5-fluorouracil(FU)-based with or without additional oxaliplatin, and treatment durations have been proposed for adjuvant treatment of stage II and stage III disease. In this chapter, we will discuss risk assessment of stage II disease, selection and duration of a chemotherapy regimen in stage II and III, nonchemotherapy-based interventions, and use of adjuvant treatment in specific populations. Finally, we will review neoadjuvant approaches and future perspectives for adjuvant therapy in colorectal cancer.

16.2 Relapse risk assessment in stage II disease

In stage II colon cancer, the benefit of adjuvant chemotherapy is relatively low. Efforts are made to identify specific populations with stage II colon cancer who are at greater risk for relapse, which might benefit from adjuvant systemic therapy (Argilés et al., 2020).

Pathological risk factors for stage II assessment that are of prognostic value are as follows:

1. less than 12 lymph nodes sampling,
2. pT4 stage including perforation,
3. high-grade tumor,
4. obstruction at the time of presentation,
5. high serum levels CEA (carcinoembryonic antigen), and
6. vascular, lymphatic, or perineural invasion (Brierley et al., 2016; Roth et al., 2012; Wells et al., 2017).

These risk factors in stage II colon cancer are in accordance with the NCCN guidelines for colon cancer (2021). However, there is no international consensus on the definition of high-risk factors in stage II colon carcinoma. In 2016 a Dutch retrospective analysis of 4940 high-risk stage II patients withheld only pT4 as a predictive marker (Verhoeff et al., 2016).

The presence of MSI, which is caused by a deficiency of the DNA MMR system, plays an important role in decision-making regarding adjuvant chemotherapy in stage II colorectal cancer. Mutations or modifications of MMR genes can result in MMR protein deficiency and MSI. The extent of instability classifies MSI as either MSI-high or MSI-low. Tumors without these mutations or modifications are classified as microsatellite stable (MSS). Specimens characterized as MSI-high are more common in stage II than in stage III and only 3.5% of stage IV tumor are classified as MSI-high (Popat et al., 2005). This suggests that colon cancers characterized by the presence of MSI-high tumor have a lower chance of encountering metastasis. In stage II disease MSI-high is a favorable prognostic factor. Currently, two standard reference methods, namely immunohistochemistry and polymerase chain reaction, are recommended for the detection of defective MMR (dMMR) and MSI status. Tumor DNA MMR deficiency testing is also important to determine potential genetic predisposition by the identification of Lynch syndrome (germline mutation MMR genes, most commonly MLH1, MSH2, or less frequently MSH6 and PMS2), which is responsible for 2%–4% of the colon cancer cases (Koopman et al., 2009; Ribic et al., 2003; Roth et al., 2010; Sargent et al., 2010; Sinicrope et al., 2011; Tejpar et al., 2011).

Currently, determination of tumor gene status for KRAS/NRAS and BRAF mutation is recommended for patients with metastatic colorectal cancer (mCRC). In contrast to the treatment of mCRC, the use of monoclonal antibodies, cetuximab, and panitumumab toward the epidermal growth factor receptor did not provide any benefit in nonmetastatic patients (Taieb et al., 2017). In clinical practice, determination of RAS and BRAF mutation in stage II and III colon cancer will not help the clinician in treatment decision-making because these parameters will not change the standard of care. Further trials may define a subset of patients who might benefit from cetuximab and panitumumab in the adjuvant setting, but at this time data is lacking and no recommendation can be made for the routine assessment of genetic markers (e.g., RAS and BRAF mutations) in localized colorectal cancer. To conclude, the decision-making for the need of adjuvant therapy in stage II and III colon cancer should be based on multiple variables: TNM classification, number of lymph nodes analyzed after surgery (<12), poor prognostic features [e.g., poorly differentiated histology (dMMR and MSI) lymphatic/vascular and perineural invasion, bowel obstruction, perforation; close, indeterminate, or positive margins], anticipated life expectancy, and the assessment of other comorbidities.

16.3 Adjuvant treatment in stage II disease

Stage II colon cancer is both clinically and biologically a heterogeneous group. Reported 5-year survival rates for stage II after surgical resection alone are 68%−83%. The risk of relapse following surgery of an MSI T3 lesion is less than 10%, whereas a patient following resection for a MMR proficient T4 tumor may have a risk of disease recurrence greater than 50% (Gill et al., 2004).

The benefit of systemic adjuvant therapy in stage II colon cancer depends on further pathological risk factors and dMMR/MSI status as discussed earlier. Pathological risk factors for stage II assessment include less than 12 lymph nodes sampling, pT4 stage including perforation, high-grade tumor, obstruction at the time of presentation, high CEA levels, and vascular, lymphatic or perineural invasion (Roth et al., 2012; Wells et al., 2017).

A low-risk subgroup is defined as patients with no pathological risk factor and the presence of dMMR/MSI. In the intermediate-risk subgroup, one or more of the risk factors must be present regardless of the dMMR/MSI status. A high-risk subgroup is defined as patients with less than 12 lymph nodes sampling, pT4, or multiple risk factors from the intermediate-risk group.

Results from a meta-analysis in 2015 showed a 5-year disease-free survival (DFS) of 81.4% in patients with stage II colon cancer who did not receive adjuvant therapy, whereas a DFS of 79.3% was seen in a patient with stage II treated with adjuvant chemotherapy. On the other hand, for patients with stage III colon cancer, the 5-year DFS was 49% and 63.6% in those treated without and with adjuvant chemotherapy, respectively (Böckelman et al., 2015).

This suggests the benefit of adjuvant therapy in high-risk patients. Patients with average risk stage II colon cancer have a good prognosis, so the possible benefit of adjuvant therapy is rather small.

The decision regarding the use of adjuvant therapy in colon cancer stage II should be based on a discussion between the patient and the physician, including the assessment of the risk factors, prognosis, adverse effects of the treatment, and the expected benefit in terms of disease-free and overall survival (OS).

Different chemotherapy regimens, mainly 5-FU-based with or without additional oxaliplatin, and treatment durations have been proposed for adjuvant treatment of stage II and stage III disease. The MOSAIC analysis showed among stage II patients that DFS at 5 years was 83.7% and 79.9% in the FOLFOX4 and LV5FU2 (bolus/infusional fluorouracil plus leucovorin) group, respectively. The DFS at 5 years in high-risk stage II patients was 82.3% and 74.6% in the FOLFOX and LV5FU2 groups, respectively (Andre et al., 2009). Updated OS data of the MOSAIC study with 10-year follow-up continues to demonstrate the OS benefit of oxaliplatin in the adjuvant treatment of patients with resected stage II to III colon cancer. Ten-year OS in the FOLFOX4 and LV5FU2 arms was 78.4% versus 79.5% for stage II and 67.1% versus 59.0% for stage III disease. There was a nonsignificant detrimental effect in the low-risk stage II group and a nonsignificant benefit in the high-risk stage II group with FOLFOX4 (André et al., 2015) (Table 16.1). Based on these results, one can conclude that in low-risk stage II colon cancer, the addition of Oxaliplatinum is not indicated and based on possible side effects is even contraindicated in this population.

TABLE 16.1 OS according to disease stage for patients in the LV5FU2 and FOLFOX4 treatment arms (André et al., 2015).

Variable	LV5FU2	FOLFOX4	HR	95% CI	P
Stage II					
No. of patients	448	451	—	—	—
OS					
No. of events	85	87	1.00	0.74–1.35	.980
5 year, % (SE)	90.0 (1.4)	89.0 (1.5)	—	—	—
10 year, % (SE)	79.5 (2.2)	78.4 (2.2)	—	—	—
Low-risk stage II					
No. of patients	223	235			
OS					
No. of events	32	39	1.02	0.68–1.53	.904
5 year, % (SE)	92.3 (1.8)	90.5 (1.9)	—	—	—
10 year, % (SE)	86.7 (2.5)	81.2 (3.0)	—	—	—
High-risk stage II					
No. of patients	222	212			
OS					
No. of events	53	48	0.89	0.6–1.32	.579
5 year, % (SE)	87.5 (2.2)	87.6 (2.3)	—	—	—
10 year, % (SE)	71.7 (3.5)	75.4 (3.3)	—	—	—
Stage III					
No. of patients	675	672	—	—	—
OS					
No. of events	250	209	0.80	0.66–0.96	.016
5 year, % (SE)	71.7 (1.8)	76.0 (1.7)	—	—	—
10 year, % (SE)	59.0 (2.1)	67.1 (2.0)	—	—	—

CI, Confidence interval; HR, hazard ratio; OS, overall survival.

It may be reasonable to accept the benefit of adjuvant therapy in stage III disease as an indirect evidence for stage II colon cancer, especially in the high-risk population.

At present, no advice can be given regarding the duration of treatment in high-risk patients with stage II colon cancer. In the absence of risk factors, follow-up without adjuvant therapy can be defended in a low-risk stage II patient. In intermediate-risk stage II with MSS treatment with fluoropyridimines monotherapy for 6 months is recommended. Patients with a high-risk stage (less than 12 lymph nodes sampling, pT4, or multiple risk factors) may be considered for 3 months of CAPOX, as the IDEA-pooled analysis showed noninferiority of 3 months of CAPOX and inferiority of 3 months of FOLFOX when compared with 6 months of FOLFOX, with all the limitations of post hoc analyses (Chang et al., 2012; Haller et al., 2011).

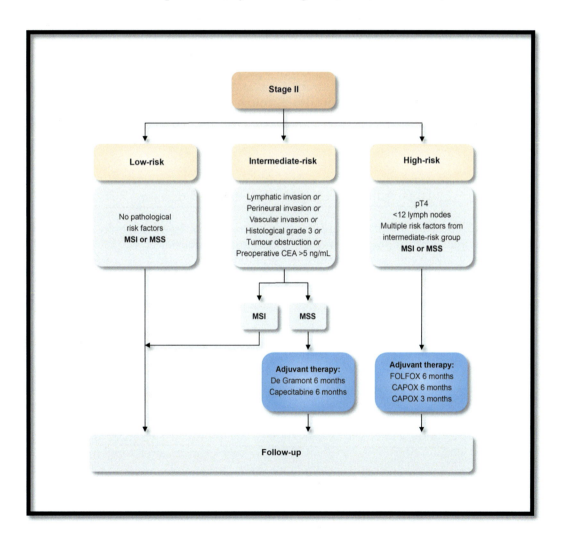

16.4 Time to treat

In general, after surgical resection, adjuvant systemic therapy should start within 2 months. A meta-analysis in 2010 examined whether a further delay of initiation of adjuvant therapy affected the OS. In this meta-analysis, data from eight different studies were included. It showed that the effect of delaying initiation or adjuvant chemotherapy after surgical resection of more than 8 weeks was associated with a significant decrease in OS [(relative risk) RR: 1.20; 95% (confidence interval) CI: 1.15–1.26] (Des Guetz et al., 2010). In 2011 a Canadian systematic review and meta-analysis showed that a 14% decrease in OS for each 4-week delay in adjuvant therapy (Biagi et al., 2011). Bos et al. (2015) published a retrospective study of 7794 stage II and II patients. A reduced survival was observed if adjuvant therapy was started after a period of 6 weeks after surgical resection. Critics like Sargent et al. already discussed the presence of multiple confounding factors in the above analyses. The presence of specific comorbidities, patient's age, and the need for emergency resection affect the prolonged hospitalization time and delay the initiation of adjuvant systemic therapy. These factors can also account for a negative effect on survival (Sargent et al., 2011). Consequently, we recommend adjuvant therapy as soon as the patient is medically able.

16.5 Adjuvant treatment in stage III disease

For several decades, fluoropyrimidine monotherapy has been the gold standard in the adjuvant treatment of stage II and III colon cancer (Moertel et al., 1990). Low-risk patients [low grade (well or moderately differentiated, or grade 1 or 2 tumors), T1/2 tumors and involvement of ≤4 lymph nodes] have a 5-year DFS of 75% when treated with adjuvant 5-FU alone in comparison with 17% in high-risk patients [>5 lymph nodes, high grade (poorly differentiated, anaplastic, and grade 3 or 4 tumors), T4 tumors] (Gill et al., 2004).

In 2009 a large international phase III clinical trial in stage II or III colon cancer was published [Multicenter International Study of Oxaliplatin/5-Fluorouracil/Leucovorin in the Adjuvant Treatment of Colon Cancer (MOSAIC)]. This randomized phase III study included 2246 patients who had undergone resection with a curative intent for either stage II ($n = 899$) or III ($n = 1347$) colon cancer. One group received a bolus plus continuous infusion FU plus leucovorin (LV5FU2) with the addition of oxaliplatin (FOLFOX). The control group received LV5FU2 alone. The results showed an improved DFS [hazard ratio (HR): 0.78, $P = .023$) and OS (HR: 0.8, $P = .005$) stage III colon cancer after receiving adjuvant oxaliplatin plus FU and leucovorin (Andre et al., 2009).

In 2015 an update of the MOSAIC trial was published after a 10-year follow-up was performed in 2246 patients with resected stage II to III colon cancer. The OS benefit of oxaliplatin-based adjuvant chemotherapy, increasing over time and with the disease severity, was confirmed at 10 years in patients with stage II to III colon cancer as discussed earlier (André et al., 2015) (Table 16.1).

The benefit of the combination of fluoropyrimidine (either oral capecitabine or bolus 5-FU) and oxaliplatin over 5-FU/leucovorin was also seen in two other landmark publications: NSABP C-07 and XELOXA (Haller et al., 2011; Kuebler et al., 2007).

The XELOXA study compared the addition of oxaliplatin to capecitabine (CAPOX) with fluorouracil in monotherapy as adjuvant therapy. In contrast to NSABP C-70 and MOSAIC, they included only stage III colon cancer. The addition of oxaliplatin showed a similar risk reduction on relapse in all three studies. A longer follow-up also showed a significant improvement in OS. FOLFOX and CAPOX are currently the gold standards for the treatment of stage III colon cancer (Andre et al., 2009; Haller et al., 2011).

The patient characteristics and preferences are not unimportant. Capecitabine (CAPOX), for example, has a higher risk of diarrhea and hand-foot syndrome compared to a bolus plus continuous infusion FU plus leucovorin (FOLFOX). Therefore CAPOX seems less appropriate for patients with ileostomy and chronic renal insufficiency. The advantage of CAPOX is that it gives less neurotoxicity in a shorter regimen and that no central venous access is needed for treatment compared to FOLFOX. Both regimens require a dose reduction of oxaliplatin following the occurrence of grade 2 neurotoxicity. Based on patient characteristics, it may be necessary to reduce the dose of capecitabine. A starting dose of 2000 mg/m^2/d can be initiated. If this dose is well tolerated, it can be further increased to the standard dose of 2500 mg/m^2/d for the second cycle (Sobrero et al., 2020).

There is currently no evidence for any benefit of irinotecan or biological agents, such as bevacizumab or cetuximab as discussed earlier, in the adjuvant setting (Taieb et al., 2017).

Initially, a treatment duration of 6 months was standard. Given the previously described sensory polyneuropathy induced by oxaliplatin, shortening of the treatment duration was desirable. This is partly due to the fact that the sensory polyneuropathy depends on the cumulatively administered dose of the drug and can manifest itself or aggravate up to months after the last administration of oxaliplatin, potentially affecting patients' daily activities for the rest of their lives.

Therefore the IDEA collaboration was designed to receive either 3 months or 6 months of adjuvant therapy with oxaliplatin and a fluoropyrimidine. It was a prospectively conducted cumulative study involving 12,834 patients with stage III colon cancer who were enrolled in six individual randomized phase III trials in the UK/Denmark/Spain/Australia/Sweden, Italy, France, US/Canada, Greece, and Japan (respectively SCOT, TOSCA, IDEA France, C80702, HORG, and ACHIEVE). The primary endpoint of the six trials was DFS, which was defined as the time from the date of randomization to the date of first relapse, the diagnosis of a secondary colorectal cancer, or death from any cause (Chang et al., 2012).

For 3 and 6 months of treatment, the 3-year DFS was similar in the overall study population (74.6% and 75.5% for 3 and 6 months, respectively). However, there was no statistical confirmation of the predefined noninferiority. The variation in the subgroups with different risk profile and treatments may explain the fact that the noninferiority was not being achieved. On the other hand, a noninferiority in 3-year DFS of treatment with CAPOX between 3 and 6 months was achieved (75.9% and 74.8% for 3 and 6 months, respectively), whereas for FOLFOX, an inferiority in DFS was seen in the 3 months regimen (73.6% and 76.0% for 3 and 6 months, respectively).

The difference in the performance between CAPOX and FOLFOX treatment regimens is remarkable. For FOLFOX, a longer duration of therapy increased the rate of DFS, particularly in high-risk population, whereas an additional 3 months of CAPOX did not affect DFS.

An explanation can be the fact that adherence and the overall dose intensity of a 6-month duration of an oral therapy might attenuate over time so that the absence of difference between 3 months and 6 months of CAPOX might be due to a reduced overall dose intensity in the 6-month therapy group.

Another hypothesis is related to the different dosing schedule of oxaliplatin (130 mg/m^2 of body-surface area every 3 weeks with CAPOX vs 85 mg/m^2 every 2 weeks with FOLFOX) and especially the fluoropyrimidine regimen a 46-h infusion of fluorouracil every 2 weeks with FOLFOX vs capecitabine twice daily for 2 of every 3 weeks with CAPOX. Long-term administration of a fluoropyrimidine with CAPOX may be more effective than the infusions with FOLFOX, which are administered only twice a month (Chang et al., 2012).

As expected, a major decrease in patients experiencing grade 2 or higher neurotoxicity was observed, 16.6% versus 47.7% with FOLFOX and 14.2% versus 44.9% CAPOX for 3 and 6 months, respectively.

In 2020 André et al. published the updated 5-year IDEA results. The update showed for DFS (HR: 0.98, 95% CI: 0.88−1.08) in the CAPOX arm between 3 months and 6 months of treatment, while in the FOLFOX arm the DFS (HR: 1.16, 95% CI: 1.07−1.26) was significantly extended at 6 months of treatment.

The 5-year OS between 3 months (82.4%) and 6 months of treatment (82.8%) was not significantly different (HR: 1.02, 95% CI: 0.95−1.11).

The absolute difference in 5-year OS rate between 3 and 6 months was only 1% (71.4% vs 72.4%, respectively) with an HR of 1.02 (95% CI: 0.89−1.20) for high-risk stage III patients treated with CAPOX, whereas for those treated with FOLFOX, the absolute difference in 5-year OS rate was 2.8% between the two treatment durations (72.5% vs 75.3%) with an HR of 1.12 (95% CI: 0.98−1.27). Given the minor differences in OS and DFS for the majority of patients, a low-risk subgroup (defined as patients with T1, T2, or T3 with N1 disease) can opt for CAPOX for 3 months. FOLFOX for 3 months can be a worthy alternative, given the minor differences in survival. In the high-risk subgroup (defined as patients with T4 or N2 disease or both), CAPOX for 3 months can also be recommended, given the minor differences in OS. When opting for FOLFOX, a therapy of 6 months remains indicated (Grothey et al., 2018).

16.5 Adjuvant treatment in stage III disease

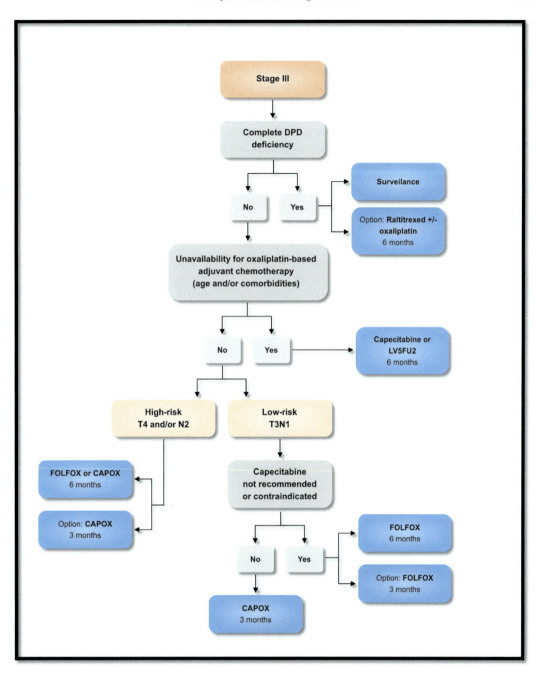

16.6 Neoadjuvant chemotherapy for locally advanced colon cancer

In contrast to rectal cancer, neoadjuvant treatment for locally advanced colon cancer is not standardized. The PRODIGE 22, a French multicenter randomized phase II trial was performed to evaluate the effect of neoadjuvant chemotherapy for locally advanced colon cancer. They included 104 patients with a high-risk, nonmetastatic locally advanced (T3–T4, and/or N2) colon cancer. As assessed by a computed tomography (CT) scan. After randomization, one group of the population received 6 months of adjuvant FOLFOX (12 cycles) after primary surgery (colectomy), the experimental group received four cycles of neoadjuvant FOLFOX before and eight cycles after surgery.

The perioperative FOLFOX was well tolerated without any increase in morbidity. Perioperative FOLFOX showed no major histological response [Tumor Regression Grade (TRG) by Ryan > 1, TRG-Ryan is based on the relationship between fibrosis and the amount of residual tumor cells] when compared to patients with no neoadjuvant therapy (Karoui et al., 2020).

The FoxTROT phase III trial included 1052 patients with a locally advanced colon tumor and compared perioperative FOLFOX with adjuvant FOLFOX after primary surgery. The perioperative group appeared to be associated with reduced therapy failure, defined as relapse or persistent disease after 2 years (13.6% in the perioperative arm vs 17.2% in the control arm). This difference, however, did not reach the target statistical significance (HR: 0.75; 95% CI: 0.55–1.04; $P = .08$). The perioperative arm showed a reduced number of incomplete resections, 5% versus 11%, respectively ($P < .05$) (Seymour & Morton, 2019).

In both trials, 33% of the population in the control group scored a higher staging based on the CT scan criteria compared to the postoperative staging. Thus low-risk patients potentially received (neo)adjuvant therapy, theoretically not requiring chemotherapy. As a result, the previously described results should be interpreted with caution. Based on these findings, perioperative chemotherapy cannot yet be considered standard of care in locally advanced colon cancer.

However, these studies may suggest considering in selected cases neoadjuvant systemic treatment in nonmetastatic locally advanced colon cancer in tumors that are a priori irresectable but may potentially become after neoadjuvant treatment.

16.7 Nutrition and lifestyle modification reduce relapse risk

The beneficial effect of *exercise* on the incidence of colon cancer has already been described.

In 2005 Samad et al. published a meta-analysis of 19 cohort studies in active males and active females which showed a 22% reduction in the risk of colon cancer. Meyerhardt et al. showed in 2006 in prospective observational study of 832 patients with stage III colon cancer enrolled in a randomized adjuvant chemotherapy trial. They observed the association of exercise on cancer recurrences and survival after surgery. The 3-year DFS for patients who engage in less than 18 metabolic equivalent of task (MET) hours per week was 75.1% and 84.5% for patients who engage in 18 or more MET-hours per week.

Physical activity seems to reduce the risk of cancer recurrence and mortality (Meyerhardt et al., 2006; Samad et al., 2005).

The effect of *nutrition* on the risk of developing colon cancer has already been described in detail. In 2007 Meyerhardt et al. suggested that following a diet after surgical resection and adjuvant systemic chemotherapy may have an effect on the DFS and OS.

Meyerhardt et al. performed an analysis of dietary intake using a semiquantitative food frequency questionnaire during and 6 months after adjuvant chemotherapy in the CALGB 89803 cohort, a prospective observational study of 1009 patients with stage III colon cancer who were enrolled in a randomized adjuvant chemotherapy trial.

They identified two major dietary patterns: (1) the prudent pattern, characterized by high intakes of fruits and vegetables, poultry, and fish; and (2) the Western pattern, characterized by high intakes of meat, fat, refined grains, and dessert. Patients in the highest quintile of the Western dietary pattern were 2.9 times more likely to recur than those in the lowest quintile. A higher overall mortality was also seen with an increasing Western dietary pattern. On the other hand, no difference in DFS and OS was objectified for the prudent dietary pattern after curative resection of stage III colon. In conclusion, a higher intake of a Western dietary pattern may be associated with a higher risk of recurrence and mortality among patients with stage III colon cancer treated with surgery and adjuvant chemotherapy (Meyerhardt et al., 2007).

16.8 Adjuvant treatment in elderly

The risk of complications secondary to adjuvant systemic therapy must always be compared to the benefits on OS. The performance status, the presence of a systemic infection, and a mono- and/or poly-organ dysfunction can be a contraindication for systemic treatment.

The presence of other comorbidities with reduced life expectancy should also be taken into account. A retrospective analysis of a Canadian database in Ontario showed an advantage of adjuvant therapy in all age groups. However, the emphasis should be on the fact that only fit elderly should be eligible for adjuvant therapy (Booth et al., 2016).

Given their poor tolerance for therapy and limited life expectancy, frail older adults, those with significant functional impairment or an Eastern Cooperative Oncology Group performance status of 3–4 are not appropriate candidates for adjuvant chemotherapy (Oken et al., 1982).

The association with oxaliplatin appears to have a greater benefit on OS and less toxicity in a younger population as shown by a pooled analysis from four randomized trials (NSAPBP-C08, XELOXA, X-ACT, and AVANT) (Schmoll et al., 2015).

16.9 Dihydropyrimidine dehydrogenase-deficient patients

The DPYD gene encodes dihydropyrimidine dehydrogenase (DPD), an enzyme that catalyzes the rate-limiting step in fluorouracil metabolism. Approximately 3%–5% of patients have deficiencies of DPD function due to genetic polymorphisms. Individuals who carry at least one copy of no function DPYD variants, such as DPYD*2A, may not be able to metabolize fluorouracil at normal rates and are at risk of potentially life-threatening fluorouracil toxicity, such as bone marrow suppression and neurotoxicity (Lee et al., 2004). The assessment of the phenotype or DPD functionality can be based on the level of uracilemia in the blood. Treatment with fluoropyrimidines is contraindicated at doses above 150 ng/mL. In the case of uracilemia between 16 and 150 ng/mL, the dose of fluoropyrimidines should be reduced by 50% (Henricks et al., 2018; Loriot et al., 2018; Meulendijks et al., 2017).

16.10 Future perspectives for (neo)adjuvant therapy in stage III colon cancer

The successful use of immune checkpoint inhibitors in MSI metastatic colon cancer has already been described (Overman et al., 2017).

In a small study of MSI localized colon cancer, 20 patients were treated with neoadjuvant immunotherapy. The combination consisted of a single dose of ipilimumab (1 mg/kg) on day 1 and nivolumab (3 mg/kg) on day 1 and 15. Among these 20 patients, 12 showed a sensational complete pathological response, clearly demonstrating the high potency of checkpoint inhibitors in MSI high colon cancer (Chalabi et al., 2020).

ATOMIC and POLEM are two trials currently running to investigate the effect of anti-PDL1 monoclonal antibodies in adjuvant setting in stage III colon cancer patients after surgery (Lau et al., 2019; Sinicrope et al., 2019).

In metastatic HER2 mutated or amplified colon cancer, HER2 inhibition (with pertuzumab and trastuzumab) is promising. In patients with TRK fusion−positive cancer, regardless of the tumor type, larotrectinib also showed antitumor activity. Therefore these active drugs deserve further exploration in the adjuvant setting in selected patients (Drilon et al., 2018; Sartore-Bianchi et al., 2016).

The role of acetylsalicylic acid (ASA) in reducing the risk of relapse has already been described in the literature. ASA may play an important role in the downregulation of the phosphatidylinositol 3-kinase (PI3K) signaling pathway, leading to a growth suppression of tumor cells.

In approximately 17% of all patients, the PIK3CA is documented. When the presence of mutated-PIK3CA tumors, the use of ASA after diagnosis was associated with significantly longer cancer-specific survival (HR: 0.18; 95% CI: 0.06−0.61; $P < .001$) and OS (HR: 0.54, 95% CI: 0.31−0.94, $P = .01$). In case of the presence of wild-type PIK3CA tumors, there was no survival benefit (Liao et al., 2012). The ASPIRIN trial is aiming to answer this question definitively. It is a randomized, double-blind, placebo-controlled study with the inclusion of stage II and III colon cancer, after R0 resection. The study assesses the effect of adjuvant ASA on recurrence.

16.11 Conclusion

Adjuvant systemic therapy in colon cancer is determined by stage and histopathological characteristics. In stage II disease, the benefit of adjuvant treatment with 6 months of 5-FU or capecitabine is restricted to the intermediate to high-risk subgroup of patients with

MRR-proficient tumors with pathological risk factors such as pT4 stage including perforation, high-grade tumor, obstruction at time of presentation, high CEA levels and vascular, lymphatic, or perineural invasion. In the high-risk stage II group, adding oxaliplatin to the adjuvant regimen (as per stage III) can be considered, irrespective of MSI status. In all patients with nodal involvement and thus stage III disease, adjuvant systemic treatment is warranted, either with CAPOX during 3 months or FOLFOX during 6 months. In high-risk stage III, FOLFOX/CAPOX during 6 months might have a minor benefit compared to CAPOX during 3 months. All adjuvant treatment should start within 2 months after resection and DPD deficiency should be ruled out before starting therapy. Adjuvant treatments have similar efficacy in the elderly and oxaliplatin might be better tolerated than in younger age groups. However, the potential gain in OS should be weighed against the toxicity in these patient groups. Lifestyle interventions such as exercise and nutrition might improve outcomes after resection of colon cancer. Novel adjuvant strategies under investigation include ASA for all patients and PDL-1 inhibition in the MSI subgroup. Currently, neoadjuvant systemic treatment in colon cancer patients is not standardized and remains investigational.

References

Andre, T., Boni, C., Navarro, M., et al. (2009). Improved overall survival with oxaliplatin, fluorouracil, and leucovorin as adjuvant treatment in stage II or III colon cancer in the MOSAIC trial. *Journal of Clinical Oncology: Official Journal of the American Society of Clinical Oncology, 27*, 3109–3116.

André, T., de Gramont, A., Vernerey, D., et al. (2015). Adjuvant fluorouracil, leucovorin, and oxaliplatin in stage II to III colon cancer: Updated 10-year survival and outcomes according to BRAF mutation and mismatch repair status of the MOSAIC study. *Journal of Clinical Oncology: Official Journal of the American Society of Clinical Oncology, 33*(35), 4176–4187.

Argilés, G., Tabernero, J., Labianca, R., et al. (2020). Localised colon cancer: ESMO Clinical Practice Guidelines for diagnosis, treatment and follow-up. *Annals of Oncology: Official Journal of the European Society for Medical Oncology/ESMO, 31*(10), 1291–1305.

Biagi, J. J., Raphael, M. J., Mackillop, W. J., Kong, W., King, W. D., & Booth, C. M. (2011). Association between time to initiation of adjuvant chemotherapy and survival in colorectal cancer: A systematic review and meta-analysis. *JAMA: the Journal of the American Medical Association, 305*(22), 2335–2342.

Böckelman, C., Engelmann, B. E., Kaprio, T., Hansen, T. F., & Glimelius, B. (2015). Risk of recurrence in patients with colon cancer stage II and III: A systematic review and meta-analysis of recent literature. *Acta Oncologica (Stockholm, Sweden), 54*(1), 5–16.

Booth, C. M., Nanji, S., Wei, X., et al. (2016). Use and effectiveness of adjuvant chemotherapy for stage III colon cancer: A population-based study. *Journal of the National Comprehensive Cancer Network: JNCCN, 14*, 47–56.

Bos, A. C., van Erning, F. N., van Gestel, Y. R., et al. (2015). Timing of adjuvant chemotherapy and its relation to survival among patients with stage III colon cancer. *European Journal of Cancer, 51*(17), 2553–2561.

Brierley, J. D., Gospodarowicz, M. K., & Wittekind, C. (Eds.), (2016). *TNM classification of malignant tumours* (8th edition). Oxford: John Wiley & Sons, Inc.

Chalabi, M., Fanchi, L. F., Dijkstra, K. K., et al. (2020). Neoadjuvant immunotherapy leads to pathological responses in MMR-proficient and MMR-deficient early-stage colon cancers. *Nature Medicine, 26*, 566–576.

Chang, H. J., Lee, K.-W., Kim, J. H., et al. (2012). Adjuvant capecitabine chemotherapy using a tailored-dose strategy in elderly patients with colon cancer. *Annals of Oncology: Official Journal of the European Society for Medical Oncology/ESMO, 23*, 911–918.

Des Guetz, G., Nicolas, P., Perret, G. Y., et al. (2010). Does delaying adjuvant chemotherapy after curative surgery for colorectal cancer impair survival? A meta-analysis. *European Journal of Cancer, 46*(6), 1049–1055.

Drilon, A., Laetsch, T. W., Kummar, S., et al. (2018). Efficacy of larotrectinib in TRK fusion-positive cancers in adults and children. *The New England Journal of Medicine, 378*, 731–739.

Gill, S., Loprinzi, C. L., Sargent, D. J., et al. (2004). Pooled analysis of fluorouracil-based adjuvant therapy for stage II and III colon cancer: Who benefits and by how much? *Journal of Clinical Oncology: Official Journal of the American Society of Clinical Oncology, 22*(10), 1797–1806.

Grothey, A., Sobrero, A. F., Shields, A. F., et al. (2018). Duration of adjuvant chemotherapy for stage III colon cancer. *The New England Journal of Medicine, 378,* 1177–1188.

Haller, D. G., Tabernero, J., Maroun, J., et al. (2011). Capecitabine plus oxaliplatin compared with fluorouracil and folinic acid as adjuvant therapy for stage III colon cancer. *Journal of Clinical Oncology: Official Journal of the American Society of Clinical Oncology, 29,* 1465–1471.

Henricks, L. M., Lunenburg, C., de Man, F. M., et al. (2018). DPYD genotype-guided dose individualisation of fluoropyrimidine therapy in patients with cancer: A prospective safety analysis. *The Lancet Oncology, 19,* 1459–1467.

Karoui, M., Rullier, A., Piessen, G., et al. (2020). Perioperative FOLFOX 4 vs FOLFOX 4 plus cetuximab vs immediate surgery for high-risk stage II and III colon cancers: A phase II multicenter randomized controlled trial (PRODIGE 22). *Annals of Surgery, 271,* 637–645.

Koopman, M., Kortman, G. A., Mekenkamp, L., et al. (2009). Deficient mismatch repair system in patients with sporadic advanced colorectal cancer. *British Journal of Cancer, 100*(2), 266–273.

Kuebler, J. P., Wieand, H. S., O'Connell, M. J., et al. (2007). Oxaliplatin combined with weekly bolus fluorouracil and leucovorin as surgical adjuvant chemotherapy for stage II and III colon cancer: results from NSABP C-07. *Journal of Clinical Oncology: Official Journal of the American Society of Clinical Oncology, 25,* 2198–2204.

Lau, D., Cunningham, D., Gillbanks, A., et al. (2019). POLEM: Avelumab plus fluoropyrimidine-based chemotherapy as adjuvant treatment for stage III dMMR or POLE exonuclease domain mutant colon cancer—A phase III randomized study. *JCO, 37,* TPS3615.2.

Lee, A., Ezzeldin, H., Fourie, J., et al. (2004). Dihydropyrimidine dehydrogenase deficiency: Impact of pharmacogenetics on 5-fluorouracil therapy. *Clinical Advances in Hematology & Oncology, 2*(8), 527–532.

Liao, X., Lochhead, P., Nishihara, R., et al. (2012). Aspirin use, tumor PIK3CA mutation, and colorectal-cancer survival. *New England Journal of Medicine, 367*(17), 1596–1606.

Loriot, M.-A., Ciccolini, J., Thomas, F., et al. (2018). Dépistage du déficit en dihydropyrimidine déshydrogénase (DPD) et sécurisation des chi- miothérapies à base de fluoropyrimidines: mise au point et recom- mandations nationales du GPCO-Unicancer et du RNPGx. *Bulletin du Cancer, 105,* 397–407.

Meulendijks, D., Henricks, L. M., Jacobs, B. A. W., et al. (2017). Pretreatment serum uracil concentration as a predictor of severe and fatal fluoropyrimidine-associated toxicity. *British Journal of Cancer, 116*(11), 1415–1424.

Meyerhardt, J. A., Heseltine, D., Niedzwiecki, D., et al. (2006). Impact of physical activity on cancer recurrence and survival in patients with stage III colon cancer: Findings from CALGB 89803. *Journal of Clinical Oncology: Official Journal of the American Society of Clinical Oncology, 24*(22), 3535–3541.

Meyerhardt, J. A., Niedzwiecki, D., Hollis, D., et al. (2007). Association of dietary patterns with cancer recurrence and survival in patients with stage III colon cancer. *JAMA: The Journal of the American Medical Association, 298*(7), 754–764.

Moertel, C. G., Fleming, T. R., Macdonald, J. S., et al. (1990). Levamisole and fluorouracil for adjuvant therapy of resected colon carcinoma. *The New England Journal of Medicine, 322,* 352–358.

Oken, M. M., Creech, R. H., Tormey, D. C., et al. (1982). Toxicity and response criteria of the Eastern Cooperative Oncology Group. *American Journal of Clinical Oncology, 5,* 649.

Overman, M. J., McDermott, R., Leach, J. L., et al. (2017). Nivolumab in patients with metastatic DNA mismatch repair-deficient or microsatellite instability-high colorectal cancer (CheckMate 142): An open-label, multicentre, phase 2 study. *The Lancet Oncology, 18,* 1182–1191.

Popat, S., Hubner, R., & Houlston, R. S. (2005). Systematic review of microsatellite instability and colorectal cancer prognosis. *Journal of Clinical Oncology: Official Journal of the American Society of Clinical Oncology, 23,* 609–618.

Ribic, C. M., Sargent, D. J., Moore, M. J., et al. (2003). Tumor microsatellite-instability status as a predictor of benefit from fluorouracil-based adjuvant chemotherapy for colon cancer. *The New England Journal of Medicine, 349,* 247–257.

Roth, A. D., Delorenzi, M., Tejpar, S., et al. (2012). Integrated analysis of molecular and clinical prognostic factors in stage II/III colon cancer. *Journal of the National Cancer Institute, 104,* 1635–1646.

Roth, A. D., Tejpar, S., Delorenzi, M., et al. (2010). Prognostic role of KRAS and BRAF in stage II and III resected colon cancer: results of the translational study on the PETACC-3, EORTC 40993, SAKK 60-00 trial. *Journal of Clinical Oncology: Official Journal of the American Society of Clinical Oncology, 28*(3), 466–474.

Samad, A. K., Taylor, R. S., Marshall, T., et al. (2005). A meta-analysis of the association of physical activity with reduced risk of colorectal cancer. *Colorectal Disease: The Official Journal of the Association of Coloproctology of Great Britain and Ireland, 7*, 204–213.

Sargent, D., Grothey, A., & Gray, R. (2011). Time to initiation of adjuvant chemotherapy and survival in colorectal cancer. *JAMA: The Journal of the American Medical Association, 306*(11), 1199–1200.

Sargent, D. J., Marsoni, S., Monges, G., et al. (2010). Defective mismatch repair as a predictive marker for lack of efficacy of fluorouracil-based adjuvant therapy in colon cancer. *Journal of Clinical Oncology: Official Journal of the American Society of Clinical Oncology, 28*, 3219–3226.

Sartore-Bianchi, A., Trusolino, L., Martino, C., et al. (2016). Dual-targeted therapy with trastuzumab and lapatinib in treatment-refractory, KRAS codon 12/13 wildtype, HER2-positive metastatic colorectal cancer (HERACLES): A proof-of-concept, multicentre, open-label, phase 2 trial. *The Lancet Oncology, 17*, 738–746.

Schmoll, H. J., Tabernero, J., Maroun, J., et al. (2015). Capecitabine plus oxaliplatin compared with fluorouracil/folinic acid as adjuvant therapy for stage III colon cancer: final results of the NO16968 randomized controlled phase III trial. *Journal of Clinical Oncology: Official Journal of the American Society of Clinical Oncology, 33*, 3733–3740.

Seymour, M. T., & Morton, D. (2019). FOxTROT: An international randomised controlled trial in 1052 patients (pts) evaluating neoadjuvant chemotherapy (NAC) for colon cancer. *JCO, 37*, 3504.

Sinicrope, F. A., Foster, N. R., Thibodeau, S. N., et al. (2011). DNA mismatch repair status and colon cancer recurrence and survival in clinical trials of 5-fluorouracil-based adjuvant therapy. *Journal of the National Cancer Institute, 103*, 863–875.

Sinicrope, F. A., Ou, F.-S., Zemla, T., et al. (2019). Randomized trial of standard chemotherapy alone or combined with atezolizumab as adjuvant therapy for patients with stage III colon cancer and deficient mismatch repair (ATOMIC, Alliance A021502). *JCO, 37*, e15169.

Sobrero, A. F., Andre, T., Meyerhardt, J. A., et al. (2020). Overall survival (OS) and long-term disease-free survival (DFS) of three vs six months of adjuvant (adj) oxaliplatin and fluoropyrimidine-based therapy for patients (pts) with stage III colon cancer (CC): Final results from the IDEA (International Duration Evaluation of Adj chemotherapy) collaboration. *JCO, 38*, 4004.

Taieb, J., Le Malicot, K., Shi, Q., et al. (2017). Prognostic value of BRAF and KRAS mutations in MSI and MSS stage III colon cancer. *Journal of the National Cancer Institute, 109*(5).

Tejpar, S., Saridaki, Z., Delorenzi, M., et al. (2011). Microsatellite instability, prognosis and drug sensitivity of stage II and III colorectal cancer: more complexity to the puzzle. *Journal of the National Cancer Institute, 103*, 841–844.

Tie, J., Cohen, J. D., Wang, Y., et al. (2019). Circulating tumor DNA analyses as markers of recurrence risk and benefit of adjuvant therapy for stage III colon cancer [published correction appears in JAMA Oncol. 2019 Dec 1;5(12):1811]. *JAMA Oncology, 5*(12), 1710–1717.

Verhoeff, S. R., van Erning, F. N., Lemmens, V. E., de Wilt, J. H., & Pruijt, J. F. (2016). Adjuvant chemotherapy is not associated with improved survival for all high-risk factors in stage II colon cancer. *International Journal of Cancer. Journal International du Cancer, 139*(1), 187–193.

Wells, K. O., Hawkins, A. T., Krishnamurthy, D. M., et al. (2017). Omission of adjuvant chemotherapy is associated with increased mortality in patients with T3N0 colon cancer with inadequate lymph node harvest. *Diseases of the Colon and Rectum, 60*, 15–21.

17

Radiotherapy for metastatic nodal disease in colorectal cancer

Melissa A. Frick, Phoebe Loo, Lucas K. Vitzthum, Erqi L. Pollom and Daniel T. Chang

Department of Radiation Oncology, Stanford University School of Medicine, Palo Alto, CA, United States

17.1 Introduction to radiation

Radiation therapy (RT), one of the main pillars of cancer treatment, works by using high-energy radiation to induce DNA damage, resulting in cellular death and mitotic failure. Ionizing radiation can act directly by causing double-stranded DNA breaks or indirectly through the formation of reactive oxygen species, which consequently generate abasic sites and single-stranded breaks of the DNA. While both cancerous and normal cells are subject to the ionizing effects of radiation, the aberrant signaling pathways of cancerous cells (that impart its malignant character) are ill-equipped to repair this radiation-induced DNA damage and are preferentially injured by radiation. Normal cells, however, have limits on the dose of radiation they can withstand before they are irreparably damaged and from which, clinically, manifest as toxicity. These normal tissue constraints often inform the maximum dose of radiation that can be safely administered within the course of radiation treatment.

A fundamental concept in radiation oncology is the therapeutic ratio, which denotes the relationship between the probability of tumor control and likelihood of normal tissue damage. The goal of RT is to maximize the dose to the tumor target while minimizing the dose to normal tissues. One of the traditional means to reduce normal tissue toxicity has been through division (fractionation) of the total radiation dose into smaller daily doses to allow normal tissues to repair radiation damage between treatments. Another method includes the use of an advanced form of three-dimensional conformal radiation therapy (3D-CRT) called intensity-modulated radiation therapy (IMRT). This modality is distinguished by the nonuniform intensity of radiation beams and computerized inverse planning, which consequently improves target conformity and enables the allocation of different dose targets (dose painting) in complex treatment plans.

Stereotactic body radiation therapy [SBRT, also known as stereotactic ablative radiation therapy (SABR)] is a radiation modality that delivers multiple radiation beams to well-defined targets in fewer but larger fractions (hypofractionation), all while greatly sparing adjacent healthy tissue. This approach has been enabled by a combination of technologies and techniques, including intensity modulation, image guidance, motion control, and stereotactic targeting, with the goal of delivering a compact ablative dose of radiation accurately and precisely to the intended target with steep dose gradients to avoid organs at risk (OARs) (Papież et al., 2003). Beyond offering advantages in dose escalation and normal tissue avoidance, SBRT is also thought to differ biologically from conventional radiation by altering vascular, stromal, and immune effects in addition to direct tumor cell killing (Kirkpatrick et al., 2008; Lee et al., 2009).

17.2 Rationale for local treatment of metastatic colorectal cancer

Our earliest evidence that suggests efficacy of local therapy for colorectal cancer (CRC) metastases derives from the surgical literature. Early in the history of surgical treatment of metastatic CRC, metastasectomy of liver-limited oligometastatic disease showed substantial curative potential with a reported median 5-year survival of ~40% for well-selected patients (Choti et al., 2002; Kanas et al., 2012; Kopetz et al., 2009; Pawlik et al., 2005; Van Cutsem et al., 2006). A randomized phase II trial demonstrated improved survival with aggressive local therapy using radiofrequency ablation (RFA) for unresectable liver metastases (Ruers et al., 2017). Using liver oligometastatic disease as a model, a similar management approach towards oligometastatic pulmonary disease has also become a standard practice (Gonzalez et al., 2015; Onaitis et al., 2009). It is from the collective results of these studies where surgical resection stands as the standard treatment recommendation for pulmonary or hepatic CRC metastases. Of note, the efficacy of pulmonary metastectomy for CRC was evaluated in the prospective randomized PulMiCC trial. While no significant difference was observed between treatment arms, this trial was underpowered and failed to accrue in large part due to the perceived lack of equipoise with omission of surgery (Treasure et al., 2019).

Yet, the assumed role of surgical resection as "standard" for nonpulmonary, extrahepatic CRC metastases remains controversial with no clear consensus guidelines. This is primarily owed to the rarity of CRC nodal metastases and consequent paucity of data. To quantify, nodal recurrence in the retroperitoneum (distinct from the bed of the primary) is estimated to occur in 5%–15% of patients following curative primary resection and if isolated, is limited to ~1% of all patients following curative surgery (Figueredo et al., 2003; Min et al., 2008; Shibata et al., 2002).

While the small set of surgical series that report on nodal metastatectomy have described favorable survival rates, these results are in part owed to a selection bias towards those candidates with resectable disease at the time of recurrence and who are considered fit enough to undergo an aggressive surgery. In these studies, 5-year survival rates have approached a maximum of 56% following complete resection, whereas incomplete resection resulted in survival rates between 0% and 7% (Bowne et al., 2005; Gwin & Sigurdson, 1993; Min et al., 2008; Shibata et al., 2002).

A significant proportion of patients with isolated nodal recurrences, however, may be unresectable. Series report between 24% and 84% of patients with isolated nodal recurrences are not resectable at presentation (Kim et al., 2020; Min et al., 2008). The conditions in which recurrences are often considered unresectable include encasement/involvement of vascular structures (i.e., aorta, superior mesenteric artery, celiac axis) as well as invasion of adjacent organs. Informed from their systematic review of management of paraaortic CRC nodal metastasis, Wong et al. (2016) suggest that surgical candidates include those with a primary tumor <5 cm (in the setting of synchronous nodal metastases), with two or less involved paraaortic lymph nodes, and a long disease-free interval in the setting of metachronous nodal metastases. Kim et al. (2020) define unresectability by multiple nodes involving both sides of the aorta and invasion of major vessels; of those patients with isolated paraaortic lymph node failure, 24% of patients were determined to be unresectable.

Even in the condition where lesions are localized and fit the technical definition of "resectable," surgical resection is not a widely accepted practice due to the relative rarity of this presentation and highly associated postoperative morbidity. Per some surgical series, postoperative morbidity reaches 30%, including abscess, phlebitis, pneumonia, intestinal obstruction, wound complications, venous thromboembolism, and bladder leakage (Min et al., 2008; Razik et al., 2014; Shibata et al., 2002; Yamada et al., 2019).

For those patients deemed medically inoperable or unresectable—in addition to those who refuse surgery—RT may be the only option of local therapy for metastatic nodal CRC. While chemotherapy may also be considered to provide systemic control, single-modality use in the recurrent setting is associated with poor outcomes, as overall survival only approaches 20 months (Kelly & Goldberg, 2005; Kopetz et al., 2009; Tepper et al., 2003).

Given the risk of morbidity, it is important to compare results of surgery with other modalities. The comparison of surgical resection versus nonsurgical management (including radiation) is limited to small, retrospective series. Min et al. (2008) reported on 38 patients with isolated paraaortic lymph node CRC recurrences. Of the six patients who underwent curative resection (note: only 16% of the evaluated cohort), the median survival after resection was 34 months versus 14 months for the 32 patients who did not undergo surgical resection. Notably, those who underwent surgical resection were more likely to have a lower carcinoembryonic antigen (CEA) at recurrence and longer disease-free interval than those who did not. Additionally, the investigators found that patients with disease located above the renal vessels were frequently unresectable due to the involvement of adjacent organs and/or vascular structures and were also observed to have worse survival outcomes.

In an additional retrospective report, Kim et al. (2009) detailed the management of isolated paraaortic lymph node recurrence for 46 patients (35 of which were classified as resectable), with 16 having undergone surgical resection with or without adjuvant chemotherapy/chemoradiotherapy (ChRT). Survival after recurrence was found to be significantly longer in patients who underwent surgical resection versus those managed without surgery (71 months vs 39 months, respectively; $P = .02$); however, the nonsurgical cohort was more likely to have aggressive tumors (in regard to rates of lymphovascular invasion and/or perineural invasion) and/or be in poorer general condition (Kim et al., 2020).

These two studies stand as examples of the innate biases in retrospective series that compare surgical interventions to nonsurgical interventions—as those patients who undergo nonsurgical management often have multiple reasons to avoid an invasive procedure (i.e., comorbidity, aggressiveness/invasiveness of tumor, location of tumor, etc.)—and why there should be caution in applying these results to clinical recommendations. The role of surgery in resectable metastatic disease has remained the standard of care, although the potential of postoperative morbidity and mortality must be taken into account when comparing its benefit to nonsurgical approaches. In such cases, the decision to proceed with surgical resection versus another modality should be made following multidisciplinary evaluation and consideration of the individual patients' personal values/goals, particularly as results of other modalities improve.

17.3 Experience with radiation

17.3.1 Radiation in the oligometastasic state

There is burgeoning evidence that supports the use of ablative radiation in oligometastatic disease, which is relevant to the discussion of RT in the treatment of CRC nodal metastases.

One of the first clinical trials that reported on the use of radiation for oligometastatic disease was a phase I dose escalation study investigating its use in those patients with five sites of extracranial metastases regardless of primary histology (~10% CRC primaries), which demonstrated a signal towards safety and potential efficacy (Salama et al., 2012). Since then, there have been multiple prospective randomized phase II studies that have reported benefits in disease control and overall survival when ablative radiation is added to standard therapy for the treatment of oligometastatic disease (Gomez et al., 2019; Iyengar et al., 2018; Palma et al., 2020). The multicenter phase II SABR-COMET trial enrolled 99 patients with one to five oligometastases from various primary histologies (18% CRC) and randomized patients to standard-of-care palliative treatments ± SBRT to all metastatic sites (Palma et al., 2020). Inclusion criteria selected for patients with excellent performance status [Eastern Cooperative Oncology Group (ECOG) 0–1], a life expectancy ≥ 6 months, and a controlled primary tumor. After a median follow-up of 51 months, there was a 22-month improvement in overall survival and a more than doubling of progression-free survival in the SBRT arm with no detriment of quality of life. Since then, a number of other prospective trials have demonstrated the benefit of aggressive local therapy for patients with oligometastatic disease for various other malignancies including prostate cancer and nasopharyngeal cancer, while more currently are underway (Phillips et al., 2020; You et al., 2020). Taken together, these data demonstrate that we may be in the midst of a paradigm shift where aggressive local therapy including RT may become the standard of care in patients with limited metastatic disease.

17.3.2 Radiation for CRC oligometastases

While none have stood at the level of evidence presented to us by SABR-COMET, there exist multiple prospective and retrospective series that report outcomes with SBRT for oligometastatic CRC. Largely speaking, these studies include unresectable or medically inoperable patients who are heavily pretreated with neoadjuvant chemotherapy and/or other local therapies (RFA, perfusion, or surgery) (Chang et al., 2011; Comito et al., 2014; Hoyer et al., 2006; McPartlin et al., 2017; van der Pool et al., 2010).

While 2-year overall survival rates have varied from 26% to 83%, it is important to note that these outcomes may depend on variations in selection criteria, many of which included multiple poor prognostic factors. For most series, SBRT was found to provide a favorable rate of local control on the order of 75%–80% at 2 years. Further, local control was often independently associated with increased total dose and/or biologic effective dose. Although a majority of these reports were largely confined to discussion of liver and/or lung CRC oligometastases, they establish a solid backbone of evidence demonstrating safety and efficacy of SBRT. These collective results have substantiated investigation of additional manifestations of metastatic CRC, including nodal disease, which lays at the crux of this discussion.

17.3.3 Radiation for colorectal nodal oligometastases

There is currently a paucity of published data on treatment outcome after delivery of radiotherapy to CRC nodal metastases. Much of the data come from reports with heterogeneous patient cohorts with various primary histologies, or from experiences derived from treatment of hepatic/pulmonary CRC metastases. In our discussion below, we include five published studies that exclusively include patients with CRC nodal metastases and report on their treatment outcomes using conventional radiation and/or ablative techniques (Table 17.1).

Yeo et al. (2010) report on their experience treating 22 patients with isolated retroperitoneal recurrence from CRC using nonablative radiotherapy techniques and concurrent chemotherapy. Of note, their treatment population was confined to patients for whom the extent of involved lymph nodes were limited to three vertebral heights. Most patients ($n = 20$) underwent treatment with 3D-CRT technique, where an elective nodal volume (5 cm above/below sites of gross disease) received 45 Gy in 25 fractions followed by a sequential boost to the involved nodes (plus a 5 mm margin) to a total of 63 Gy in 35 fractions ($n = 12$) or 55.8 Gy in 31 fractions ($n = 8$). There were two additional patients who received treatment with helical tomotherapy; the total dose to elective nodes was 44 Gy with an integrated boost of 60 Gy to the expanded involved node volume over 20 fractions. All patients received concurrent chemotherapy with the majority (%) receiving capecitabine, and the remaining receiving fluorouracil (5-FU) plus leucovorin, capecitabine plus oxaliplatin, or capecitabine plus irinotecan. Following ChRT, adjuvant chemotherapy was given to 73% of patients. Recurrence was reported in 15 (68%) patients although only 4 (%) patients experienced local (defined as infield relapse/progression) ± distant recurrence suggesting efficacy of local control with ChRT. Multivariable analysis identified infrarenal sites of nodal recurrence and normal CEA levels to be favorable prognostic factors for recurrence-free survival, the former of which is consistent with Min et al.'s surgical series (Min et al., 2008). Responses to ChRT and adjuvant chemotherapy were associated with better overall survival. While common terminology criteria for adverse events (CTCAE) grade 1–2 acute toxicity rates were over 80%, this was likely due to the less conformal nature of their external beam radiation therapy techniques, their generous volume definitions, and/or the addition of chemotherapy; no severe toxicities were reported.

TABLE 17.1 Comparison of treatment techniques and outcomes for nodal recurrence from colorectal cancer treated by radiotherapy.

Study	Design	N	Primary site	Failure site	Notable inclusion criteria	Median DFI	Treatment/ technique	RT dose	Median survival	Median PFS	Median local failure	Survival rate (years)	PFS rate (years)	Local control rate (years)	Reported toxicities
Kim et al. (2008)	Retrospective	23	All rectal	Pelvis	Progressed through ≥1 lines of ChT	32 months	SBRT, EBRT + SBRT boost	36–51 Gy (median 39 Gy) in 3 fractions (SBRT alone); 45 Gy in 25 fractions (EBRT) plus 16 Gy SBRT boost	37 months	55 months	23 months[a]	25% (4); 23% (5)	51% (4)	LPFS[a] 74% (4)	Grade 1/2 (39%); grade 4 (4%)
Kim et al. (2009); Choi et al. (2009)	Retrospective	7	All rectal	PALN	Progressed through ≥1 lines of ChT; 1–3 isolated paraaortic lesions	21 months	SBRT	36–51 Gy (median 48 Gy) in 3 fractions	37 months	NA	13 months[a]	71% (3)	NA	NA	Grade 1 (29%); grade 4 (14%)
Lee et al. (2015)	Retrospective	52	NA	RP	—	13 months	EBRT + ChT	31–88 EQD2 Gy (median 54 EQD2 Gy)	41 months	13 months	15 months[b]	70% (2)	38% (2)	LPFS[b] 37% (2); infield failure-free survival 69% (2)	Acute grade 1/2 (40%); late grade 1/2 (12%)
Yeo et al. (2010)	Phase II	22	Colon (41%); rectum (59%)	RP	Disease involvement limited to 3 vertebral heights	15 months	EBRT + ChT	63 Gy in 35 fractions; 55.8 Gy in 31 fractions; 60 Gy in 20 fractions	21 months	20 months	NA	65% (3); 36% (5)	34% (3); 26% (5)	LPFS[a] 76% (3)	Grade 1/2 (82%)
Franzese et al. (2017)	Retrospective	35 pts, 38 treatment courses, 47 lesions	Colon (74%); rectum (26%)	PALN, pelvis, other abdominal LN sites	—	45 months	SBRT	45 Gy in 6 fractions (majority); 36 Gy in 6 fractions (large minority)	Not reached	16 months	11 months[a]	100% (1); 81% (2); 81% (3)	70% (1); 33% (2); 19% (3)	LC[a] 85% (1); 75% (2); 75% (3)	Grade ½ (13%, per treatment course)

[a] Local relapse defined as progression/recurrence within the radiation treatment field.
[b] Local relapse defined as progression/recurrence in the abdominopelvic cavity, including both in-field and out-of-field failures.

Notes: ChT, chemotherapy; CRC, colorectal cancer; DFI, disease-free interval; EBRT, external beam radiation therapy; LN, lymph node(s); LC, local control; LPFS, local progression-free survival; NA, not assessed; PALN, paraaortic lymph node(s); pts, patients; RP, retroperitoneum; SBRT, stereotactic body radiotherapy.

Lee et al. (2015) also provide an informative report on their use of three-dimensional conventional radiation techniques in the treatment of 52 patients with isolated CRC nodal recurrence, 25 of which received upfront RT and 27 having deferred RT until after initiation of salvage chemotherapy. Radiation dose and fractionation were quite heterogeneous: while most patients received conventionally fractionated RT (79%), the minority received short-course RT (17%) or SBRT (2%). Local recurrence was the predominant pattern of relapse and was associated with a 2-Gy fraction equivalent total RT dose of ≤ 54 Gy. The authors, interestingly, defined local recurrence by progression within the abdominopelvic cavity, including both in-field and out-of-field locations. When examining in-field recurrence, failure-free survival was 69% at 2 years. Notably, out-of-field recurrences most commonly recurred in the cranial direction within 1.5 vertebral bodies and were more likely to occur when the tumor was located above the renal vein. These observations not only support the need for adequate high-dose radiation towards the involved nodes but also the role of generous margins when delineating elective nodal fields and particularly for tumors above the renal vein. Repeated radiation in the event of further out-of-field recurrence successfully salvaged a number of these patients, further emphasizing the importance of intensifying local treatment for isolated nodal recurrence. Timing of RT (upfront vs deferred) and response to salvage chemotherapy did not influence overall survival, which suggests that the decision to defer salvage RT until after salvage chemotherapy may be clinically judicious in those patients who are at high risk of failure [i.e., large gross tumor volume (GTV) >30 cc, disease-free interval <12 months] and/or whose tumor directly abuts bowel, so as to avoid radiation-related toxicity. Of note, acute and late toxicities were limited to grade 1–2 gastrointestinal complaints in 40% and 12% of patients, respectively.

Both reports from Lee et al. (2015) and Yeo et al. (2010) discuss outcomes largely pertaining to use of 3D-CRT techniques. While, overall, their outcomes are good, we must acknowledge the innate limitations of 3D-CRT and conventionally fractionated treatment schedules, as well as some unique features of CRC nodal metastases that favor treatment with stereotactic approaches.

First, there is increasing evidence to suggest that CRC metastases are more radioresistant than other cancer histologies. Takeda et al., (2011) compared clinical outcomes between the local control of oligometastatic lung lesions (21 of 44 lesions from CRC primaries) compared to primary lung cancers following treatment with SBRT, establishing a 1-year local control rate of 80%, 94%, and 97% ($P < .05$) for CRC primaries, oligometastases from other primaries, and lung primaries, respectively. A phase II study and European database report also support the importance of tumor genotype, demonstrating poorer local control rates ($\sim 60\%$ at 1 year) of CRC hepatic metastases versus other histologies following treatment with SBRT (Andratschke et al., 2018; Hong et al., 2017). Using a multigene expression index for tumor radiosensitivity, Ahmed et al. showed that metastatic colorectal adenocarcinomas scored as more radioresistant than other histologies such as anal squamous cell cancer, breast adenocarcinoma, and lung adenocarcinoma. These calculated differences corresponded to worse local control rates of liver metastases of CRC origin treated with SBRT at 12 and 24 months (79% and 59%, respectively), compared to 100% at both time points for non-CRC lesions ($P = .19$) (Ahmed et al., 2016), respectively, compared to 100% at both time points for non-CRC lesions ($P = .19$) (Ahmed et al., 2016). Among CRC metastases, those with a KRAS oncogene mutation have been shown to be particularly radioresistant (Hong et al., 2017).

This evidence of radioresistance of CRC metastases suggests the need for dose escalation as higher radiation doses have been associated with improved local control of radioresistant histologies (Ahmed et al., 2016; McCammon et al., 2009). Studies of SBRT to CRC hepatic and pulmonary metastases have demonstrated that increased dose leads to improved local control rates. However, the ability to deliver ablative radiation doses within the abdominopelvic region—the invariable location of many CRC nodal metastases—poses unique dosimetric challenges due to the proximity of radiosensitive luminal GI organs and other critical structures (spinal cord, liver, kidneys, etc.), all of which may limit dose escalation.

Previous reports using novel radiation techniques including IMRT and helical tomotherapy in the paraaortic region have demonstrated superiority in overcoming problems with critical organ radiation tolerance and dose uniformity (Hermesse et al., 2005; Pezner et al., 2006). Additional series have investigated the use of SBRT of nodal metastases of various primary histologies, including CRC. These, too, demonstrate both safety and efficacy of nodal irradiation in the abdominopelvic cavity. Choi et al. (2009) used SBRT in patients with isolated paraaortic lymph node metastasis from uterine cervical and corpus cancer, delivering a dose of 33–45 Gy over three fractions. Only one patient developed a late grade 3 radiation-related toxicity manifesting in ureteral stricture 20 months following treatment. Shahi et al. (2020) reported on 51 patients with 58 nonliver abdominopelvic oligometastases. A dose of 25–40 Gy was delivered in five fractions. One patient required hospitalization for grade 3 nausea and one late grade 1 vertebral compression of T12 was reported, however, was more likely related to an adjacent lymph node who had underwent SBRT 1 year previously. Finally, Rwigema et al. (2015) evaluated outcomes of 38 patients with 44 oligometastases in the abdomen/pelvis treated with 24–50 Gy in four to five fractions using SBRT. Two late toxicities were observed including colovesicular fistula in the setting of local tumor progression and a large volume of bowel (26.9 cc) receiving greater than 20 Gy; the second late toxicity was a grade 2 proctitis. This group concluded that SBRT doses of 40–50 Gy in five every-other-day fractions are efficacious and associated with minimal toxicity, if constraining the volume of bowel receiving 20 Gy to <20 cc.

Some early evidence regarding the utility of ablative radiation for CRC nodal metastases comes from the Korean experience, in which 23 patients with isolated pelvic nodal recurrence from rectal cancer were treated with SBRT (Kim et al., 2008). All patients had progressed through at least one line of salvage chemotherapy. Five patients with separate lymph nodes were treated with external beam RT (45 Gy in 25 fractions) followed by a single-fraction SBRT boost of 16 Gy. Eighteen patients were treated with SBRT alone, ranging from a total dose of 36–51 Gy (median 39 Gy), all in three fractions. Although these patients were enrolled on a dose escalation protocol with the intent to increase dose by 1 Gy/fraction, if at least five patients did not show grade 4 toxicity, the protocol was not applied strictly as the applied dose was influenced by irradiated volume. In follow-up, one patient who had received 51 Gy developed a rectal perforation (grade 4) that was managed surgically. Grade 1 or 2 acute toxicity (mostly nausea, vomiting, and/or pain) was reported in nine (39%) patients; no acute grade 3 toxicities nor late complications were reported. No complications occurred in four reirradiated patients who had initially received adjuvant concurrent ChRT. Overall median survival was 37 months, with 23% of patients alive 5 years following SBRT. Local progression-free survival was 74% at 4 years, while nearly half of patients remained free of any disease progression at 4 years. This preliminary study suggests that SBRT can be considered as a noninvasive local modality for isolated pelvic metastases in recurrence rectal cancer.

The same group published on their experienced of SBRT for isolated paraaortic lymph node recurrence from rectal cancer with a very limited case series of seven patients (Kim et al., 2009). All patients had progressed through at least one line of salvage chemotherapy for recurrence and had between one and three isolated paraaortic lymph node lesions. They received between 36 and 51 Gy (median 48 Gy) in three fractions. Overall median survival was 37 months, with 71% of patients alive 3 years following SBRT. Although the absolute recurrence rate was high (five of seven patients), only one patient recurred in the treatment field 13 months following SBRT. There was a low rate of mild acute toxicity (two of seven patients, 29%) and one patient experienced grade 4 toxicity relating to bowel obstruction requiring bypass surgery, which was attributed to the volume of normal bowel receiving high-dose radiation (3.6 cc received >45 Gy) and a high maximum point intestinal dose (53 Gy).

The most contemporary report detailing radiotherapy for CRC nodal metastases comes from an Italian experience, published by Franzese et al. in 2017, which included a cohort of 35 patients treated for 47 lesions over 38 treatment courses, involving either the paraaortic (51%), pelvic (26%), or other nodal sites within the abdomen (23%) (Franzese et al., 2017). There was no elective coverage of at-risk nodes. A majority of treatment courses (74%) were delivered by 45 Gy in six fractions with a larger minority receiving 36 Gy (13%) in six fractions. Overall survival was relatively high (81%) at 3 years. The rate of local in-field control was 75% at 3 years and was observed to have a median time to local relapse of 11 months. In-field relapse was observed in only six patients, while out-of-field lymph node progression was observed in 10 patients. Disease-free survival, however, was relatively low compared to other series (19%, 3 years), likely attributed to a preponderance of distant metastases (23 of 39 relapse events) that developed after a median of 7 months. Treatment was well tolerated with acute grade 1/2 events reported in 5 of 38 treatment courses, all in cases after paraaortic node treatments.

Although endpoints such as local control and overall survival are often highlighted in the SBRT literature, other relevant outcomes such as chemotherapy-free survival, treatment-intensification-free survival, and quality of life data are frequently underreported. Patients who are able to delay initiation or intensification of systemic therapy by receiving metastasis-directed therapy with SBRT may best preserve their current quality of life by forestalling additional chemotherapy-related morbidity. This endpoint remains even more true in those patients who poorly tolerate chemotherapy and/or targeted therapy.

Franzese et al. (2020) reported on their experience of a large, single-center cohort of 418 nodal oligometastases from 278 total patients (with various primary histologies) treated with SBRT and include in their results the endpoint of free-from treatment intensification (FFTI), calculated as the time from SBRT to initiation of intensification of systemic therapy. One- and 2-year actuarial FFTI was 83% and 75%, respectively. In those patients who reported an intensification of systemic therapy, median time to intensification was 8.4 months. For those patients who experienced progression in which retreatment with SBRT was feasible, 13% underwent repeat SBRT (some up to five times, until polymetastatic diffusion) and for whom median FFTI was extended to 14.6 months.

Other authors have also suggested the efficacy of additional SBRT in select cases of local-regional and/or out-of-field recurrence. Following SBRT for treatment of oligometastatic lymph node metastases, Loi et al. (2018) observed that—of those patients who progressed—more than half did so with limited involved sites (three metastases), allowing for continuation of local treatment in 41% of these selected patients. Shahi et al. (2020) also report that nearly half of patients with oligometastatic relapse/progression went on to receive a second course of SBRT, with 2- and 4-year chemotherapy-free survival rates of 47% and 37%, respectively. These experiences additionally support the role of SBRT in a strategy incorporating repeated use of local treatments for limited-extent relapse and prolonged chemotherapy-free survival.

Noting the above studies, these experiences make apparent that many patients with oligometastatic disease have undetectable micrometastases at the time of ablative treatment. As such, an important question to be answered is whether or not elective nodal regions adjacent to involved lymph nodes should be included in the treatment target volumes to prevent future regional recurrences. Detailed reports on patterns of recurrence following local treatment of lymph node metastases are needed.

17.3.4 Neoadjuvant radiotherapy

The use of radiotherapy, often considered in combination with concurrent chemotherapy, in the neoadjuvant setting is not well defined. Neoadjuvant chemoradiation was delivered to 16% of patients in the Shibata series and in 50% of Min et al.'s 2008 series (Min et al., 2008; Shibata et al., 2002). The indication for neoadjuvant therapy was not delineated in their reports, nor was its effect on subsequent surgical management. In their systematic review, Wong et al. (2016) suggest neoadjuvant ChRT should be reserved only for borderline resectable or unresectable patients in the setting of metachronous lymph node metastases. Potential benefits may include reduction in GTV and improving margin-negative resection. Further clarification regarding selection criteria, treatment parameters, and duration of therapy is needed.

There exists a singular experience from the Mayo Clinic that reports on the use of radiation within a trimodality treatment approach of isolated abdominal lymph node CRC metastases (Johnson et al., 2018). In this cohort of 65 patients, most patients received preoperative RT (median dose of 50.4 Gy) in conjunction with radiosensitizing chemotherapy followed by maximal tumor resection and intraoperative radiotherapy (IORT, median dose 12.5 Gy). About one-third of patients therein received adjuvant chemotherapy. Overall, 31% of patients were initiated on standard-of-care systemic therapy prior to being deemed candidates for trimodality therapy. Many patients were able to achieve sustainable long-term survival with a median overall survival of 55.4 months, an estimated 5-year survival rate of 45%, and recurrence-free survival of 40%. Additionally, trimodality therapy was well tolerated without any reports of grade 3 or 4 toxicity; radiation-related toxicity included three reports of IORT-related neuropathy and one of experienced ureteral fibrosis. We note this was a highly selected group of patients, as all patients were ECOG ≤1 (89% ECOG 0) and most of whom proceeded with curative-intent trimodality therapy at the time of isolated relapse. Conversely, it should be noted that most of the aforementioned RT studies included patients who had progressed on salvage chemotherapies and many of which were deemed inoperable due to invasive nature of relapse and/or medical comorbidities.

17.3.5 Prognostic factors

There have been multiple efforts to define which patient subgroups are expected to derive greatest benefit from local therapies and, conversely, for which patients the benefit of intervention is limited.

Early surgical series in the setting of hepatic and pulmonary CRC metastases have consistently identified several parameters to portend inferior outcomes, including multiple metastatic sites vs. a single site; a short disease-free interval from primary tumor resection; and elevated preoperative CEA (Casiraghi et al., 2011; Fong et al., 1999; Gonzalez et al., 2015; Pastorino et al., 1997; Rees et al., 2008). Although derived from surgical experiences, the aforementioned parameters remain applicable in informing which patients are expected to benefit most from radiation, as it too is considered a local therapy.

In regard to radiation-specific experiences, we must mostly rely on studies that report on the treatment of nodal metastases from various histologies. We, however, do note that the prognostic factors identified in the two CRC-specific radiation series corroborate the larger groups' findings (Table 17.2). A few trends become apparent. First, patients with long disease-free interval, small tumors, and solitary tumors appear to derive the largest treatment benefit in local control, disease-free survival, and overall survival with RT. Second, dose escalation is imperative to optimize local control and may contribute to a benefit in overall survival. Third, a short disease-free interval, large tumor size, elevated CEA, and suprarenal location are associated with poor overall and disease-free survival overall regardless of treatment. The presence of any of these risk factors may be a harbinger for micrometastatic disease and warrant close surveillance and/or addition of systemic therapy rather than aggressive local therapy. Finally, a history of oligoprogression, poor response to radiation, and/or prior receipt of chemotherapy are associated with poor local control, distant progression-free survival, and overall survival, which may potentially correlate to a resistance acquired from previous treatments, either via radiotherapy or chemotherapy.

17.3.6 Future directions

Multiple opportunities exist to expand and advance our collective understanding in the use of radiation for treatment of nodal CRC metastases.

First and foremost, we need prospectively enrolled patient cohorts to demonstrate safety and efficacy. As we recognize that CRC nodal metastases represents a rare phenomenon, multiinstitutional collaboration should be considered as a means to ensure adequate patient enrollment and robust conclusions. These cohorts will also be used to better define dose constraints and optimized dose regimen, particularly relevant in this setting due to the radioresistant nature of CRC metastases (Ahmed et al., 2016; McCammon et al., 2009; Takeda et al., 2011).

TABLE 17.2 Selected prognostic factors related to clinical outcomes following radiotherapy for nodal metastases.

Prognostic factor	Stratification	Comparative statistic	P-value	Reference
Overall survival				
Tumor size	GTV ≥ 30 mL vs <30 mL	aHR 3.49[a]	.03	Lee et al. (2015)
	Continuous variable	HR 1.01[a]	.02	Franzese et al. (2020)
Location	Above vs below renal vessels	aHR 4.61[a]	<.01	Lee et al. (2015)
Number of metastases	Solitary vs nonsolitary oligometastases	73% vs 51% (2 years)	.01	Ito et al. (2020)
Metastasis status	Metastasis in other organs vs isolated nodal metastasis	HR 1.91[a]	<.01	Franzese et al. (2020)
Mode of progression	Oligoprogression vs oligorecurrence	HR 2.15[b]	<.01	Franzese et al. (2020)
DFI	<12 months vs ≥ 12 months	aHR 2.79[a]	.01	Lee et al. (2015)
	≥8.5 months vs <8.5 months	73% vs 53% (2 years)	.04	Ito et al. (2020)
RT dose	≥75 Gy BED vs <75 Gy BED	HR 0.59[a]	.01	Franzese et al. (2020)
Response to RT	CR vs PR vs SD	92% vs 50% vs 0% (3 years)	.03	Yeo et al. (2010)
	Local failure vs local control	HR 3.06[a]	.01	Loi et al. (2018)
Progression/disease-free survival				
Tumor size	Per GTV increase of 1 cc	1% increase in progression probability	<.01	Jereczek-Fossa et al. (2014)
	Diameter ≥ 30 mm vs <30 mm	HR 2.59[b]	.05	Loi et al. (2018)
	PTV ≤17 cc vs >17 cc	65% vs 26% (4 years)	.04	Choi et al. (2009)
Location	Above vs below renal vessels	aHR 2.00[b]	.04	Lee et al. (2015)
	Above vs below renal vessels	0% vs 40% (3 years)	.04	Yeo et al. (2010)
Number of metastases	Solitary vs nonsolitary oligometastases	42% vs 0% (2 years)	<.01	Scorsetti et al. (2012)
CEA (ng/dL) before RT	≤5 vs >5	66.7% vs 0% (3 years)	<.01	Yeo et al. (2010)

DFI	<12 months vs ≥12 months	aHR 1.96[b]	.03	Lee et al. (2015)
	≥8.5 months vs <8.5 months	25% vs 14% (2 years)	.01	Ito et al. (2020)
Response to RT	CR vs PR vs SD	50% vs 17% vs 0% (3 years)	.01	Yeo et al. (2010)
Local (in-field) recurrence-free survival				
Tumor size	Diameter ≥30 mm vs <30 mm	HR 4.59[b]	.01	Loi et al. (2018)
	PTV ≤17 cc vs >17 cc	100% vs 34% (4 years)	<.01	Choi et al. (2009)
Number of metastases	Solitary vs nonsolitary oligometastases	70% vs 48% (2 years)	.01	Ito et al. (2020)
Mode of progression	Oligoprogression vs oligorecurrence	HR 2.10[b]	.04	Franzese et al. (2020)
DFI	<12 months vs ≥12 months	aHR 3.40[b]	.03	Lee et al. (2015)
RT dose	≤54 EQD2 Gy vs >54 EQD2 Gy	aHR 3.57[b]	.03	Lee et al. (2015)
	≥75 Gy BED vs <75 Gy BED	HR 0.46[b]	<.01	Franzese et al. (2020)
	≥60 Gy EQD2 vs <60 Gy EQD2	75% vs 45% (2 years)	<.01	Ito et al. (2020)
Distant progression-free survival				
Tumor size	GTV ≥30 mL vs <30 mL	aHR 2.79[b]	.01	Lee et al. (2015)
Mode of progression	Oligoprogression vs oligorecurrence	HR 2.24[b]	<.01	Franzese et al. (2020)
Previous chemotherapy	Received before SBRT vs chemonaive	HR 1.95[b]	<.01	Franzese et al. (2020)
	Lines of chemotherapy pre-SBRT	HR 1.15[b]	<.01	Franzese et al. (2020)
RT response	Nonresponder vs responder	aHR 2.46[b]	.03	Ito et al. (2020)

[a]*Calculated for relative risk of death.*
[b]*Calculated for relative risk of recurrence.*

Notes: aHR, Adjusted hazard ratio; BED, biologically effective dose; CR, complete response; DFI, disease-free interval; EQD2, equivalent dose in 2 Gy fractions; GTV, gross tumor volume; HR, hazard ratio; PR, partial response; PTV, planning treatment volume; RT, radiation therapy; SBRT, stereotactic body radiation therapy; SD, stable disease.

Second, particle therapy (i.e., proton or carbon ions) has yet to be explored in the treatment of nodal metastases. Compared with photon-based therapies, heavy-ion therapy may potentially increase effectiveness and safety due to its favorable depth−dose profile characterized by a low entry dose, a localized peak of energy deposition with rapid dose fall-off (the "Bragg peak"), and a consequent lack of exit dose. This could offer a dosimetric advantage by sparing adjacent anterior structures in the abdomen and pelvis including large and small bowel. In addition, carbon ions have a greater relative biological effectiveness that is promising for tumors, like CRC, traditionally considered radioresistant.

While a majority of experiences using particle therapy has been within the context of protracted fractionation, these modalities are capable of significant geometric sparing and are, in theory, ideal for delivering SBRT. Indeed, a meta-analysis of 74 conventional photon SBRT clinical studies and 9 hypofractionated particle beam therapy studies on non-small cell lung cancer suggests clinical benefit in using particle-based SBRT. In their multivariable analysis, particle-based SBRT/hypofractionated regimens were associated with improved local control and lower overall incidence of grade 3−5 toxicities (Chi et al., 2017). A case report on the first application of proton SBRT to treat synchronous bilateral renal cell carcinomas offers evidence that proton SBRT within the retroperitoneal space—like many cases of nodal CRC metastases—is feasible, efficacious, and associated with minimal toxicities (Frick et al., 2017).

Third, recent technologic innovations in radiation oncology offer ample opportunity to overcome the unique dilemmas of irradiating abdominopelvic metastases. Early evidence suggests that stereotactic magnetic resonance image (MRI)-guided adaptive RT can facilitate safe delivery of ablative radiation dose to metastases in the abdominopelvic region by utilizing on-table adaptive replanning based on daily changes in normal anatomy, with particular attention to gastrointestinal luminal structures. In a small retrospective analysis of 28 abdominopelvic oligometastases (32% CRC primary) treated on a MR linear accelerator with adaptive replanning, the authors were able to deliver a median prescribed dose of 50 Gy in 5 fractions (range, 30−50 Gy in 5−10 fractions) despite proximity to stomach and/or bowel with high rates of local control (96% at median follow-up of 6 months) and no severe toxicity (Chuong et al., 2020).

Biology-guided radiation therapy is an additional novel modality that uses real-time positron emission tomography (PET) radiotracer emissions for direct tumor tracking during radiation delivery. While this technology has not yet been used in clinical practice, preclinical studies have shown feasibility and improved dosimetric outcomes by means of reduced treatment volume, escalated dose delivery to gross tumor, and improved dose sparing of normal tissues (Fan et al., 2013; Liang et al., 2019; Yang et al., 2014). We eagerly await the application of these technologies towards the treatment of CRC nodal metastases.

17.4 Practical considerations in radiation and radiation techniques

We offer some practical considerations when evaluating patients for radiotherapy, as well as creating and implementing a radiation treatment plan. These recommendations are made through curation of the literature's reported experience and our institutional practice.

TABLE 17.3 Recommendations for patient selection.

Inclusion:	
	Control of primary site, by:
	Definitive therapy, OR
	≥3 months of systemic therapy
	Oligometastatic or oligoprogressive disease
	Gross disease burden enables safe radiation delivery
	Consider parameters of nodal size, volume, number, and location to critical structures
	Extranodal sites of active metastatic disease allowed, if amenable to ablative treatments/surgery or stable with systemic therapy
	Eastern Cooperative Oncology Group ≤2
Exclusion:	
	Time from curative treatment to recurrence ≤6 months
	Previous radiation treatment at site of progressive nodal disease
	Peritoneal disease/carcinomatosis

17.4.1 Patient selection

It is crucial to select appropriate patients for radiotherapy so to maximize benefit and minimize toxicity (Table 17.3). We first suggest that the primary site of disease be controlled either by initial definitive therapy or at least 3 months of systemic agents. The patient should have either oligometastatic or oligoprogressive disease, recognizing that the definition of oligometastatic disease has be variously defined in the literature with upper limits of 1, 3, 5, or 10 metastases. In regard to tumor size/volume, limited reports include recommendations of a single nodal conglomerate, two to three nodes in close proximity (<1 cm), and a maximum tumor diameter of 5 or 8 cm (Franzese et al., 2017; Kim et al., 2009). Our institutional practice does not place an explicit limit on tumor size/volume; rather, we consider the multiple parameters of nodal size, volume, number, and location to critical structures in evaluating whether a radiation may ultimately be delivered safely. If there are extranodal sites of active metastatic disease, patients can still be considered for radiotherapy if these other sites are amenable to ablative treatments/surgery or if these sites have remained stable while on systemic therapy. Patients with peritoneal disease likely have more extensive carcinomatosis and may not be good candidates for aggressive radiotherapy. Patients should have an adequate performance status which suggests more favorable prognosis. We generally suggest an ECOG score of ≤2.

Radiotherapy would not be encouraged in the case where the time from curative treatment to recurrence is less than 6 months, as this suggests an aggressive tumor biology and increased likelihood of micrometastatic disease elsewhere. Systemic therapy would be preferred in this situation. Previous radiation treatment in the area of progressive nodal disease would also exclude further consideration of radiotherapy. While a history of inflammatory bowel disease has been historically considered an exclusion criterion for consideration of radiation, modern evidence using newer radiation techniques suggest patients with IBD may safely undergo abdominal and pelvic RT (White et al., 2015).

The decision to continue systemic therapy through and beyond RT should be made in conjunction with medical oncologist, and we generally would recommend that all cases should be discussed at a multidisciplinary tumor board prior to initiating local therapy.

17.4.2 Simulation

Patients should undergo high-quality planning scans, ideally with coregistration of recent PET, MRI, or other diagnostic imaging scans to aide in target delineation. As abdominal tumors are often subject to respiratory motion, thorough assessment of the magnitude and direction of tumor motion is recommended. At our institution, we obtain computed tomography (CT) at during deep-inspiration breath hold (with contrast), at end expiration, and throughout the full respiratory cycle (four-dimensional CT) at the time of simulation to inform our target volumes and respiratory motion management. Other mechanical techniques such as abdominal compression can also be used to limit tumor motion. Immobilization devices (i.e., Wing Board, Alpha Gradle, Vac-Loc, etc.) should also be used to minimize setup errors and efficiency for treatment setup.

17.4.3 Treatment planning

The first step in defining treatment volumes is delineating the GTV on cross-sectional imaging. If treating the patient using breath-hold technique, the GTV is assumed as the clinical target volume (CTV). If using a respiratory-gated technique, an internal target volume is defined to encompass the target volume of selected expiratory respiratory phases. An additional isotropic expansion of 5–7 mm is added to define a planning treatment volume (PTV) to account for setup uncertainty. At our institution, we also define an elective nodal volume covering at-risk micrometastatic nodal spread, which generally includes the adjacent nodal regions or stations. In the case of the paraaortic lymph node metastases, the elective nodal target includes at least 1–2 vertebral levels superior and inferior to the involved lymph nodes.

The prescribed dose should not be uniform for all patients and depends on the target characteristics such as number of lesions, volume, and proximity to OARs. We recommend that the prescribed dose to the involved node PTV exceed a biologically effective dose (BED$_{10}$: a calculated measure of the true biological dose delivered influenced both by dose per fraction and total dose) of 100 Gy (i.e., 50 Gy in five fractions), delivered as an integrated boost to the elective nodal volume receiving a BED$_{10}$ of ~35–40 Gy. Our dose is prescribed to the periphery of the tumor with where we aim for 95% coverage of the PTV. Dose heterogeneity can exceed 15%–20% in the setting of both strict dose constraints and high conformality. Fractions should be separated with intervening days to decrease toxicity.

Plans are generated with two to three coplanar volumetric-modulated arcs using 6–10 mV flattening free filter (FFF) beams. The use of flattening filter-free (FFF) mode increases the dose rate and reduces the time of treatment, not only making treatment more convenient but also reducing the risk of patient movement due to discomfort or distraction (Scorsetti et al., 2012).

17.4.4 Treatment delivery

Verification of tumor position and movement at the time of treatment is also important. On-board kilovoltage imaging and cone-beam CT (CBCT) may be used although often provide poor soft-tissue resolution. Gated CBCT technique improves image resolution by reducing motion artifact (Kincaid et al., 2013). Fiducial markers can aid in target alignment and for real-time tracking of motion during treatment delivery. For this reason, fiducial placement should be requested if a patient is undergoing biopsy confirmation of nodal metastasis and there is any potential consideration for future radiotherapy.

17.4.5 Treatment-related toxicity/dose constraints

Reported early and late bowel SBRT-related toxicities include nausea, abdominal discomfort, loose stools, stricture, obstruction, ulceration, and perforation (Franzese et al., 2017; Kim et al., 2008; Lee et al., 2015; Yeo et al., 2010). Treatment-related toxicity is highly dependent upon which organs lay in adjacent to the nodal metastases, often including: spinal cord, stomach, small bowel, large bowel, liver, kidney, and/or rectum. In particular, the tolerance of the intestine is an important dose-limiting factor in SBRT for abdominopelvic targets. Respecting dose constraints should take precedence over maximizing PTV coverage.

17.4.6 Follow-up

Patients should have interval follow-up every 3 months for 2 years and every 6 months thereafter with physical examination, CEA measurement, and CT chest/abdomen/pelvis.

17.5 Case examples

17.5.1 Case 1

This is a 49-year-old man who was diagnosed 4 years prior with de novo metastatic moderately differentiated rectal adenocarcinoma (cT4N2M1) involving pelvic, retroperitoneal, upper mediastinal, and right supraclavicular lymph nodes, in addition to the liver and lung. He initially was treated with capecitabine (Xeloda) and oxaliplatin (XELOX) chemotherapy with tumor marker and radiographic response before transitioning to maintenance Xeloda plus Avastin. After completing 20 cycles of overall chemotherapy and having maintained a radiologic complete response for 3 months, he was placed on a chemotherapy break. Sixteen months later, he began to experience rectal discomfort. A subsequent colonoscopy demonstrated a partially obstructing circumferential mass 15 cm from the anal verge. Shortly thereafter, an MRI of the abdomen/pelvis demonstrated locally recurrent rectal cancer in the rectosigmoid junction invading the peritoneal reflection and one left perirectal lymph node. There was no evidence of further distant disease on CT restaging scans; he was now fully restaged as cT4N1M0. He underwent one cycle of XELOX before proceeding with chemoradiation with concurrent Xeloda, delivering 45 Gy to the mesorectum and pelvic nodes with sequential boost to 50.4 Gy to the mesorectum

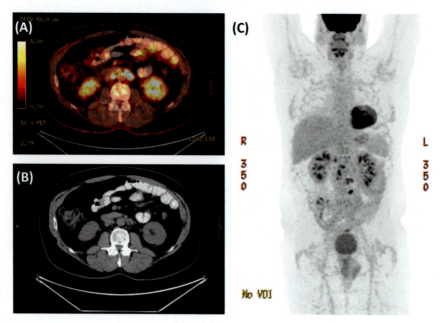

FIGURE 17.1 Pretreatment (A) axial positron emission tomography, (B) axial computed tomography fusion, and (C) maximum intensity projection for case 1.

and 54 Gy to the gross tumor, all in 1.8 Gy per fraction. He then underwent low anterior resection with pathology demonstrating a 3.0 cm adenocarcinoma invading through the muscularis propria with negative margins and 1 of 12 lymph nodes positive for disease, yielding ypT3N1a disease. Postsurgical imaging was without evidence of disease and further chemotherapy was not recommended.

Eight months later, a CT of the abdomen/pelvis showed increased size of aortocaval and left periaortic lymph nodes. Further work-up with a PET scan showed multiple new hypermetabolic aortocaval and left periaortic lymph nodes; there was no evidence of local recurrence. As he was found to be in an oligorecurrent state whereby no systemic disease was present and his pelvic disease was well controlled 4 years after his initial diagnosis, and SABR was recommended (Fig. 17.1).

At the time of simulation, the patient was placed in the supine position with arms positioned above his head. A Wing Board and Vac-Lok cushion were used as immobilization devices to minimize interfractional variability in patient positioning. A CT with intravenous contrast was obtained using inspiratory breath-hold technique. His most recent PET scan was fused to the CT simulation images to aid in delineating his target volumes using specialized computer software. GTV was defined as PET-avid nodal disease and was considered to be identical to the CTV. A 3 mm margin was added to account for small deviations in organ movement and interfractional position, defining the final PTV. An elective nodal volume was defined and covered the nodal basins surrounding the inferior vena cava and aorta, superiorly bounded by the T12/L1 vertebral interspace and inferiorly bounded by the aortic bifurcation. An isotropic 7 mm margin was added to this, again, to

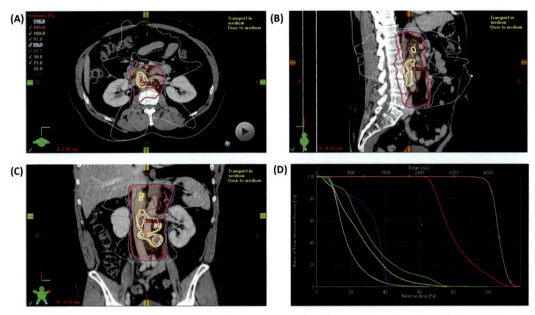

FIGURE 17.2 Stereotactic body radiation therapy treatment plan for case 1 with axial (A), sagittal (B), and coronal (C) views demonstrating treatment volumes and isodose lines. Treatment volumes: gross tumor volume nodes (*white*), planning treatment volume (PTV) prescribed 40 Gy (*pink*), PTV prescribed 25 Gy (*red*). Isodose lines: 42 Gy (*orange*), 40 Gy (*yellow*), 38 Gy (*green*), 36 Gy (*blue*), 25 Gy (*magenta*), 20 Gy (*cyan*), 10 Gy (*light green*). Dose–volume histogram is shown in (D), demonstrating adequate coverage of PTV40 (*pink*) and PTV25 (*red*) with dose sparing of spinal cord (*purple*), small bowel (*green*), stomach (*yellow*), right kidney (*cyan*), and left kidney (*orange*).

account for inter- and intrafractional positioning deviations. Plans were generated with two coplanar volumetric-modulated arcs using a 10 MV FFF beam. The plan was normalized so that 94% of the target volume received 100% of the prescribed dose. He was prescribed 25 Gy in five fractions to the elective lymph node volume with an integrated boost of 40 Gy to five paraaortic nodal conglomerates. Fractions were delivered every other day over the course of 1.5 weeks. Patient was treated with inspiratory breath hold with audiovisual biofeedback. In designing his radiation treatment plan, special attention was taken to avoid nearby OARs, principally the small bowel, kidneys, spinal cord, and stomach. His treatment volumes and dose constraints are demonstrated by Fig. 17.2. Daily orthogonal oblique X-ray and CBCT images were used to align the patient to nodal disease and ensure accurate position verification. He experienced grade 1 nausea managed by Zofran. No treatment breaks were necessary, nor did he report any weight loss.

Three months following SBRT, a PET/CT demonstrated overall decreased size and fluorodeoxyglucose (FDG) avidity of the previously radiated retroperitoneal lymph nodes, however, with increased size and avidity of a retrocrural lymph node and at least three new FDG-avid liver lesions concerning for new sites of metastatic disease. Given disease progression, he was initiated on XELOX. He is currently alive 6 months after SBRT. This case points to the risk of additional lymphatic disease as highlighted by Lee et al. (2015), which highlighted the risk of further regional failures superiorly along the paraaortic lymphatic chain.

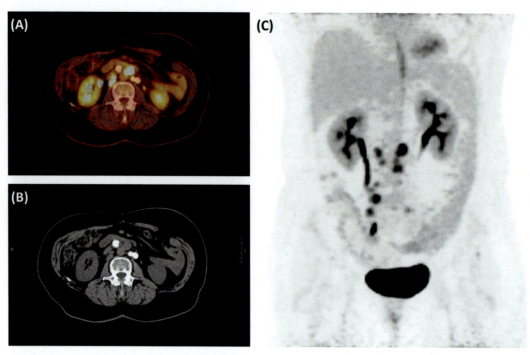

FIGURE 17.3 Pretreatment (A) axial positron emission tomography, (B) axial computed tomography fusion, and (C) maximum intensity projection for case 2.

17.5.2 Case 2

This is a 62-year-old woman who was diagnosed 3 years prior with de novo metastatic sigmoid colon adenocarcinoma (pT4bN2aM1b) involving retroperitoneal, mesenteric, right retrocrural, and paraesophageal lymph nodes, in addition to direct extension into the left psoas and iliacus muscles and liver metastases. She was initially taken to surgery for debulking of tumor following bowel perforation, undergoing open sigmoid colectomy with colostomy, small bowel resection and en bloc removal of left ovary and fallopian tube. She soon thereafter initiated leucovorin (folinic acid), fluorouracil, and oxaliplatin (FOLFOX) with initial partial response. Panitumumab was added to her chemotherapy regimen at cycle 6 due to KRAS wild-type status. Following 12 cycles of FOLFOX plus panitumumab with stable scans, she was transitioned to maintenance chemotherapy with panitumumab plus 5-FU. She remained on maintenance chemotherapy for nearly 2 years before she developed oligoprogressive retroperitoneal and pelvic lymphadenopathy. Multidisciplinary consensus did not recommend surgery but rather SBRT with the goal of allowing her to remain on her current well-tolerated systemic therapy while undergoing definitive local treatment with ablative radiation (Fig. 17.3).

At simulation, the patient was positioned in the supine position with arms positioned above her head. A Wing Board and Vac-Lok cushion were used as immobilization devices to minimize positional variation between treatment fractions. A CT with IV contrast and

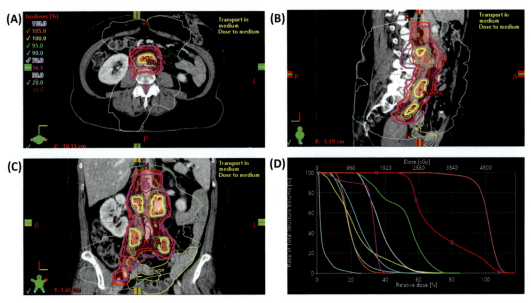

FIGURE 17.4 Stereotactic body radiation therapy treatment plan for case 1 with axial (A), sagittal (B), and coronal (C) views demonstrating treatment volumes and isodose lines. Treatment volumes: gross tumor volume nodes (*white*), planning treatment volume (PTV) prescribed 48 Gy (*pink*), PTV prescribed 27 Gy (*red*). Isodose lines: 50.4 Gy (*orange*), 48 Gy (*yellow*), 45.6 Gy (*green*), 43.2 (*cyan*) 33.6 Gy (*blue*), 25 Gy (*magenta*), 10 Gy (*light green*). Dose–volume histogram is shown in (D), demonstrating adequate coverage of PTV48 (*pink*) and PTV27 (*red*) with dose sparing of spinal cord (*magenta*), duodenum (*green*), lower small bowel (*yellow*), upper small bowel (*white*), large bowel (*cyan*), right kidney (*light green*), and left kidney (*orange*).

PET were obtained using inspiratory breath-hold technique and assisted in delineating her tumor treatment volumes. Her GTV was defined as PET-avid nodal disease; there was no marginal expansion for a clinical treatment volume. A 3 mm margin was added to account for small deviations in organ movement and interfractional position, defining the final PTV. A clinical elective nodal volume covered at-risk nodal basins. It was superiorly bounded 3 cm above the cranial-most PET-avid lymph node and extended inferiorly into her pelvis, bounded asymmetrically 3 cm below the inferior-most involved lymph nodes on either side of her iliac vessels. A 5 mm margin was added to create her elective nodal PTV. Her treatment plan was generated with the use of three coplanar volumetric-modulated arcs and a 10 MV FFF beam. She was prescribed 27 Gy in six fractions to the elective lymph node volume with an integrated boost of 48 Gy to her paraaortic and pelvic nodal conglomerates. The plan was normalized so that 65% of the target volume received 100% of the prescribed dose so to meet dose constraints of nearby OARs, principally the bowel, liver, kidneys, and spinal cord (Fig. 17.4). Fractions were delivered every other day over the course of 2 weeks. She was treated while maintaining a voluntary deep inspiratory breath hold, aided with audiovisual biofeedback. Daily orthogonal oblique X-ray and CBCT images were used to align the patient to nodal disease and ensure accurate position verification. During her treatment course, she experienced grade 2 nausea managed by Zofran. She did not require any treatment breaks.

This patient remains without further toxicity at 1 month following completion of radiotherapy. She continues with her maintenance chemotherapy regimen of panitumumab plus 5-FU, having just receiving her 67th overall chemotherapy cycle. She will be undergoing PET and a CT of the chest, abdomen, and pelvis at 3 months after completion of radiotherapy to assess response to RT and overall disease status.

References

Ahmed, K. A., Caudell, J. J., El-Haddad, G., Berglund, A. E., Welsh, E. A., Yue, B., ... Torres-Roca, J. (2016). Radiosensitivity differences between liver metastases based on primary histology suggest implications for clinical outcomes after stereotactic body radiation therapy. *International Journal of Radiation Oncology, Biology, Physics*, 95(5), 1399–1404. Available from https://doi.org/10.1016/j.ijrobp.2016.03.050.

Andratschke, N., Alheid, H., Allgäuer, M., Becker, G., Blanck, O., Boda-Heggemann, J., ... Habermehl, D. (2018). The SBRT database initiative of the German Society for Radiation Oncology (DEGRO): Patterns of care and outcome analysis of stereotactic body radiotherapy (SBRT) for liver oligometastases in 474 patients with 623 metastases. *BMC Cancer*, 18(1), 283. Available from https://doi.org/10.1186/s12885-018-4191-2.

Bowne, W. B., Lee, B., Wong, W. D., Ben-Porat, L., Shia, J., Cohen, A. M., ... Weiser, M. R. (2005). Operative salvage for locoregional recurrent colon cancer after curative resection: An analysis of 100 cases. *Diseases of the Colon and Rectum*, 48(5), 897–909. Available from https://doi.org/10.1007/s10350-004-0881-8.

Casiraghi, M., De Pas, T., Maisonneuve, P., Brambilla, D., Ciprandi, B., Galetta, D., ... Spaggiari, L. (2011). A 10-year single-center experience on 708 lung metastasectomies: The evidence of the "international registry of lung metastases". *Journal of Thoracic Oncology*, 6(8), 1373–1378. Available from https://doi.org/10.1097/JTO.0b013e3182208e58.

Chang, D. T., Swaminath, A., Kozak, M., Weintraub, J., Koong, A. C., Kim, J., ... Schefter, T. E. (2011). Stereotactic body radiotherapy for colorectal liver metastases: A pooled analysis. *Cancer*, 117(17), 4060–4069. Available from https://doi.org/10.1002/cncr.25997.

Chi, A., Chen, H., Wen, S., Yan, H., & Liao, Z. (2017). Comparison of particle beam therapy and stereotactic body radiotherapy for early stage non-small cell lung cancer: A systematic review and hypothesis-generating meta-analysis. *Radiotherapy & Oncology*, 123(3), 346–354. Available from https://doi.org/10.1016/j.radonc.2017.05.007.

Choi, C. W., Cho, C. K., Yoo, S. Y., Kim, M. S., Yang, K. M., Yoo, H. J., ... Kim, B. J. (2009). Image-guided stereotactic body radiation therapy in patients with isolated para-aortic lymph node metastases from uterine cervical and corpus cancer. *International Journal of Radiation Oncology, Biology, Physics*, 74(1), 147–153. Available from https://doi.org/10.1016/j.ijrobp.2008.07.020.

Choti, M. A., Sitzmann, J. V., Tiburi, M. F., Sumetchotimetha, W., Rangsin, R., Schulick, R. D., ... Cameron, J. L. (2002). Trends in long-term survival following liver resection for hepatic colorectal metastases. *Annals of Surgery*, 235(6), 759–766. Available from https://doi.org/10.1097/00000658-200206000-00002.

Chuong, M. D., Herrera, R., Contreras, J., Kotecha, R., Kalman, N. S., Garcia, J., ... Hall, M. D. (2020). Ablative dose prescribed to oligometastases near gastrointestinal luminal structures is well tolerated using stereotactic magnetic resonance image-guided adaptive radiation therapy (SMART). *International Journal of Radiation Oncology, Biology, Physics*, 108(3), e181. Available from https://doi.org/10.1016/j.ijrobp.2020.07.1393.

Comito, T., Cozzi, L., Clerici, E., Campisi, M. C., Liardo, R. L. E, Navarria, P., ... Scorsetti, M. (2014). Stereotactic ablative radiotherapy (SABR) in inoperable oligometastatic disease from colorectal cancer: A safe and effective approach. *BMC Cancer*, 14(1), 619. Available from https://doi.org/10.1186/1471-2407-14-619.

Fan, Q., Nanduri, A., Yang, J., Yamamoto, T., Loo, B., Graves, E., ... Mazin, S. (2013). Toward a planning scheme for emission guided radiation therapy (EGRT): FDG based tumor tracking in a metastatic breast cancer patient. *Medical Physics*, 40(8), 081708. Available from https://doi.org/10.1118/1.4812427.

Figueredo, A., Rumble, R. B., Maroun, J., Earle, C. C., Cummings, B., McLeod, R., ... Zwaal, C. (2003). Follow-up of patients with curatively resected colorectal cancer: A practice guideline. *BMC Cancer*, 3, 26. Available from https://doi.org/10.1186/1471-2407-3-26.

Fong, Y., Fortner, J., Sun, R. L., Brennan, M. F., & Blumgart, L. H. (1999). Clinical score for predicting recurrence after hepatic resection for metastatic colorectal cancer: Analysis of 1001 consecutive cases. *Annals of Surgery*, *230*(3), 309–318. Available from https://doi.org/10.1097/00000658-199909000-00004, discussion 318–321.

Franzese, C., Badalamenti, M., Comito, T., Franceschini, D., Clerici, E., Navarria, P., ... Scorsetti, M. (2020). Assessing the role of Stereotactic Body Radiation Therapy in a large cohort of patients with lymph node oligometastases: Does it affect systemic treatment's intensification? *Radiotherapy & Oncology*, *150*, 184–190. Available from https://doi.org/10.1016/j.radonc.2020.06.029.

Franzese, C., Fogliata, A., Comito, T., Tozzi, A., Iftode, C., Clerici, E., & Scorsetti, M. (2017). Stereotactic/hypofractionated body radiation therapy as an effective treatment for lymph node metastases from colorectal cancer: An institutional retrospective analysis. *The British Journal of Radiology*, *90*(1079), 20170422. Available from https://doi.org/10.1259/bjr.20170422.

Frick, M. A., Chhabra, A. M., Lin, L., & Simone, C. B. (2017). First ever use of proton stereotactic body radiation therapy delivered with curative intent to bilateral synchronous primary renal cell carcinomas. *Cureus*, *9*(10), e1799. Available from https://doi.org/10.7759/cureus.1799.

Gomez, D. R., Tang, C., Zhang, J., Blumenschein, G. R., Hernandez, M., Lee, J. J., & Heymach, J. V. (2019). Local consolidative therapy vs. maintenance therapy or observation for patients with oligometastatic non-small-cell lung cancer: Long-term results of a multi-institutional, phase II, randomized study. *Journal of Clinical Oncology*, *37*(18), 1558–1565. Available from https://doi.org/10.1200/JCO.19.00201.

Gonzalez, M., Poncet, A., Combescure, C., Robert, J., Beat Ris, H., & Gervaz, P. (2015). Risk factors for survival after lung metastasectomy in colorectal cancer patients: Systematic review and meta-analysis. *Annals of Surgical Oncology*, *11*(2 Suppl.), 31–33. Available from https://doi.org/10.2217/fon.14.259.

Gwin, J. L., & Sigurdson, E. R. (1993). Surgical considerations in nonhepatic intra-abdominal recurrence of carcinoma of the colon. *Seminars in Oncology*, *20*(5), 520–527.

Hermesse, J., Devillers, M., Deneufbourg, J.-M., & Nickers, P. (2005). Can intensity-modulated radiation therapy of the paraaortic region overcome the problems of critical organ tolerance? *Strahlentherapie und Onkologie*, *181*(3), 185–190. Available from https://doi.org/10.1007/s00066-005-1324-8.

Hong, T. S., Wo, J. Y., Borger, D. R., Yeap, B. Y., McDonnell, E. I., Willers, H., ... Zhu, A. X. (2017). Phase II study of proton-based stereotactic body radiation therapy for liver metastases: Importance of tumor genotype. *Journal of the National Cancer Institute*, *109*(9). Available from https://doi.org/10.1093/jnci/djx031.

Hoyer, M., Roed, H., Hansen, A. T., Ohlhuis, L., Petersen, J., Nellemann, H., ... Von der Maase, H. (2006). Phase II study on stereotactic body radiotherapy of colorectal metastases. *Acta Oncologica*, *45*(7), 823–830. Available from https://doi.org/10.1080/02841860600904854.

Ito, M., Kodaira, T., Koide, Y., Okuda, T., Mizumatsu, S., Oshima, Y., ... Suzuki, K. (2020). Role of high-dose salvage radiotherapy for oligometastases of the localised abdominal/pelvic lymph nodes: A retrospective study. *BMC Cancer*, *20*(1), 540. Available from https://doi.org/10.1186/s12885-020-07033-7.

Iyengar, P., Wardak, Z., Gerber, D. E., Tumati, V., Ahn, C., Hughes, R. S., ... Timmerman, R. D. (2018). Consolidative radiotherapy for limited metastatic non-small-cell lung cancer: A phase 2 randomized clinical trial. *JAMA Oncology*, *4*(1), e173501. Available from https://doi.org/10.1001/jamaoncol.2017.3501.

Jereczek-Fossa, B. A., Piperno, G., Ronchi, S., Catalano, G., Fodor, C., Cambria, R., ... Orecchia, R. (2014). Linac-based stereotactic body radiotherapy for oligometastatic patients with single abdominal lymph node recurrent cancer. *American Journal of Clinical Oncology*, *37*(3), 227–233. Available from https://doi.org/10.1097/COC.0b013e3182610878.

Johnson, B., Jin, Z., Haddock, M. G., Hallemeier, C. L., Martenson, J. A., Smoot, R. L., ... Grothey, A. (2018). A curative-intent trimodality approach for isolated abdominal nodal metastases in metastatic colorectal cancer: Update of a single-institutional experience. *The Oncologist*, *23*(6), 679–685. Available from https://doi.org/10.1634/theoncologist.2017-0456.

Kanas, G. P., Taylor, A., Primrose, J. N., Langeberg, W. J., Kelsh, M. A., Mowat, F. S., ... Poston, G. (2012). Survival after liver resection in metastatic colorectal cancer: Review and meta-analysis of prognostic factors. *Clinical Epidemiology*, *4*, 283–301. Available from https://doi.org/10.2147/CLEP.S34285.

Kelly, H., & Goldberg, R. M. (2005). Systemic therapy for metastatic colorectal cancer: Current options, current evidence. *Journal of Clinical Oncology*, *23*(20), 4553–4560. Available from https://doi.org/10.1200/JCO.2005.17.749.

Kim, M.-S., Cho, C. K., Yang, K. M., Lee, D. H., Moon, S. M., & Shin, Y. J. (2009). Stereotactic body radiotherapy for isolated paraaortic lymph node recurrence from colorectal cancer. *World Journal of Gastroenterology*, *15*(48), 6091–6095. Available from https://doi.org/10.3748/wjg.15.6091.

Kim, M.-S., Choi, C., Yoo, S., Cho, C., Seo, Y., Ji, Y., ... Kang, H. (2008). Stereotactic body radiation therapy in patients with pelvic recurrence from rectal carcinoma. *Japanese Journal of Clinical Oncology, 38*(10), 695−700. Available from https://doi.org/10.1093/jjco/hyn083.

Kim, Y. I., Park, I. J., Park, J.-H., Ro, J-S., Lim, S-B., Yu, C. S., & Kim, J. C. (2020). Management of isolated para-aortic lymph node recurrence after surgery for colorectal cancer. *Annals of Surgical Treatment and Research, 98*(3), 130−138. Available from https://doi.org/10.4174/astr.2020.98.3.130.

Kincaid, R. E., Yorke, E. D., Goodman, K. A., Rimner, A., Wu, A. J., & Mageras, G. S. (2013). Investigation of gated cone-beam CT to reduce respiratory motion blurring. *Medical Physics, 40*(4), 041717. Available from https://doi.org/10.1118/1.4795336.

Kirkpatrick, J. P., Meyer, J. J., & Marks, L. B. (2008). The linear-quadratic model is inappropriate to model high dose per fraction effects in radiosurgery. *Seminars in Radiation Oncology, 18*(4), 240−243. Available from https://doi.org/10.1016/j.semradonc.2008.04.005.

Kopetz, S., Chang, G. J., Overman, M. J., Eng, C., Sargent, D. J., Larson, D. W., ... McWilliams, R. R. (2009). Improved survival in metastatic colorectal cancer is associated with adoption of hepatic resection and improved chemotherapy. *Journal of Clinical Oncology, 27*(22), 3677−3683. Available from https://doi.org/10.1200/JCO.2008.20.5278.

Lee, J., Chang, J. S., Shin, S. J., Lim, J. S., Keum, K. C., Kim, N. K., ... Koom, W. S. (2015). Incorporation of radiotherapy in the multidisciplinary treatment of isolated retroperitoneal lymph node recurrence from colorectal cancer. *Annals of Surgical Oncology, 22*(5), 1520−1526. Available from https://doi.org/10.1245/s10434-014-4363-5.

Lee, Y., Auh, S. L., Wang, Y., Burnette, B., Wang, Y., Meng, Y., ... Fu, Y-X. (2009). Therapeutic effects of ablative radiation on local tumor require CD8+ T cells: Changing strategies for cancer treatment. *Blood, 114*(3), 589−595. Available from https://doi.org/10.1182/blood-2009-02-206870.

Liang, J., Silva, A. D., Han, C., Neylon, J., Amini, A., Sampath, S., ... Wong, J. Y. C. (2019). Biology-guided radiotherapy for lung sbrt reduces planning target volume and organs at risk doses. *International Journal of Radiation Oncology, Biology, Physics, 105*(1), S254. Available from https://doi.org/10.1016/j.ijrobp.2019.06.2468.

Loi, M., Frelinghuysen, M., Klass, N. D., Oomen-De Hoop, E., Granton, P. V., Aerts, J., & Nuyteens, J. (2018). Locoregional control and survival after lymph node SBRT in oligometastatic disease. *Clinical & Experimental Metastasis, 35*(7), 625−633. Available from https://doi.org/10.1007/s10585-018-9922-x.

McCammon, R., Schefter, T. E., Gaspar, L. E., Zaemisch, R., Gravdahl, D., & Kavanagh, B. (2009). Observation of a dose-control relationship for lung and liver tumors after stereotactic body radiation therapy. *International Journal of Radiation Oncology, Biology, Physics, 73*(1), 112−118. Available from https://doi.org/10.1016/j.ijrobp.2008.03.062.

McPartlin, A., Swaminath, A., Wang, R., Pintilie, M., Brierley, J., Kim, J., ... Dawson, L. A. (2017). Long-term outcomes of phase 1 and 2 studies of SBRT for hepatic colorectal metastases. *International Journal of Radiation Oncology, Biology, Physics, 99*(2), 388−395. Available from https://doi.org/10.1016/j.ijrobp.2017.04.010.

Min, B. S., Kim, N. K., Sohn, S. K., Cho, C. H., Lee, K. Y., & Baik, S. H. (2008). Isolated paraaortic lymph-node recurrence after the curative resection of colorectal carcinoma. *Journal of Surgical Oncology, 97*(2), 136−140. Available from https://doi.org/10.1002/jso.20926.

Onaitis, M. W., Petersen, R. P., Haney, J. C., Saltz, L., Park, B., Flores, R., ... Downey, R. (2009). Prognostic factors for recurrence after pulmonary resection of colorectal cancer metastases. *The Annals of Thoracic Surgery, 87*(6), 1684−1688. Available from https://doi.org/10.1016/j.athoracsur.2009.03.034.

Palma, D. A., Olson, R., Harrow, S., Gaede, S., Louie, A. V., Haasbeek, C., ... Senan, S. (2020). Stereotactic ablative radiotherapy for the comprehensive treatment of oligometastatic cancers: Long-term results of the SABR-COMET phase II randomized trial. *Journal of Clinical Oncology, 38*(25), 2830−2838. Available from https://doi.org/10.1200/JCO.20.00818.

Papież, L., Timmerman, R., Desrosiers, C., & Randall, M. (2003). Extracranial stereotactic radioablation physical principles. *Acta Oncologica, 42*(8), 882−894. Available from https://doi.org/10.1080/02841860310013490.

Pastorino, U., Buyse, M., Friedel, G., Ginsberg, R. J., Gerard, P., Goldstraw, P., ... Putnam Jr, J. B. (1997). Long-term results of lung metastasectomy: Prognostic analyses based on 5206 cases. *The Journal of Thoracic and Cardiovascular Surgery, 113*(1), 37−49. Available from https://doi.org/10.1016/s0022-5223(97)70397-0.

Pawlik, T. M., Scoggins, C. R., Zorzi, D., Abdalla, E. K., Andres, A., Eng, C., ... Vauthey, J-N. (2005). Effect of surgical margin status on survival and site of recurrence after hepatic resection for colorectal metastases. *Annals of Surgery*, *241*(5), 715−722. Available from https://doi.org/10.1097/01.sla.0000160703.75808.7d, discussion 722−724.

Pezner, R. D., Liu, A., Han, C., Chen, Y-J., Schultheiss, T. E., & Wong, J. Y. C (2006). Dosimetric comparison of helical tomotherapy treatment and step-and-shoot intensity-modulated radiotherapy of retroperitoneal sarcoma. *Radiotherapy & Oncology*, *81*(1), 81−87. Available from https://doi.org/10.1016/j.radonc.2006.08.025.

Phillips, R., Shi, W. Y., Deek, M., Radwan, N., Lim, S. J., Antonarakis, E. S., ... Tran, P. T. (2020). Outcomes of observation vs stereotactic ablative radiation for oligometastatic prostate cancer: The ORIOLE phase 2 randomized clinical trial. *JAMA Oncology*, *6*(5), 650−659. Available from https://doi.org/10.1001/jamaoncol.2020.0147.

Razik, R., Zih, F. S. W., Haase, E., Mathieson, A., Sandhu, L., Cummings, B., ... Swallow, C. J. (2014). Long-term outcomes following resection of retroperitoneal recurrence of colorectal cancer. *European Journal of Surgical Oncology*, *40*(6), 739−746. Available from https://doi.org/10.1016/j.ejso.2013.10.008.

Rees, M., Tekkis, P. P., Welsh, F. K. S., O'Rourke, T., & John, T. G. (2008). Evaluation of long-term survival after hepatic resection for metastatic colorectal cancer: A multifactorial model of 929 patients. *Annals of Surgery*, *247*(1), 125−135. Available from https://doi.org/10.1097/SLA.0b013e31815aa2c2.

Ruers, T., Van Coevorden, F., Punt, C. J. A., Pierie, J-P. E. N., Borel-Rinkes, I., Ledermann, J. A., ... Nordlinger, B. European Organisation for Research of Treatment of Cancer Gastro-Intestinal Tract Cancer Group. Arbeitsgruppe Lebermetastasen und tumoren in der Chirurgischen Arbeitsgemeinschaft Onkologie. National Cancer Research Institute Colorectal Clinical Study Group. (2017). Local treatment of unresectable colorectal liver metastases: Results of a randomized phase II trial. *Journal of the National Cancer Institute*, *109*(djx015). Available from https://doi.org/10.1093/jnci/djx015.

Rwigema, J.-C. M., King, C., Wang, P.-C., Kamrava, M., Kupelian, P., Steinberg, M. L., & Lee, P. (2015). Stereotactic body radiation therapy for abdominal and pelvic oligometastases: Dosimetric targets for safe and effective local control. *Practical Radiation Oncology*, *5*(3), e183−e191. Available from https://doi.org/10.1016/j.prro.2014.09.006.

Salama, J. K., Hasselle, M. D., Chmura, S. J., Malik, R., Mehta, N., Yenice, K. M., ... Weichselbaum, R. R. (2012). Stereotactic body radiotherapy for multisite extracranial oligometastases: Final report of a dose escalation trial in patients with 1 to 5 sites of metastatic disease. *Cancer*, *118*(11), 2962−2970. Available from https://doi.org/10.1002/cncr.26611.

Scorsetti, M., Alongi, F., Fogliata, A., Pentimalli, S., Navarria, P., Lobefalo, F., ... Tinterri, C. (2012). Phase I-II study of hypofractionated simultaneous integrated boost using volumetric modulated arc therapy for adjuvant radiation therapy in breast cancer patients: A report of feasibility and early toxicity results in the first 50 treatments. *Radiation Oncology*, *7*, 145. Available from https://doi.org/10.1186/1748-717X-7-145.

Shahi, J., Peng, J., Donovan, E., Vansantvoort, J., Wong, R., Tsakiridis, T., ... Swaminath, A. (2020). Overall and chemotherapy-free survival following stereotactic body radiation therapy for abdominopelvic oligometastases. *Journal of Medical Imaging and Radiation Oncology*, *64*(4), 563−569. Available from https://doi.org/10.1111/1754-9485.13057.

Shibata, D., Paty, P. B., Guillem, J. G., Wong, W. D., & Cohen, A. M. (2002). Surgical management of isolated retroperitoneal recurrences of colorectal carcinoma. *Diseases of the Colon and Rectum*, *45*(6), 795−801. Available from https://doi.org/10.1007/s10350-004-6300-3.

Takeda, A., Kunieda, E., Ohashi, T., Aoki, Y., Koike, N., & Takeda, T. (2011). Stereotactic body radiotherapy (SBRT) for oligometastatic lung tumors from colorectal cancer and other primary cancers in comparison with primary lung cancer. *Radiotherapy & Oncology*, *101*(2), 255−259. Available from https://doi.org/10.1016/j.radonc.2011.05.033.

Tepper, J. E., O'Connell, M., Hollis, D., Niedzwiecki, D., Cooke, E., & Mayer, R. L. Intergroup Study 0114. (2003). Analysis of surgical salvage after failure of primary therapy in rectal cancer: Results from Intergroup Study 0114. *Journal of Clinical Oncology*, *21*(19), 3623−3628. Available from https://doi.org/10.1200/JCO.2003.03.018.

Treasure, T., Farewell, V., Macbeth, F., Monson, K., Williams, N. R., & Brew-Graves, C., ... PulMiCC Trial Group. (2019). Pulmonary Metastasectomy vs Continued Active Monitoring in Colorectal Cancer (PulMiCC): A multicentre randomised clinical trial. *Trials*, *20*(1), 718. Available from https://doi.org/10.1186/s13063-019-3837-y.

Van Cutsem, E., Nordlinger, B., Adam, R., Kohne, C-H., Pozzo, C., Poston, G., ... Rougier, P. (2006). Towards a pan-European consensus on the treatment of patients with colorectal liver metastases. *European Journal of Cancer*, *42*(14), 2212−2221. Available from https://doi.org/10.1016/j.ejca.2006.04.012.

van der Pool, A. E. M., Méndez Romero, A., Wunderink, W., Heijmen, B. J., Levendag, P. C., Verhoef, C., & Ijzermans, J. N. M. (2010). Stereotactic body radiation therapy for colorectal liver metastases. *The British Journal of Surgery*, 97(3), 377−382. Available from https://doi.org/10.1002/bjs.6895.

White, E. C., Murphy, J. D., Chang, D. T., & Koong, A. C. (2015). Low toxicity in inflammatory bowel disease patients treated with abdominal and pelvic radiation therapy. *American Journal of Clinical Oncology*, 38(6), 564−569. Available from https://doi.org/10.1097/COC.0000000000000010.

Wong, J. S. M., Tan, G. H. C., & Teo, M. C. C. (2016). Management of para-aortic lymph node metastasis in colorectal patients: A systemic review. *Surgical Oncology*, 25(4), 411−418. Available from https://doi.org/10.1016/j.suronc.2016.09.008.

Yamada, K., Tsukamoto, S., Ochiai, H., Shida, D., & Kanemitsu, Y. (2019). Improving selection for resection of synchronous para-aortic lymph node metastases in colorectal cancer. *Digestive Surgery*, 36(5), 369−375. Available from https://doi.org/10.1159/000491100.

Yang, J., Yamamoto, T., Mazin, S. R., Graves, E. E., & Keall, P. J. (2014). The potential of positron emission tomography for intratreatment dynamic lung tumor tracking: A phantom study. *Medical Physics*, 41(2), 021718. Available from https://doi.org/10.1118/1.4861816.

Yeo, S.-G., Kim, D. Y., Kim, T. H., Jung, K. H., Hong, Y. S., Kim, S. Y., ... Oh, J. H. (2010). Curative chemoradiotherapy for isolated retroperitoneal lymph node recurrence of colorectal cancer. *Radiotherapy & Oncology*, 97(2), 307−311. Available from https://doi.org/10.1016/j.radonc.2010.05.021.

You, R., Liu, Y.-P., Huang, P.-Y., Zou, X., Sun, R., He, Y.-X. , ... Chen, M.-Y. (2020). Efficacy and safety of locoregional radiotherapy with chemotherapy vs chemotherapy alone in de novo metastatic nasopharyngeal carcinoma: A multicenter phase 3 randomized clinical trial. *JAMA Oncology*, 6(9), 1345−1352. Available from https://doi.org/10.1001/jamaoncol.2020.1808.

Index

Note: Page numbers followed by "*f*" and "*t*" refer to figures and tables, respectively.

A
Acetylsalicylic acid (ASA), 268
Adenomatous polyposis *coli* (Apc) genes, 5
American Joint Committee on Cancer (AJCC) system, 45
Anatomy of lymphatic system, 59–61
    collecting lymphatic vessels and lymph nodes, 60
    lymphatic capillaries, 59–60
    lymphatic trunks and ducts, 61
Anatomy of lymphatic system of colon and rectum, 61–69
    lymphatic drainage, 63–69
        of colon, 63–65, 67t
        lumbar and intestinal lymph trunks and cisterna chyli, 66–68
        preaortic and lumbar lymph nodes, 66
        of rectum, 65–66, 67t
        thoracic duct, 68–69
    macroscopic anatomy of colon and rectum, 61–62
    mesocolon, lymphatic vessel anatomy in, 62–63
    mesorectum, lymphatic vessel anatomy in, 63
    vascularization of colon and rectum, 62
Anoikis, 8–9
Arc of Riolan, 168
Artificial intelligence (AI) models, 76
Artificial neural network (ANN), 84
Ascending colon cancer, 200–202
ASPIRIN trial, 268

B
Biomechanics of lymphatic metastasis, 30–34
    cancer tissue, biomechanical environment of, 30–31
    future perspectives, 34–35
    interstitial and lymphatic transport in cancer tissue, 33
    lymphangiogenesis in cancer, 31–33
    mechanisms of cancer cell invasion of initial lymphatics, 33–34
    metastatic lymph node, 34
Biomechanics of normal lymphatic system, 21–30
    collecting lymphatics and lymphangions, 26–29
    future perspectives, 34–35
    interstitial flow and lymph formation, 22–24
    lymphatic endothelial cells and initial lymphatics, 24–26
    lymph nodes, 29–30
    mechanical properties of interstitium, 21–22
Biot's theory of poroelasticity, 28–29
Blood vessels, 63, 66, 69–70
Body mass index (BMI), 174–175
Bowel resected, length of, 181

C
Cancer, lymphatic system in, 31, 32f
Cancer-associated fibroblasts (CAFs), 30
Cancer-associated lymphatic system
    biomechanics of lymphatic metastasis, 30–34
        cancer tissue, biomechanical environment of, 30–31
        interstitial and lymphatic transport in cancer tissue, 33
        lymphangiogenesis in cancer, 31–33
        mechanisms of cancer cell invasion of initial lymphatics, 33–34
        metastatic lymph node, 34
    biomechanics of normal lymphatic system, 21–30
        collecting lymphatics and lymphangions, 26–29
        interstitial flow and lymph formation, 22–24
        lymphatic endothelial cells and initial lymphatics, 24–26
        lymph nodes, 29–30
        mechanical properties of interstitium, 21–22
    future perspectives, 34–35
Cancer cell invasion of initial lymphatics, 33–34
Cancer cells (CCs), 4–5
Cancer Genome Atlas study, 50
Cancer tissue, biomechanical environment of, 30–31
Capecitabine (CAPOX), 263–264
Capillaries, lymphatic, 59–60
Cardinal veins, 58
CD8 + T cells, 44–45
CD44, 30
CD151 expression, 7
Cecum cancer, 200
    ascending, 200–202
    descending, 206

Cecum cancer (*Continued*)
  sigmoid, 207
  transverse, 202–206
Central (apical) lymph node resection, 137
Central anastomotic mesenteric artery, 168
Central vascular ligation (CVL), 111, 167–168, 200, 210–211
Chemoradiotherapy (CRT), 156
Chemotherapy, neoadjuvant, 266
Circulating tumor cells (CTCs), 30, 133
Circulating tumor DNA (ctDNA), 133
Clinical target volume (CTV), 288
Clonal sweep, 46–47
Collecting lymphatics and lymphangions, 26–29
Colon, lymphatic drainage of, 63–65, 67t
Colon cancer, tumor deposits in, 98–99
Colorectal cancer spread, molecular markers of, 50
Colorectal cancer staging and treatment, 45–46
Colorectal cancer treatment, 45–46
  lymph node metastasis, evolutionary mechanisms of, 46–50
Colorectal nodal oligometastases, radiation for, 277–282
Colorectal polyps segmentation, 77–78
Colorectal tumors segmentation, 78–79
Complete mesocolic excision (CME), 167–168, 193, 210–211
  and D3 dissection versus standard resection, 137–138
    importance of pathological quality control in CME surgery, 176–185
      bowel resected, length of, 181
      distance between tumor and central arterial ligation point, 179–181
      lymph node yield, 182–183
      mesocolon, integrity of, 177–179
      specimen photography, 183–185
  Japanese D3 versus, 111
    oncological benefits of, 171–174
    patients benefitted from, 172–174
    potential limitations of, 174–176
      complications, 174
      technical difficulties, 174–175
    principles of CME surgery, 169–170
      minimally invasive CME surgery, role of, 170
    in transverse colon cancer, 175–176
Computational fluid dynamics (CFD) modeling, 28–29
Cone-beam CT (CBCT), 289
Convolutional neural network (CNN), 76, 84
Coronavirus disease, 179–181
COVID-19 pandemic, 179–181
Cytokeratin, 116–117
Cytokeratin 19 (CK19) mRNA, 113
Cytokeratin 20 (CK20) antibodies, 116–117

D

D3 lymphadenectomy versus D2 lymphadenectomy, 208–209
Darcy's law, 22–23, 28–29
Darwinian evolutionary principles of diversification, 133
Descending colon cancer, 206
Dihydropyrimidine dehydrogenase (DPD)-deficient patients, 267–268
Disease-free survival, 172–174, 183
Distal rectal cancers, 227, 241
DNA breaks, 273
DNA repair mechanisms, 48
Dose-limiting factor, 289
Dose painting, 273

E

Eastern and Western surgeons, 228–231
eIF-3, 50–51
Embryology of lymphatic system, 58–59
  lymph sacs, 58–59
  molecular mechanisms, 58
EMT hybrid state, 6–7
Endothelium, lymphatic, 57
Epithelial–mesenchymal transition (EMT), 5–8
European CME with CVL, Japanese D3 dissection and, 210–211
Excitation light, 246, 247f
External stress, 30–31
Extracellular matrix (ECM), 21–22
Extra-mesenteric lymph node dissection, 136–137

F

Fatty tissue, 89, 96
Flattening free filter (FFF) beams, 288
Fluorescence-guided sentinel lymph node detection in colorectal cancer, 247–249
  fluorescence-guided surgery, 246–247, 247f
  future perspectives, 249–252
  sentinel lymph node (SLN) mapping, concept of, 245–246
Fluorescence-guided surgery, 246–247, 247f
Fluorophore, 247f, 251–252
Fluoropyrimidine, 263
Fluorouracil, and oxaliplatin (FOLFOX), 263–264, 266
FOXTROT trial, 162, 177–179
Fractionation, 273, 286
Free-from treatment intensification (FFTI), 281

G

Glycoproteins, 24
Glycosaminoglycans, 24
Gross tumor volume (GTV), 282, 288

H

Hemicolectomy, segmental resection versus, 136
Hepatocyte growth factor (HGF), 7
HSA800, 249
Hyaluronic acid (HA), 30–31
Hydraulic conductivity, 23
Hypofractionation, 274
Hypoxia, 4
  and metastatic dissemination, 5–9
    anoikis, mitochondrial hyperpolarization, and lymphatic dissemination, 8–9
    EMT, cell motility, and invasion, 6–8
    lymphatic intravasation and extravasation, 8
    lymphatic niche formation and clonal expansion, 9
    therapeutic perspectives, 10
Hypoxia-inducible factors (HIFs), 4–5
Hypoxic signaling in tumor microenvironment, 4–5

I

Immuno-engineering, 34–35
Immunohistochemistry (IHC), 115–116, 117f
Indocyanine green (ICG), 246–247
Inferior mesenteric artery (IMA)
  high versus low ligation of, 137
Integrins, 7
Intensity-modulated radiation therapy (IMRT), 273, 280
Interobserver agreement regarding tumor deposits
  in previous TNM editions, 96
  in TNM 8th edition, 96–97
Interobserver variation improvement regarding tumor deposits, 97–98
Interstitial and lymphatic transport in cancer tissue, 33
Interstitial flow and lymph formation, 22–24
Interstitial fluid pressure (IFP), 4–5, 21–22, 31f
Interstitium, 21–22
  mechanical properties of, 21–22
Intratumor heterogeneity, 48
Isolated lymph node recurrence, extended lymphadenectomy on, 138–139
Isolated tumor cells (ITCs), 115–117, 117f, 122

J

Japanese Classification of Colorectal, Appendiceal, and Anal Carcinoma (JCCRC) system, 107, 112–113
Japanese D3 dissection in colon cancer, 193
  ascending colon cancer, 200–202
  cecum cancer, 200
  D3 lymphadenectomy versus D2 lymphadenectomy, 208–209
  descending colon cancer, 206
  and European CME with CVL, 210–211
  future perspective, 212
  left colic artery (LCA) preservation for left-sided colon cancer, 207–208
  lymphadenectomy, history of, 193–194
    Japanese classification and Japanese guidelines, 193–194
  lymph node dissection
    basic principles of, 194
    changes in the recommended area of, 195–196
    classifications of, 196–197, 200
  lymph node groups and station numbers, 198
  lymph node metastases
    classification of, 198–199
    current classifications of, 196–197
    recommendations of Japanese guidelines, 2019, 209–210
  regional lymph node classification, basic principles of, 200
  sigmoid colon cancer, 207
  technique of, 200
  transverse colon cancer, 202–206
Japanese D3 lymphadenectomy, 107–111, 109f
  versus complete mesocolic excision (CME), 111
  concept of, 108–109
  Japanese lymph node classification, 107–108
  lateral lymph node dissection, 110–111
Japanese guidelines, 2019
  current recommendations of, 209–210
Japanese lymph node classification, tumor node metastasis versus, 112–113
Japanese Society for Cancer of the Colon and Rectum (JSCCR), 169, 194
JK-1 gene, 51–52

L

Lacteals, 24–25
Lateral lymph node dissection (LLND) in rectal cancer, 227–228, 235–239
  Eastern and Western surgeons, 228–231
  functional outcomes after, 240t
  future, 234–235
  lateral nodal disease, defining, 231–234
  procedural variation, 239–241
  risks, 239
Lateral Nodal Recurrence in Rectal Cancer (LaNoReC) study, 235
Left colic artery (LCA) preservation for left-sided colon cancer, 207–208
Leucovorin, 263
Linear models, 46–50
Localized colorectal cancer, systemic treatment of, 257
  adjuvant treatment
    in elderly, 267
    in stage II disease, 259–261

Localized colorectal cancer, systemic treatment of (*Continued*)
  in stage III disease, 262–265
  dihydropyrimidine dehydrogenase (DPD)-deficient patients, 267–268
  future perspectives for (neo)adjuvant therapy in stage III colon cancer, 268
  locally advanced colon cancer, neoadjuvant chemotherapy for, 266
  nutrition and lifestyle modification reducing relapse risk, 266–267
  relapse risk assessment in stage II disease, 257–259
  time to treat, 262
Locally advanced colon cancer, neoadjuvant chemotherapy for, 266
Lumbar and intestinal lymph trunks and cisterna chyli, 66–68
Lymphadenectomy, 138–139, 173–175
  for colon cancer in Japan, 193–194
  Japanese classification and Japanese guidelines, 193–194
Lymphangiogenesis, 24
  in cancer, 31–33
Lymphatic capillaries, 24–25
Lymphatic dissemination, 8–9
Lymphatic drainage, 63–69
  of colon, 63–65, 67t
  lumbar and intestinal lymph trunks and cisterna chyli, 66–68
  preaortic and lumbar lymph nodes, 66
  of rectum, 65–66, 67t
  thoracic duct, 68–69
Lymphatic endothelial cells (LECs), 8, 24–25
  and initial lymphatics, 24–26
Lymphatic intravasation and extravasation, 8
Lymphatic niche formation and clonal expansion, 9
Lymphatic pathways, 144
Lymphatic spread in colon cancer, 43, 132, 134f
  colorectal cancer staging and treatment, 45–46
  colorectal cancer treatment, 45–46
    evolutionary mechanisms of lymph node metastasis, 46–50
    intratumor heterogeneity, 48
    linear models, 46–50
  mechanisms of lymphatic system, 43–44
    lymph formation and movement, 43–44
  molecular markers of colorectal cancer spread, 50
  molecular markers of metastases, 50–52
  parallel model, 49
  tumor microenvironment, 44–45
Lymphatic vessel anatomy
  in mesocolon, 62–63
  in mesorectum, 63

Lymphatic vessel hyaluronan receptor-1 (LYVE-1), 58
Lymph formation and movement, 43–44
Lymph node classification in colorectal cancer
  Japanese D3 lymphadenectomy, 107–111, 109f
    versus complete mesocolic excision (CME), 111
  concept of Japanese lymphadenectomy, 108–109
  Japanese lymph node classification, 107–108
  lateral lymph node dissection, 110–111
  tumor node metastasis versus Japanese lymph node classification, 112–113
Lymph node count (LNC), 136, 140, 141t
Lymph node dissection
  classification of, 200
  in Japan, 194
    changes in the recommended area of, 195–196
    current classifications of, 196–197
Lymph node groups
  and nodal station numbers at colon, 198t
  and station number at colon, 198, 199f
Lymph node metastases (LNMs), 43, 75
  classification of, 198–199
  current classifications of LNM in Japan, 196–197
    regional lymph node classification, basic principles of, 200
  evolutionary mechanisms of, 46–50
  models of, 47f
Lymph node ratio (LNR), 140, 160–161
Lymph nodes (LNs), 29–30, 66, 131–133, 135–136
Lymph node yield, 157–160, 182–183
  importance of, 156–157
Lymphovascular niches, 9
Lymph sacs, 58–59

M
Macroscopic anatomy of colon and rectum, 61–62
Matrix metalloproteinases (MMPs), 7
Mesocolon
  integrity of, 177–179
  lymphatic vessel anatomy in, 62–63
Mesorectum, lymphatic vessel anatomy in, 63
Metachronous isolated lymph node metastasis, 217–218
Metachronous metastases, 80, 82–83
Metastasis
  anatomical patterns of, 141–143
  molecular markers of, 50–52
Metastasis, temporal patterns of, 132–140
  autopsy findings in metastatic colorectal cancer, evidence from, 136
  circulating and tissue biomarkers, evidence from, 133
  clinical studies, evidence from, 136–140
    central (apical) lymph node resection, 137

complete mesocolic excision (CME) and D3
 dissection versus standard resection, 137–138
extended lymphadenectomy on isolated lymph
 node recurrence, 138–139
extra-mesenteric lymph node dissection, 136–137
high versus low ligation of inferior mesenteric
 artery (IMA), 137
lymph node counts and outcome, 140
segmental resection versus hemicolectomy, 136
genomic and phylogenetic studies, evidence from,
 133–135
growth rate of primary and metastatic colorectal
 tumors, evidence from, 135
implications for research, 144
Metastasis classification and prediction, 79–83
pathological classification, 81–82
radiological lymph node metastase (LNM)
 classification, 82–83
metachronous metastasis prediction, 82–83
Metastatic colorectal cancer (mCRC), 258–259
local treatment of, 274–276
Metastatic dissemination, hypoxia and, 5–9
anoikis, mitochondrial hyperpolarization, and
 lymphatic dissemination, 8–9
epithelial–mesenchymal transition (EMT), cell
 motility, and invasion, 6–8
lymphatic intravasation and extravasation, 8
lymphatic niche formation and clonal expansion, 9
therapeutic perspectives, 10
Metastatic lymph node, 34
Methylene blue, 246–247
Micrometastases and isolated tumor cells
in colorectal cancer, 119–123
 implications, 119–120
 occult disease in lymph nodes, 120–122
 standardized histopathological analysis of lymph
  nodes, 123
definition of, 115–119
 AJCC/UICC definition, 115–116
 biological significance, 118
 methods of detection, 116–117
 reporting, 119
differential diagnosis of, 127
after neoadjuvant therapy, 126–127
sentinel lymph-node biopsy in colorectal cancer,
 124–126
sentinel lymph-node mapping with ultra-staging,
 123–124
Minimally invasive CME surgery, role of, 170
Mitochondrial hyperpolarization, 8–9
Molecular markers
of colorectal cancer spread, 50
of metastases, 50–52

Molecular mechanisms, 58
Monoclonal anticytokeratin antibodies, 116–117
Multidisciplinary team meetings (MDTs), 176
Multiscale network architecture, 81
Myeloid-derived suppressor cell (MDSC), 9

N
National Cancer Database (NCDB), 159–160
Near-infrared (NIR) fluorescence imaging, 246
Neoadjuvant chemoradiotherapy (NCRT), 94–95
Neoadjuvant chemotherapy for locally advanced colon
 cancer, 266
Neoadjuvant radiotherapy, 282–283
Neoadjuvant therapy, 155–156, 228–233
lymph node ratio, 160–161
lymph node yield, 157–160
 importance of, 156–157
micrometastases and isolated tumor cells after,
 126–127
nodal involvement with complete clinical response
 after, 161
total neoadjuvant therapy, effects of, 162
tumor deposits and, 94–96
Nodal disease, colorectal, 75–76
future directions, 85
metastasis classification and prediction, 79–83
 pathological classification, 81–82
 radiological LNM classification, 82–83
segmentation and endoscopic detection, 77–79
 colorectal polyps segmentation, 77–78
 colorectal tumors segmentation, 78–79
staging, 76–77
treatment response, recurrence, and survival, 83–84

O
Obturator nerve, 236
Occult disease in lymph nodes in colorectal cancer,
 120–122
Oligometastases, radiation for, 276–277
Oligometastasic state, radiation in, 276
One-step nucleic acid amplification (OSNA), 113
ORCHESTRA trial, 222
Oxaliplatin, 263
Oxaliplatinum, 259–260

P
Para-aortic nodal disease in colon cancer, 215–216
chemotherapy, role of, 221–222
future perspectives, 222
imaging and implications for prognosis, 216
metachronous isolated lymph node metastasis,
 217–218

Para-aortic nodal disease in colon cancer (*Continued*)
  right-sided and left-sided para-aortic node involvement, 216–217
  surgery, morbidity of, 220–221
  synchronous metastases, 218–220
Parallel model, 49
Pericolic lymph nodes, 197f
Persistent homology profiles (PHP), 81
Planning treatment volume (PTV), 288
Poroelasticity, Biot's theory of, 28–29
Preaortic and lumbar lymph nodes, 66
PRODIGE 23 trial, 162
Programmed death ligand 1 (PD-L1), 44–45
Prospero homeobox 1 (PROX1), 58
PulMiCC trial, 274

R
Radiation, 273–274
  for colorectal nodal oligometastases, 277–282
  for CRC oligometastases, 276–277
  neoadjuvant radiotherapy, 282–283
  in oligometastasic state, 276
  prognostic factors, 283
Radiation and radiation techniques, practical considerations in, 286–289
  follow-up, 289
  patient selection, 287–288
  simulation, 288
  treatment delivery, 289
  treatment planning, 288
  treatment-related toxicity/dose constraints, 289
Radiation therapy (RT), 273
Radiocolloid tracer, 246
Radiofrequency ablation (RFA), 274
Radiological lymph node metastase (LNM) classification, 82–83
  metachronous metastasis prediction, 82–83
Radiology, future of tumor deposits in, 100
Radiotherapy (RT), 158
Radiotherapy for metastatic nodal disease in colorectal cancer
  case examples, 289–294
  experience with radiation, 273–274, 276–286
    for colorectal nodal oligometastases, 277–282
    for CRC oligometastases, 276–277
    future directions, 283–286
    neoadjuvant radiotherapy, 282–283
    in oligometastasic state, 276
    prognostic factors, 283
  metastatic colorectal cancer, local treatment of, 274–276
Rectal cancer, tumor deposits in, 99–100

Rectum, lymphatic drainage of, 65–66, 67t
Regional lymph node classification, basic principles of, 200
Regional lymph nodes and resection line in cecum cancer, 201f
Residual stress, 30–31
Reynolds number, 23

S
Segmental resection versus hemicolectomy, 136
Sentinel lymph node (SLN), 119, 245–246
Sentinel lymph-node biopsy
  in colorectal cancer, 124–126
    micrometastases and isolated tumor cells in, 123–126
Sentinel lymph-node mapping with ultra-staging, 123–124
Sigmoid colon, 61–62
Sigmoid colon cancer, 207
Skip metastases, 251
Solid stress, 30–31
Specimen photography, 183–185
Stage II disease
  adjuvant treatment in, 259–261
  relapse risk assessment in, 257–259
Stage III disease, adjuvant treatment in, 262–265
Staging systems, 92, 99
Starling equation, 23
Starling principle, 23
Stereotactic ablative radiation therapy (SABR), 274
Stereotactic body radiation therapy (SBRT), 274, 280–281
Stokes shift, 246
Subclones, 46–47
Superior mesenteric vein (SMV), 174
Surgical resection, 257
Surveillance, epidemiology, and end results (SEER) database, 94–95, 156–157
Sustained proliferation, 4–5
Swelling stress, 30–31
Synchronous metastases, 218–220
Systemic adjuvant therapy, 257, 259

T
Thoracic duct, 68–69
Tissue fluid homeostasis, 22
Toldt's fascia, 62–63
Total mesorectal excision (TME), 110, 156, 167–168, 185, 227
Total neoadjuvant therapy (TNT), 162
Transforming growth factor β (TGF-β), 23–24
Transverse colon cancer, 175–176, 202–206

Tumor and central arterial ligation point, distance between, 179–181
Tumor cells, 48
Tumor deposits (TDs), 75–76, 89
  biological mechanisms underlying development of, 92
  in colon cancer, 98–99
  definition of, 89–90
  future of, in radiology, 100
  histology of, 90f
  interobserver agreement regarding
    in previous TNM editions, 96
    in TNM 8th edition, 96–97
  interobserver variation improvement regarding, 97–98
  and neoadjuvant therapies, 94–96
  origins of, 91–92, 91f, 95f
  pathological assessment of, 96–98
  prognostic value of, 93–94
  radiological assessment of, 98–100
  radiological concept regarding the origin of, 100
  in rectal cancer, 99–100
  staging of, 92–93
Tumor-draining lymph nodes (TDLNs), 33
Tumor microenvironment, 44–45
  hypoxic signaling in, 4–5
Tumor node metastasis versus Japanese lymph node classification, 112–113
Tumour nodes metastasis (TNM) staging, 45, 50–51, 173

U
Ureter, medial retraction of, 235–236, 235f

V
Vascular endothelial growth factor receptor 3 (VEGFR-3), 58
Vascularization of colon and rectum, 62
VEGFR-3 expression, 58
Von Hippel–Lindau (VHL) tumor suppressor, 5

X
XELOXA study, 263

Y
Young's modulus E, 21–22

Z
ZEB2 expression, 50–51

Printed in the United States
by Baker & Taylor Publisher Services